Introductory Phycology

Introductory Phycology

F. R. TRAINOR
University of Connecticut

JOHN WILEY & SONS
New York   Santa Barbara   Chichester   Brisbane   Toronto

To Peg and Harold

Library of Congress Cataloging in Publication Data:

Trainor, Francis Rice.
  Introductory phycology.

  Includes bibliographies and index.
    1. Algology.   I.   Title.
QK566.T7   589'.3   77-26663
ISBN 0-471-88190-2

Printed in the United States of America

10 9 8 7 6 5 4 3 2 1

Preface

Interest in the algae, whether marine or freshwater organisms, algal components of the plankton or benthic organisms, has steadily increased over the last decade. In the past most students enrolling in introductory courses in phycology (the term applied to the study of the algae) were either botany majors (undergraduate or graduate students) or potential phycologists. Today a different interest group is emerging in many colleges and universities. Students majoring in marine biology, fisheries, population biology, and several areas in engineering now need and want to know something about the components of the algal flora and about algae as primary producers or as troublesome organisms in ecological systems.

This book was developed over a period of several years while I taught an undergraduate course to such a diverse group. It is apparent that many such students would be unable to enroll in a course on the algae if the prerequisites were excessive, or even approached what some of us would consider ideal. One advisor or another might want to justify prerequisites such as plant morphology, plant physiology, taxonomy, or ecology. In teaching this course, however, my prerequisites have been an introductory course in biology and three other courses in the sciences. The average student is a junior.

With such a group, through the years, I have found that some general background material should be made available to the nonbiologists, along with a small library and reading list. Such students can then learn quickly some of the facts and concepts essential for this course. Keeping these students in mind, as well as those biologists who had thought little about plants, I assembled the material in Chapter 2. Much of this information, which is very general in nature, has proven valuable to at least some of the students each year. It is not unusual to find an undergraduate quite surprised to learn that the occurrence of meiosis is not limited to those nuclear divisions just preceding the formation of gametes!

Some phycologists feel strongly that we do the student an injustice by omitting an ongoing treatment of algal classification, with characteristics of divisions, classes, orders, and even some families. It is held that the student needs this material to erect a framework for study and discussion of the organisms. I prefer to place little emphasis on these categories in the early chapters dealing with organisms. Instead, I treat classification as a separate subject in Chapter 16.

Experience has taught me that it is possible to talk about the major groups of algae, that is, the algal classes, and to discuss the organisms themselves. Whenever possible in this book, even species names are avoided. This approach has proved successful—the students become familiar with the structure and reproduction of a number of algae, and it frees some lecture time as well. Additional lectures can then deal with subjects covered in the later chapters of this book. For those who might want to emphasize classification, during lecture time a framework can easily be added to the material in the first 14 chapters.

The approach taken here, aimed primarily at the undergraduate, would need modification for those whose first contact with the algae comes at the graduate level. Additional readings, either certain of the chapter references, or recent publications in *Phycologia,* the *Journal of Phycology, Revue Algologique,* the *British Phycological Journal,* or in numerous other journals can be assigned. *Science,* the *American Journal of Botany, Limnology and Oceanography,* the *Canadian Journal of Botany,* the *Journal of Protozoology, Archives of Microbiology, Protoplasma, Botanica Marina,* among others, often have interesting papers on algae. Other journals are listed in Appendix I.

Often I have attempted to avoid technical terms. It is possible to establish a fact or explain a process or principle without the use of specialized terminology. Later, during lectures, many terms are gradually introduced. In taking this approach, a glossary is essential; therefore, important terms are defined in the Glossary at the end of the book. The serious student, or one who will deal with the algae in detail at a later date, will of necessity learn all the vocabulary so essential for understanding the literature or material in reference volumes.

The later chapters deal with subjects that not only attract student interest but also are timely. Numerous additions—especially in physiological topics—or substitutions come to mind. Selection of these topics has been the most difficult part of developing this book. Material now treated well in recent reference volumes might provide information for an additional lecture or two, as individual courses are developed.

I thank the following colleagues who aided in the development of one or more chapters: H. Andrews, E. Bonneau, J. Cain, P. Cook, E. Cox, P. Edwards, G. Floyd, J. Foerster, J. Grochowski, M. Harlin, D. Hibberd, W. Johanson, L. Klotz, N. Lang, K. Mattox, R. McLean, T. Monahan, M. Neushul, J. Page, N. Proctor, L. Shubert, E. Swift, D. Vance, P. Walne, B. Whitton, R. Wilce, R. Wood, C. Yarish, my graduate students, and especially S. Van Valkenburg and R. Hilton. Those who granted the use of photographs are acknowledged elsewhere; they have made this a more attractive volume. I apologize for any inadvertent omission of an acknowledgment. The typing of the manuscript through various stages of development by E. DeCarli and M. J. Trainor was done with accuracy and efficiency. I thank M. Hubbard for many of the diagrams, J. Grochowski for assistance with the proof, and finally, the entire Wiley staff.

It is my hope that suggestions for improving this volume will be sent to me at the University of Connecticut, BSG, Botany, Storrs, CT 06268.

F. R. Trainor

To the Student

In this book I have compiled information that I feel is central to the study of the algae. One of the main objectives in writing such a volume is to make the study of the algae pleasant for the student. This is not a simple task since each of you, with different backgrounds and interests, could fall into one of several categories. Thus, if you are primarily interested in the microalgae, or if you are planning a career in marine biology, you might wish that there were more information, more examples, and considerable detail in just these areas. I have attempted to provide a complete introduction and at the same time to keep the volume of moderate size. Thus there is more that could be said in all areas.

In any introductory course the instructor must work to develop a balance between what is known and what is yet to be learned. If everything is presented as fact, material that some undergraduates find easier to learn, the student could get the false impression that there is not much more to be learned about the algae. On the other hand, if it is pointed out too often that we are quite uncertain about the complete life history with one organism, and that we need a lot more research to develop basic information concerning the ultrastructure of another group, the student might get the impression that we have few fundamental and basic principles.

As will become apparent as the student works his (her) way through this book, we do have considerable information about this heterogeneous group of organisms. Often when we cannot make a statement that covers all the algae, it is because we are dealing with this diversity. We have a vast fund of knowledge, but we need more researchers not only to gather new data but also to confirm findings demonstrated only once, or with only one species in a very large genus. In at least a general way the student should learn the true state of the art in this discipline.

The student is encouraged to consult chapters in reference volumes and to become familiar with at least a few papers in the recent literature. Only then will it become apparent how factual material in textbooks is developed and how new areas of interest become popular. The student should learn how new information is added and older facts and interpretations are modified or discarded. It is a tedious, but exciting process.

The ultimate source of information is the organism and the experimentation with it. The student should use the laboratory sessions to examine and experiment with living material. Cultures of many algae are available and most organisms can be collected locally. In many cases the student can collect much of his (her) own study material.

I hope you enjoy your introduction to the algae.

F. R. Trainor

Contents

Introduction

A DEFINITION

What are algae? A technical definition is fairly easy: Algae are photosynthetic, nonvascular plants that contain chlorophyll a and have simple reproductive structures. However, this definition does not reflect the fact that algae exhibit a remarkable diversity of form and size, and exist in nearly every environment. The brown kelps up to 70 m in length, the flagellated swimming green cells as minute as 1 micron in diameter, rose-red, feathery fronds in deep, clear marine coastal waters, green scum on a quiet pond, sargasso weed in the massive Atlantic gyre, and organisms responsible for coloration on mountain snow are all algae.

Some algae resemble animals in that they actually ingest particulate food; others resemble higher plants because they have organs that *superficially* resemble roots, stems, and leaves. A further complication is that some nonpigmented (and hence nonphotosynthetic) organisms must be considered with algae, for reasons which will become clear as our discussion proceeds.

It is evident then that these "simple" plants are not simple at all. It is hoped that you will find them fascinating, as we do.

CLASSIFICATION OF THE ALGAE

The information about the major groups of algae that is briefly examined here will provide a framework for later discussions. Because distinctions between different systems of classification are often subtle, this topic will be covered in more detail later (Chapter 16) after we have studied many different organisms.

Algae, as well as other organisms, can be classified both for convenience and to show relationships. In the first case it is convenient to group similar things in a particular category; it is simple and orderly. If we wished to show that certain types originated separately from others, we might erect a different system, one that would tell us something about algal evolution. Thus organisms grouped together as brown algae, or Phaeophyceae, have evolved independently of the other classes of algae.

Linnaeus in 1754 first applied the term algae to lower plants, but his grouping included many organisms now considered more complex than true algae. Since that time, dozens of schemes for classifying algae have been proposed. Some of these were based almost solely on morphological similarities, others on supposed evolutionary relationships. Modern schemes are based almost entirely on the latter, and are bound to change as new evidence on evolutionary relationships comes to light. Morphological characteristics, physiological data, ultrastructural evidence, and so forth must now be considered, as well as evidence from organisms other than the algae. Where at one time organisms were placed in only two kingdoms, either as plants or animals, there are now suggestions for erecting up to eight kingdoms. In all cases it is now felt that bacteria and blue-green algae should be grouped together as the prokaryotic types.

Evidence for the different approaches comes from ultrastructural studies, pigment analyses of representatives of particular groups, biochemical data, and morphological and physiological studies of living organisms. As these data accumulate, modifications of a previously accepted classification of the algae are proposed.

Beginning biologists and workers in interdisciplinary fields may be perplexed by the various classification schemes (see Chapter 16), and some may believe that the entire discipline is in chaos. However, this is a rapidly developing field; we simply do not yet have agreement on the significance of some recent discoveries, and, therefore, we will find various interpretations.

The Categories Used in Classification

All living things are first placed in kingdoms. The animal kingdom is next subdivided into phyla, whereas the plant kingdom is divided into equivalent categories called divisions. Further subgroups, in descending order, are: class, order, family, genus, and species. (With some well-known groups of organisms, these categories may also be subdivided.) In this book, classes and genera are emphasized. Frequently terms with the ending -phyte, will be used (Table 1-1). This is a *common term,* and does not indicate a particular level of classification. All algal division names end in -phyta and classes in -phyceae (Table 1-2). Later (Chapter 16) we see how the algal classes can be grouped, indicating relationships.

Distinguishing among the Groups of Algae

Originally, organisms of somewhat similar form and color were classified by the common names green, brown, and red algae. Early investigators might have called a greenish form a green alga, and tentatively placed it in the Chlorophyceae, but today it is wise to be cautious. The organism

Table 1-1 DIVISIONS OF ALGAE, AND THE COMMON NAMES, PROPOSED BY G. M. SMITH (1950)[a]

Division	Common Name
Cyanophyta	Blue-green algae
Chlorophyta	Green algae
Rhodophyta	Red algae
Phaeophyta	Brown algae
Chrysophyta	Chrysophytes
Pyrrophyta	Dinoflagellates
Euglenophyta	Euglenoids

[a] Smith used an additional category for miscellaneous forms. Some of the latter were imperfectly known, whereas others did not conveniently fit into the known divisions.

Table 1-2   THE DIVISIONS AND CLASSES OF ALGAE TREATED IN
THIS VOLUME[a]

Division	Classes within the Division	Common Name
Cyanophyta	Cyanophyceae	Blue-green algae
Chlorophyta	Chlorophyceae	Green algae
	(Charophyceae)	(Charophytes)
	Euglenophyceae	Euglenoids
Rhodophyta	Rhodophyceae	Red algae
Chromophyta	Phaeophyceae	Brown algae
	Bacillariophyceae	Diatoms
	Xanthophyceae	Yellow-green algae
	Chrysophyceae	Golden algae
	Haptophyceae	Haptophytes
	Eustigmatophyceae	Eustigmatophytes
	Dinophyceae	Dinoflagellates
Cryptophyta	Cryptophyceae	Cryptophytes

[a] The separation of the Chlorophyta into Chlorophyceae and Charophyceae is along new lines discussed in Chapter 4. Further details on classification will be found in Chapter 16.

might, for example, be a red alga in which the red pigment had been destroyed or was masked by overproduction of a second pigment.

In order to distinguish the various classes of algae we now rely on the following characteristics.

1. Pigmentation. The various kinds of pigments, rather than simply the color of the alga, are determined. All of the groups contain chlorophyll and several carotenoids. The carotenoids include carotenes and xanthophylls. In one or more divisions one can find chlorophylls *a* and *b,* alpha or beta carotene, and a selection from a group of more than a dozen xanthophylls. In addition to the above pigments, which are soluble in organic solvents, there are the water soluble phycobiliproteins (sometimes called phycobilins), which are found in blue-green algae, red algae, and a small group of flagellates. We will discuss the entire group of pigments in more detail with each group of algae.
2. The stored reserve of photosynthesis. Reserve food material is usually stored within the cell and frequently within the plastid in which photosynthesis occurred. Starch, starchlike compounds, fats, or oils are the most common forms. It now appears that some

organisms release a portion of their excess material into the environment, and, in a sense, might be using the environment as their storage area. The released material might then be brought back into the cell at a later date.

3.   Motility. Some organisms are motile during much of their lives, whereas other genera lack motility, or any motile reproductive stages. Most algae do not move about by active means when they are adults, but often have some reproductive stages that are motile. As examples we might think of certain benthic algae such as brown and green seaweed. The number of flagella, their position of insertion, and flagellar surface structure (i.e., the presence or absence of scales or hairs) have taxonomic value. Some filamentous blue-green algae and certain pennate diatoms move in a steady gliding manner *without* flagella (Chapters 3 and 8).

4.   Wall composition. Although most algae have a conspicuous cell wall, some genera and certain reproductive cells do not. The cell wall may be a simple outer covering around the protoplast or an elaborately ornamented structure. There are marked differences in wall composition, and, at times, structure, among the various groups of algae. Among the materials isolated from algal walls are cellulose, xylans, mannans, sulphated polysaccharides, alginic acid, protein, silicon dioxide, and calcium carbonate. A typical algal wall is not constructed of one compound but instead is a matrix of one material interlaced with another or is formed by layers of different materials.

   Beginning students of phycology do not usually examine cell wall chemistry. Determination of wall chemistry requires sophisticated techniques, including chemical analyses or use of an electron microscope.

5.   Gross structure and plant body types. After examining a number of plants, the student will see that groups of algae can be distinguished by morphology. In this book this topic is usually presented first in order to show the range of form in each algal class. In addition, certain ultrastructural details, such as arrangement of thylakoids in plastids or nuclear and cell division phenomena, are important or at least strengthen the determinations already made.

Which of the five characteristics do you feel would be the most practical, or easiest to use? Which characteristics might provide the most valuable information for determining evolutionary relationships? Possible answers will develop as you learn about the individual algal groups.

DIVISIONS AND CLASSES USED IN THIS BOOK

If we gather data as outlined, and use new information as it becomes available, at certain times we will want to revise old systems of classification or erect completely new ones. A number of phycologists have now drifted away from that system proposed by Smith in 1950 (Table 1-1). Some now think of more divisions, others fewer. Despite these differences of opinion, the class has remained a rather stable unit of algal classification. Therefore, this book will emphasize classes for the most part, because at that level it is often possible to make clear distinctions among various groups of organisms (Table 1-2). When necessary, the student can refer to the divisions of algae proposed by Christensen (Table 1-2).

REFERENCES

Christensen, T. 1962. Alger. In: T. Bücher, M. Lange, and T. Sorensen (Eds.) *Botanik. Systematisk Botanik.* Vol. 2, No. 2. Munksgaard, Copenhagen. pp. 128–46.
Smith, G. 1950. *The fresh-water algae of the United States.* McGraw-Hill, New York. 719 p.

Algal cells, cell structure, and reproduction

INTRODUCTION

This chapter is an introduction to algal cytology and reproduction. An understanding of features common to many algae can prepare the student with little background for the material covered in later chapters. Some students will need to be alerted to the great diversity of cell types, morphology, and reproduction in the algae; others may wish to take a comparative approach to some subjects covered in the book. In a few cases, additional detailed cytological information is presented just once, when dealing with a specific organism later in the text.

More than 200 years ago Robert Hooke examined a small piece of cork under the primitive microscope and saw the outlines of simple plant cells. Since that first glimpse, there has been intense interest in the structure and function of the cell. Few organisms provide the cytologist with more diversity and interest than do the algae.

In this brief introduction we will see how ultrastructural cytology is providing information basic to our understanding of structure, function, and classification of the algae. Any student not familiar with recent advances in our understanding of both cell structure and function should consult some general works in this area.

ALGAL CELLS

In the algae a broad range of cell types can be observed. Some algae are unicells, for example, *Porphyridium* cells, *Chlorella,* euglenoids, many diatoms, and many kinds of flagellates. But most algae are more complex; that is, they may exist as colonies, as filaments, or as plant bodies composed of intertwining filaments. Other complex forms have at least simple tissues. Most multicellular algae have unicellular reproductive cells. In both unicellular and complex forms we refer to vegetative cells, i.e., those that are photosynthesizing and growing, and cells involved in reproduction, such as gametes.

Some examples of macroscopic cells, even if just visible to the naked eye, include certain centric diatoms, an occasional green alga or euglenoid, and the well-known *Acetabularia.* Most algal cells, whether the individual organism or a single cell of a multicellular organism, are microscopic. We have been able to learn a great deal about cell structure using the light microscope. With white light and the best of objectives, magnifications that could resolve individual objects to a minimum distance of 0.2 to 0.3 μm were possible.

The diagrams shown in Fig. 2-1 illustrate our knowledge of simple prokaryotic and eukaryotic cells as seen with the light microscope.

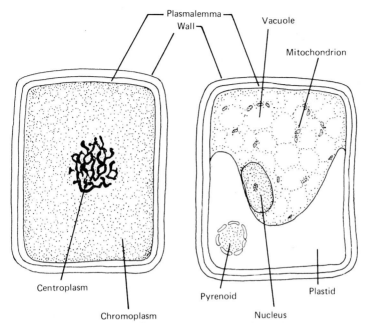

Figure 2-1. The structure of both a prokaryotic cell and a eukaryotic cell, based on observations made with the light microscope. Some structures could be best observed after the material was stained.

Cytology

The many advances in our understanding of cell structure and function in the last two decades have overshadowed earlier discoveries. In the early half of this century we had considerable knowledge of algal cytology. The nucleus was understood to hold hereditary information. Examination of mitosis and meiosis in scores of organisms, along with genetic studies of selected algae, provided basic information in algal cytology. Considerable data on the various algal life histories were gathered. At the same time, physiological studies revealed the workings of the chloroplast and the photosynthetic apparatus. Algae have been, and still are, favored material for these investigations. Algal plastid structure was shown to be quite variable in groups such as the spherical green unicells. Most algal cells were known to have a wall and large central vacuoles. Contractile vacuoles and stigmata, or red eye spots, were described for some cells.

Higher magnifications were made possible with the development of the electron microscope (Figs. 2-2 and 2-3). A stream of electrons, focused

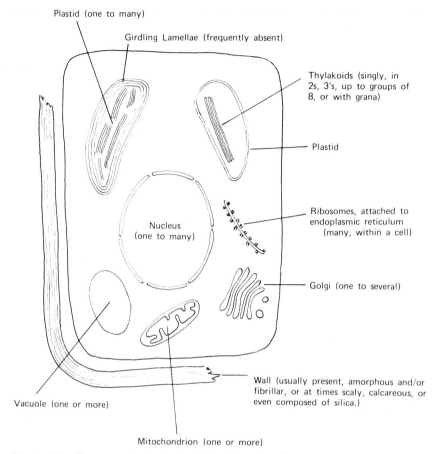

Figure 2-2. Diagrammatic presentation of a eukaryotic cell as seen with the electron microscope, along with the possible distribution of the organelles in each cell. Girdling lamellae are figured, but most plastids do not have them.

electromagnetically, is used in place of light passing through glass lenses. Resolution is extended down to the range of 0.2 to 0.3 nm, and thus cell detail can be magnified tens of thousands of times. In earlier studies, resolution was not always good, but with better technical approaches, and some knowledge of the chemistry of fixation and embedding, great strides were made. A whole new world of cellular detail was then discovered with electron microscopy, and even those structures previously studied yielded new and valuable information. Thin sections of specimens are examined in a transmission electron

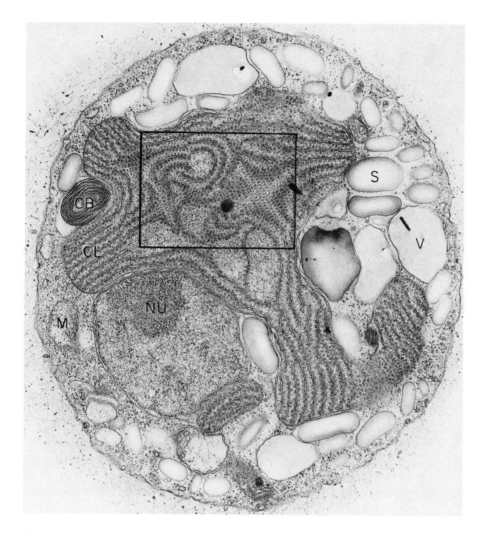

Figure 2-3. Cell of the red alga *Porphyridium* with the membrane-bound nucleus, plastid, and mitochondrion. Rows of phycobilisomes are visible in the boxed area. 31,000 X. Courtesy of Gantt and Conti and Journal of Cell Biology.
S = starch, V = vacuoles, Nu = nucleus, M = mitochondrion, Cl = chloroplast.

microscope (TEM). The scanning electron microscope (SEM) is used to examine surfaces, such as entire cells or even small organisms.

With this additional cytological detail, it was possible to make clear distinctions between prokaryotic and eukaryotic forms (Figs. 2-3 and 2-

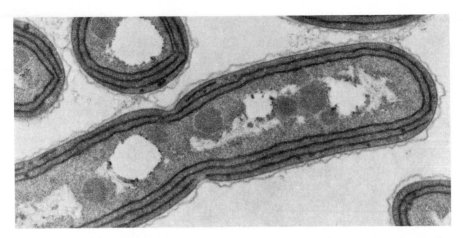

Figure 2-4. Longitudinal and cross sections of a blue-green alga. Membrane-bound organelles are lacking. Note thylakoids toward the periphery. 28,000 X. Courtesy of N. Lang.

4). The close alignment of the bacteria and blue-green algae became clear, for these prokaryotic forms lack *membrane-bound* organelles. Blue-green algae do possess single thylakoids, membranes associated with photosynthesis.

The structure of the chloroplast, and its association with other cellular organelles, proved to be most informative. All chloroplasts are bounded by a double membrane. In addition, many chloroplasts are enclosed by cytoplasmic endoplasmic reticulum, which may be continuous with the outer membrane surrounding the nucleus. The nucleus and chloroplast may be enclosed together in one sac. Inasmuch as the plastid may be red, green, brown, and so forth, some researchers prefer the term plastid rather than chloroplast.

Within the plastid the pigmentation is associated with membranes, the thylakoids (Figs. 2-2 and 2-3). These appear to be closely appressed lines when seen in the two dimensions of an ultrastructural figure. Actually, the individual thylakoid can be considered to be a membraned sac. If the sac is then flattened so that the opposite membranes come into close proximity, one has the typical plant thylakoid. In certain groups of algae, thylakoids are frequently found in groups of three (Table 2-1). In addition, more extensive stacking of portions of thylakoids produces the granum. Numbers and arrangements of thylakoids in chloroplasts are significant when distinguishing among groups of algae (Table 2-1).

Additional thylakoids, the girdling lamellae, may encircle the entire chloroplast, just inside the chloroplast envelope (Fig. 2-2, Table 2-1).

Table 2-1   THE ARRANGEMENT OF THYLAKOIDS IN PLASTIDS
OF VARIOUS CLASSES OF ALGAE

Thylakoids present, but not surrounded by a chloroplast envelope	Cyanophyceae
Thylakoids present, surrounded by a chloroplast envelope	
A. Thylakoids single with a girdling thylakoid in most	Rhodophyceae
B. Thylakoids in groups of two with no girdling thylakoids	Cryptophyceae
C. Thylakoids in groups of three	
1. With girdling thylakoids	Phaeophyceae Chrysophyceae Xanthophyceae Bacillariophyceae
2. Lacking girdling thylakoids	Haptophyceae Eustigmatophyceae Dinophyceae
D. Thylakoids never single, but groupings variable, from two to many; lacking girdling thylakoids. Some stacking into grana possible in certain genera.	Chlorophyceae Charophyceae Euglenophyceae

The girdling lamellae are present in many red algae, in the diatoms, golden and brown algae, and some yellow-green algae, but are generally lacking in other groups. When phycobiliproteins are present, they are found in phycobilisomes, which are distinct granules attached to thylakoids (in red and blue-green algae), or between thylakoids (in cryptophytes).

Pyrenoids are present in the plastids of most algal classes. Whether actually embedded in the plastid or almost extruded from it, the pyrenoid is surrounded by the chloroplast envelope. The pyrenoid is a granular, or sometimes crystalline, proteinaceous body. It is sometimes associated with production of reserve foods, such as starch or other carbohydrates. Some pyrenoids are stalked, many have thylakoids within their structure, and some have invaginations of other cell constituents.

In eukaryotic forms it is apparent that the nucleus is bound by a membrane (Fig. 2-2). Because of the chemistry of the hereditary material and the properties of fixatives used, little detail was seen in many early studies. Today ultrastructural studies of organisms in several algal classes are revealing information on chromosomal structure and behavior as well as cytokinetic phenomena.

Mitochondria (Fig. 2-2) are spherical or cylindrical membrane-bound bodies; the convoluted inner membrane provides increased surface area. The invaginations are called cristae. Mitochondria are directly involved in respiratory processes within the cell, and are typically found in most plant and animal cells. There may be fewer than 100 per cell or several thousand.

Ribosomes, involved in protein synthesis in cells and cell organelles, are found in the algae. Different types of ribosomes are recognized because of their size and sedimentation coefficients (S). The latter are determined with ultracentrifugation. The 70 S, or relatively small ribosomes, are found in the blue-green algae and in the matrices of chloroplasts and mitochondria of eukaryotic algae. The 80 S ribosomes are located in the cytoplasm of eukaryotic algae, as well as in higher plants and animals. Ribosomes found outside organelles are either free or associated with endoplasmic reticulum.

The Golgi apparatus, a system of closely associated, variously sized, often collapsed sacs, is present in most eukaryotic cells. Wall components, such as scales on *Chrysochromulina* or on a coccolithophorid, are assembled within Golgi sacs or vesicles. The vesicles migrate toward the cell periphery and eventually reach the plasmalemma. At the point of contact, the membrane is disrupted and the contents of the vesicles are released through the plasmalemma. Other components of the walls of marine and freshwater algae, such as microfibrils and amorphous materials, are released from within vesicles in a similar manner. Vesicles may vary greatly in size. Many wall components are assembled outside the plasmalemma.

Endoplasmic reticulum (ER) is a system of membranes within the cytoplasm that is sometimes associated with cell organelles. The membranes of the ER are paired, may have ribosomes associated with them (rough ER) (Fig. 2-2), or appear without attached ribosomes (smooth ER).

In the vegetative state many algae are flagellated and motile; reproductive cells in a variety of algal classes are also motile. There can be one or more flagella per cell, inserted at either pole, or at times on a lateral surface. Those at the anterior end may be at the apex or displaced slightly to one side. When two flagella are present, they may be of equal length, slightly unequal in length, or markedly so (Fig. 2-5). The flagellum surface is smooth or covered with hairs or scales (Table 2-2). If flagella are unequal in length and each has a different surface composition, the shorter flagellum, or the trailing flagellum, is usually smooth.

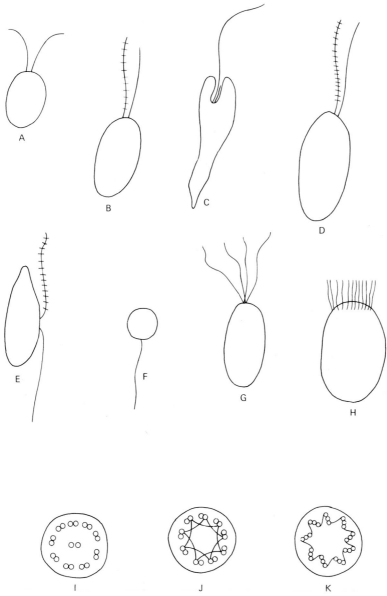

Figure 2-5.   Arrangement of flagella, both the hairy and smooth types, on algal cells in the apical (A, B, C) subapical (D), lateral (E), and posterior positions (F). Uniflagellated, biflagellated, quadriflagellated, and multiflagellated types are known. Three "typical" cross sections of flagella as seen with the electron microscope (TEM), are presented, one along the free portion (I), another near the point of attachment, the transition region (J), and a third (K) within the cell.

Table 2-2   THE FLAGELLAR TYPES FOUND IN VARIOUS
ORGANISMS IN THE DIFFERENT CLASSES OF ALGAE[a]

Algal Class	Number of Flagella	Flagellum Smooth	Hairy	Rows of Hairs	Comments
Cyanophyceae	none				
Chlorophyceae	2 or more	most	some	two, in some	some scaly flagella
Euglenophyceae	2 or more	one	one	single	small flagellum smooth
Rhodophyceae	none				
Phaeophyceae	two	one	one	two	
Bacillariophyceae	one	none	one	two	
Xanthophyceae	two	one	one	two	
Chrysophyceae	two	one	one	two	
Haptophyceae	two	two	none	none	
Eustigmatophyceae	one	one	none	none	
Dinophyceae	two	one	one	single	
Cryptophyceae	two	none	two	two	

[a] Some flagella are smooth, whereas others have either scales or projecting hairs. There can be hairs located along the flagellar length in either one or two rows. Other arrangements of hairs, as well as further flagellar complexity, are not presented or discussed here. For those organisms which have both a smooth (whiplash) and a hairy (tinsel) flagellum, the shorter or the trailing flagellum is normally the smooth type. Exceptions to this summary information can be found in several algal classes.

In cross section most flagella are identical, and have a 9 + 2 microtubule arrangement (Figs. 2-5 and 2-6). The figure 9 refers to the 9 fused doublets that are found in a ring. The central microtubules (the 2 of the 9 + 2) are not fused and terminate near the flagellum base, before it enters the cell surface. In that transition region, a stellate pattern, with connections between the 9 doublets, is observed in cross sections of flagella (Fig. 2-5 J). Inside the cell the flagellum base is called the basal body. One then sees 9 peripheral triplets of microtubules with interconnections (Fig. 2-5 K).

Figure 2-6.   Cross section of green algal flagellum, showing microtubules in the 9 + 2 arrangement. 90,000 X. Courtesy of Laurendi and McLean.

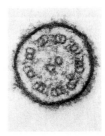

Table 2-3 DIVERSITY IN CELL COVERINGS WITHIN THE ALGAE[a]

A. Forms that possess only a plasmalemma:
    1. Spermatia (reproductive cells) of red algae.
    2. The green flagellate, *Dunaliella.*
    3. Some naked dinoflagellates.
B. Organisms that have additional covering material *inside* the plasmalemma:
    1. The euglenoids have a pellicle.
    2. Dinoflagellates may possess thecae.
    3. Cryptophycean forms have a pellicle.
C. Forms that have a cell covering *outside* the plasmalemma:
    1. Organisms with scales of organic matter, calcite, or silica, such as coccolithophorids and others.
    2. Cell walls that are:
        a. Incomplete, as with a lorica of *Dinobryon.* There is an opening at one end.
        b. Complete, composed of microfibrillar and/or amorphous material, organic matter, or silica. May be impregnated with metals. Most algae, including diatoms, fall into this category.
        A number of organisms, especially some marine red and green algae, can be highly calcified.

[a] After Dodge.

Although there is a great deal of variety in the types and arrangements of flagella, there is some uniformity within an algal class. Some exceptions will always be noted.

Certain microscopic algae and reproductive cells in other groups may lack a cell wall. In these cases the plasmalemma is in direct contact with the environment. Table 2-3 presents examples of the diverse cell coverings found in the algae. In certain algae details of wall composition and wall ornamentation have been shown to be of taxonomic value, even at the level of class.

Wall formation can be an exciting area of study. Not only are there interesting cytological and morphological observations at the light microscope level, but also the subject has attracted the attention of the electron microscopist. Studies during cell division, including nuclear division, cytoplasmic division, and wall formation, can provide clues to the evolution of both the green algae and certain higher plants.

CELL DIVISION

It is not surprising that most algae have typical nuclei and that they divide either by mitosis, passing on the same chromosome number to daughter nuclei, or by meiosis (reduction division), a process that reduces the chromosome number. Cytokinesis, or cytoplasmic division, follows. During and after these events new cell wall material must be deposited.

First consider cell division in the algal prokaryotes, the blue-green algae. During cell division we do not find the typical figures of prophase, metaphase, anaphase, and telophase. The genetic material of blue-green algae is not organized into visible chromosomes, and thus one cannot observe these stages of division. However, some changes in the centroplasm can be seen during division. Blue-green algae have been stable forms for a few billion years, so obviously they replicate their DNA, even though we do not see chromosomes in movement.

In blue-green algae replication of DNA is eventually followed by cytoplasmic cleavage and wall formation, which proceed in a centripetal direction, from the margin toward the center of the cell. A new cleavage and formation of another cross wall in each of the two "progeny" can be initiated before the initial cleavage is complete.

Division of the nucleus can occur in the cells of most eukaryotic algae, but it will not take place in some cells, such as gametes. In plants, normally we think of cell division as a mitotic division followed by cleavage of the cytoplasm. This cytoplasmic division occurs in a variety of ways:

1. A cleavage furrow may appear at one point at the cell surface and develop across the cell.
2. There may be centripetal growth of a furrow, initiated around the circumference of the cell.
3. Fusion of vesicles is first observed between the recently formed nuclei, and this process continues until it has proceeded to the cell margins.
4. A cell plate can be formed between nuclei. As in higher plants it develops in a centrifugal direction.
5. There are other possibilities, including combinations of some of the above.

Nuclear and cytoplasmic divisions take place in both haploid and diploid cells. Thus some organisms have an extensive free-living haploid phase, as well as a free-living diploid phase. In other organisms one portion of the plant body may be haploid, with the remainder diploid. Some interesting life histories are then possible.

You already know that a mitotic nuclear division is not always followed by cytoplasmic division. In fact, the exceptions are quite common. If cytokinesis always immediately followed mitosis, how would multinucleate cells be formed?

Certain algae always produce new individuals or reproductive cells within the original cell wall, and they do not incorporate the old wall, per se, into new cells. New individuals are released when the parent wall breaks or is lysed.

Meiosis or Reduction Division

In the two successive nuclear divisions of meiosis the chromosome number is halved. Although there are usually four products of this reduction division, as in the red algal tetraspore formation, this is not always the case. In the more advanced brown algae (e.g., *Sargassum*) reduction division in the female structure may result in only one egg. Studies with animals show that this is accomplished when one product of each division in meiosis fails to mature. Diatom female gametes are also formed in this way. In the algae when male gametes are formed the four products of meiosis survive.

It is not uncommon to observe one or more additional divisions (mitotic) of the four nuclei resulting from meiosis. Thus in a brown algal sporangium in which meiosis has occurred, further mitotic divisions produce a few dozen reproductive cells. The products of meiosis have increased chances of survival because of genetic recombination.

The divisions that result in gamete formation, or spore formation, or the germination of a zygote, can be meiotic (Fig. 2-13). Because meiosis can occur at these points in algal life histories, these plants are challenging to the student first attempting to understand their reproduction. In some simple red algae as well as in many other organisms, we do not know where meiosis occurs. In certain algae, meiosis apparently does not occur.

Ultrastructural Studies of Cell Division

Recent studies of cell division have produced considerable evidence for two evolutionary lines in the green algae. This has led to a reconsideration of their classification, as well as of the origin of the vascular plants. Primitive organisms, such as *Chlamydomonas* and colonial relatives, are not directly linked to the evolution of terrestrial plants.

This ultrastructural approach to classification was based initially on studies of algae that did not develop a cell plate or phragmoplast. A few green algae, even filamentous forms, do have a phragmoplast as found in higher plants. In these organisms the longitudinal spindle is persistent and daughter nuclei, formed after chromosomes separate, remain quite distant.

Other green algae do not have the persistent interzonal spindle. There is early collapse of the spindle at telophase. Daughter nuclei are then positioned closer to each other. Most of these organisms have microtubules that lie parallel to the line of cytokinesis. This system has been called a phycoplast (Fig. 2-7) (see Chapter 4).

Ultrastructural studies have shown other features; for example the development of connections between cells, such as the pit connections in the red algae, or the formation of the precise patterns of silica in the diatom wall. We have good research data on nuclear organization and division not only in eukaryotes, but also in the mesokaryotic nucleus of the dinoflagellates. These and other studies have shown us some interesting phenomena and essential information not only about the

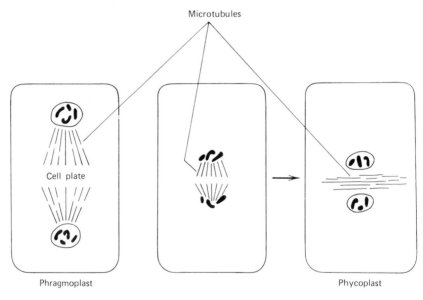

Figure 2-7. Diagrams of a phragmoplast (left only) and a phycoplast (right), including the stage prior to phycoplast formation (center). A cell plate develops at the center of the persistent spindle of the phragmoplast and nuclei remain some distance from each other. In those organisms with phycoplast formation, because of an early collapse of the spindle, newly formed nuclei remain near each other (right). Microtubules, arranged parallel to the cell equator, are characteristic of a phycoplast.

individual organism, but also its relationship to other algae. Information on cell division of a variety of organisms in all algal classes is needed.

REPRODUCTION AND REPRODUCTIVE CELLS

Those unfamiliar with microorganisms will be surprised to learn that many unicellular microorganisms reproduce only by cell division (Fig. 2-8a). In these organisms no specialized types of reproductive cells are known. Furthermore, some simple filaments and colonies, which increase their cell number by division, reproduce only when the alga fragments. Organisms that have a more sophisticated method of reproduction might reproduce by fragmentation in some environments, but by the alternate method in other conditions.

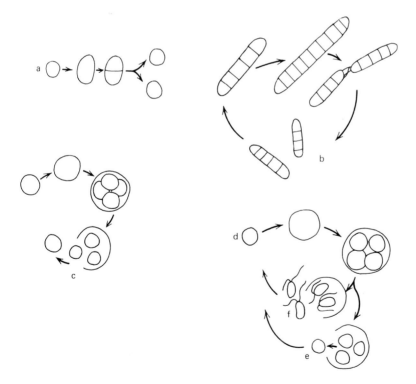

Figure 2-8. (a) Reproduction in algae by cell division (A), fragmentation (B) and spore production (C, D). Planospores (E) and aplanospores (F) are shown.

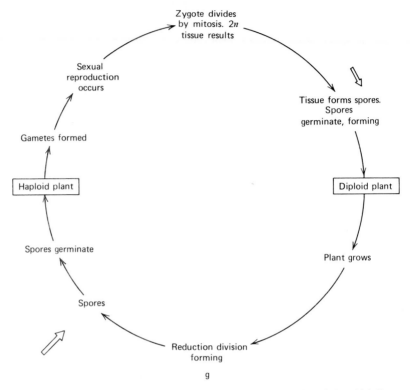

Figure 2-8.
Figure 2-8. (b) Certain algae reproduce by means of a sexual cycle in which there is *also* spore production (large arrows).

Spores

The majority of algae have specialized reproductive cells, most of which could be called either spores or gametes. Although the asexual reproductive cells are typically called spores, the term usually has a prefix. The variety of spore types is staggering, ranging from auxospores to zoospores. In addition, there are specialized terms for sporelike structures in certain algal groups, such as the akinete of blue-green algae, which are discussed later in the book. Some of them have a broad use, whereas others are restricted to just a few organisms, as are the androspores found in *Oedogonium*. Spores are formed by any cell in certain genera, whereas other organisms have specialized cells involved in spore production.

Some organisms have the ability to produce nonmotile spores (Fig. 2-8); others form only motile types. With certain algae lacking motile cells, there is evidence for a loss of motility during their evolution. The terms planospore and aplanospore are used with those algae which have both motile (planospore) and nonmotile reproductive cells (aplanospores) (Fig. 2-8). The planospore is found only when conditions in the environment are suitable. The term most commonly used for a motile, flagellated asexual reproductive cell is zoospore (Fig. 2-9). A zoospore of a macroscopic plant can be almost identical in appearance to a primitive alga in the same group.

Although higher plant spores, such as megaspores and microspores of flowering plants, are haploid, algal spores can be either n or $2n$. Some organisms can produce either haploid or diploid spores in different phases of their life histories (e.g., certain red algae) (Fig. 2-8*g*). This fact, as well as the occurrence of reduction division at different times in the life histories of different algae, is one reason for some interesting, but often complex, life histories in the algae.

Spores not only provide a means of reproduction for algae, but also can have additional functions. Some spores that are quite ornamented or thick walled are said to aid survival in unfavorable environmental conditions, even over winter. Most of the evidence for any protective role of the wall with regard to temperature changes or desiccation is morphological, but recent experimental data support such a concept of spore function.

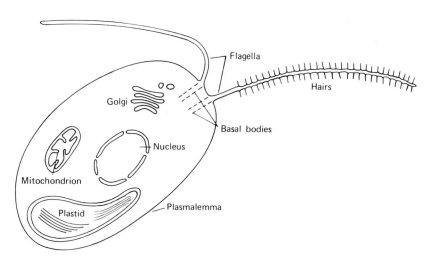

Figure 2-9. Diagrammatic presentation of a motile spore as seen with the electron microscope. The flagella may be smooth or bear lateral hairs. There may be material outside the plasmalemma (scales or a wall) or inside (a pellicle or thecae). Each cell will have most of the typical cell organelles. Contractile vacuoles and a stigma are often present.

Sex Cells

Algae in most classes exhibit a sexual stage. However, there are many different types, sizes, and shapes of gametes, and even the chambers in which they are produced may vary. In addition, gametes in one genus may be produced by a haploid plant, whereas a genus in the same class might form gametes when cells of a diploid plant undergo meiosis. In some algal groups, the gametes are motile and morphologically similar to zoospores (Figs. 2-5 and 2-9).

In primitive types the sex cells are isogamous, that is, both cells are motile and similar in size and shape (Fig. 2-10). Other forms evolved from this stock may be heterogamous (one cell is larger) or oogamous (at least one cell has lost motility). The larger cell is referred to as female. In several groups of related organisms one can see a gradual change or graded series from isogamy to oogamy.

A homothallic organism can produce both male and female cells on one individual. If the sexes are separate, the species is called heterothallic. With isogamous heterothallic populations frequently one cell is arbitrarily called plus and the other minus.

The blue-green algae lack the sex cells and reproductive mechanisms discussed thus far, but in the last decade several investigators have pointed out that genetic material *can* be exchanged between organisms.

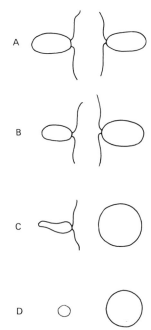

Figure 2-10. The grades of sexuality in the algae. After observing whether the gametes are motile and if they are the same size, one can distinguish isogamy (A), heterogamy (B), and oogamy (C, D). If the organism is capable of producing motile male cells the type of oogamy as figured in C is observed. However, in certain oogamous organisms, such as the red algae, the male cells lack motility (D).

The mechanism in this class of algae is similar to that reported for bacteria. Oogamy is the rule in the red algae, but it is a special case since nonflagellated male cells are carried only passively to the female cell.

Sex cells may be produced singly or in groups within a common wall. The structure containing the gametes is termed the gametangium. In the algae, unlike vascular plants, the gametangium is always unicellular in origin and no sterile cells are produced from the single cell during its formation. However, in *Chara* sterile cells *with an independent origin* are sometimes associated with gametangia. Antheridia and archegonia, found in green land plants, have sterile cells surrounding reproductive cells.

ALGAL LIFE HISTORIES

You are now aware that some algae are unicellular, others exist as colonies or filaments, and the most complex are composed of simple tissues. As we discuss individual organisms, we shall see how these algae develop, beginning with the single cell stage or juvenile plant and terminating with a complex, reproducing plant.

A great variety of life histories occur within the classes of algae. However, here we are concerned only with the basic types common to many organisms, and with how an algal life history develops. This will introduce the detailed material that appears later in the book.

Life Histories of Organisms Lacking a Sexual Stage

Certain organisms lack sex cells but are very successful in their reproduction by cell division, spore production, and fragmentation (Figs. 2-8 and 2-11). In fact, they can become dominant organisms in a water bloom, attached to aquatic life, living in sediments, or thriving on a solid substrate. These are by far the simplest of life histories possible in the algae.

Life Histories of Organisms Possessing a Sexual Stage

At first glance, a study of life histories of organisms possessing a sexual stage may appear to be a simple undertaking. Like mammals, plants would be diploid, would have male and female sex cells produced by separate plants, and would reproduce, forming individuals similar to themselves. Perhaps in some cases male and female cells would be produced by a single individual. Reduction division (RD) would take place during formation of gametes. Fusion of the gametes would restore the diploid number. The zygote develops into a new individual. A model of the life history is presented in Fig. 2-12, where a circle represents a plant. Also diagrammed is the alternate cycle resulting when the sexes

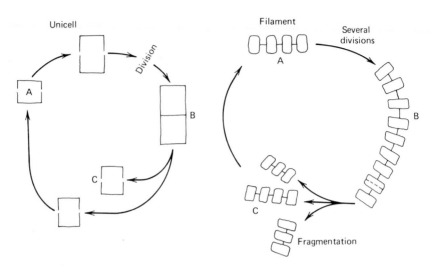

Figure 2-11.   Algal life histories for some organisms that lack a sexual stage, as in unicellular and filamentous diatoms. The young plant is figured in A. Photosynthesis and growth take place, followed by cell division (B), and separation of the division products (C in the unicell) or fragmentation (C in the filament).

are separate. Since the life histories are basically similar (Fig. 2-12), with only free-living diploid plants and with reduction division taking place in the formation of gametes (gametic meiosis), further discussions in this chapter will not be complicated by the diagrams with both male and female individuals.

This life history does exist in the algae, but it is not the way that most algae reproduce, nor is it the most primitive type of life history. The point at which meiosis takes place is critical in the discussion of a life history. In an individual species that reproduces sexually, meiosis is fixed at a certain point in the life history. However, in all the algae there are normally *three possible locations for reduction division*. It can take place:

1.   during the formation of gametes, that is, gametic meiosis.
2.   when there is formation of spores, that is, sporic meiosis.
3.   when the zygote germinates, that is, zygotic meiosis.

Diagrams indicating how the positioning of reduction division can affect the life history are shown in Fig. 2-13.

Gametic Meiosis
If meiosis takes place in the formation of gametes, there is just a diploid, free-living phase. Inasmuch as gametes are the only haploid cells, and their fate is fusion with the opposite type, no haploid, vegetative phase exists (Fig. 2-13).

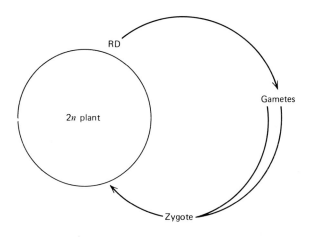

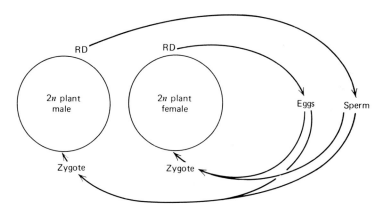

Figure 2-12.   Diagrams of life histories of algae with gametic meiosis; reduction division (RD) occurs in the formation of the gametes. Gametes may be produced by one individual (upper), or the sexes may be separate (lower).

Zygotic Meiosis

Many algae have only a haploid vegetative phase. Mitosis and cytokinesis take place in haploid cells just as easily as when the chromosome number is double. At maturity certain cells or gametangia produce gametes; meiosis *does not* take place as the gametes are formed. The gametes fuse, forming a zygote. However, the zygote can divide only by meiosis. In this case, only a haploid, free-living phase is possible (Fig. 2-13).

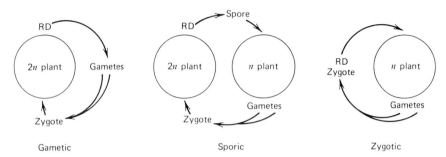

Figure 2-13. Diagrammatic presentations of life histories in the algae when there is gametic meiosis (left), sporic meiosis (center), and zygotic meiosis (right).

Sporic Meiosis

Free-living haploid (n) and diploid ($2n$) plants are detected in any organism in which reduction division takes place during the *formation of spores* (Fig. 2-13). The spore develops into a haploid plant that can produce gametes. The gametes fuse, forming a zygote, which germinates and produces a diploid plant. The divisions of the zygote and the resulting cells are mitotic. Free-living haploid and diploid plants of the same species exist!

In organisms possessing sporic meiosis, the haploid plant is sometimes morphologically similar to the diploid one. Such organisms are characterized by an alternation of *isomorphic* plants. When the n and $2n$ stages are dissimilar, the organism exhibits a *heteromorphic* alternation. In some cases the n plant is the larger of the two, but the reverse is frequently true (Fig. 2-14). Thus there are really three different categories in the life histories of the algae (Fig. 2-13), with some additional variation when meiosis is sporic (Fig. 2-14).

Additional Comments on Life Histories

Life histories become more complicated, and difficult to diagram simply, when one considers the possibility of variability. In many algae, completion of the life history is not obligatory (Fig. 2-15). (Therefore many researchers avoid using the term life cycle.) Either the organism has *an asexual means* of duplicating itself, or a different route may be followed. For example, certain gametes are not obligately gametes. If fusion does not occur, the gamete itself can germinate parthenogenetically and form an adult plant (Fig. 2-15). Temperature, day length, and nutrition affect the organism, and can sometimes dictate the pathway followed. These variations will become clear when the details of individual cycles are presented.

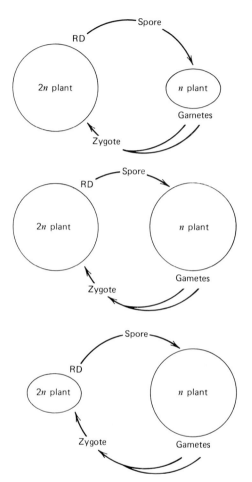

Figure 2-14. Diagrams of the life histories found in algae that possess sporic meiosis. The plant is represented by the circle; the area indicates the size the organism may achieve. The haploid and diploid generations may be isomorphic (center) or heteromorphic (upper and lower).

In other organisms the gametes, spores, or the zygote are not liberated from the producing organism. In the red algae, retention of the zygote and its germination products on the female plant results in an additional stage, the carposporophyte. A more complex life history results. In some organisms a thallus may be part haploid and part diploid. Thus in the red alga, *Lemanea,* the vegetative plant is diploid below and haploid above. A life history of this type does not conveniently fit into *any* of those diagrammed (Figs. 2-13 and 2-14). We discuss these details further when we study individual organisms.

In many land plants the haploid phase (in ferns a prostrate, frequently heart-shaped, green plant body) produces gametes. This phase is

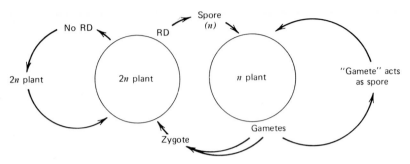

Figure 2-15. Detailed diagram of the life history of an organism with sporic meiosis. Gametes which do not fuse can behave as asexually reproducing cells. In other cases a diploid plant can reproduce by means of spores, and form another diploid plant. With these several pathways the life history might not be "completed" in the shortest possible manner.

termed the gametophyte. The diploid phase (the leafy plant we *recognize as a fern*) is the spore producing plant. In some algae it is possible to use the terms gametophyte and sporophyte for haploid and diploid plants. However, when there is gametic meiosis the plant forming gametes is diploid. And many haploid plants (e.g., a filamentous *Ulothrix*) can produce spores or gametes. Great caution must be exhibited when attempting to use such terms broadly in the algae.

In the 1960s and 1970s our understanding of life histories advanced markedly. Previously, investigators looked at an organism, described and named it, and then attempted to interpret its position in one of the categories describing life histories. At times this procedure was essential for correct taxonomic placement. However, we should consider how one determines whether a plant is haploid or diploid. How would you make a chromosome count? Would it be easy to obtain cells in division, in order that chromosomes could be counted? Would you be able to observe fusion of gametes and then follow the development of the zygote? If you observed spore release in some recently collected plants, how would you determine whether reduction division took place prior to their release? If you observed certain of these stages in culture, could you be certain that you are not dealing with artifacts? And if we observed a phenomenon once, is that sufficient?

I ask these questions merely to indicate the complexity of an analysis of a life history. In certain cases we have correctly speculated about some events in the development of an organism. We have also been wrong. Thus the student will sometimes see an organism with one name for diploid phase and *another* for the alternate phase. Previously we identified organisms on the basis of their morphology. We had no reason to consider some types as alternate stages of the same plant, but more recent and more complete research has shown this to be the case.

Using the materials presented above as an outline, in most instances you will be able to place each organism you study in one of the categories. This will make a comparative study of all groups of algae much easier.

REFERENCES

Bisalputra, T. and C. Lembi 1971. Bibliography on ultrastructure. In J. Rosowski and B. Parker (Eds.) *Selected Papers in Phycology.* Univ. Nebraska Press, Lincoln. pp. 298–304.

Bold, H. C. 1973. *Morphology of Plants.* Harper and Row, New York. 668 p.

Dodge, J. 1973. *The Fine Structure of Algal Cells.* Academic Press, London. 261 p.

Fritsch, F. 1935–45. *The Structure and Reproduction of the Algae.* The University Press, Cambridge. 2 vol.

Hilton, R. and R. McLean 1971. Bibliography on morphology and life histories. In J. Rosowski and B. Parker (Eds.) *Selected Papers in Phycology.* Univ. Nebraska Press, Lincoln. pp. 159–63.

Hoshaw, R. and J. West 1971. Morphology and life histories. In J. Rosowski and B. Parker (Eds.) *Selected Papers in Phycology.* Univ. Nebraska Press, Lincoln. pp. 153–8.

Pickett-Heaps, J. 1975. *Green Algae: Structure, Reproduction and Evolution in Selected Genera.* Sinauer Assoc., Inc., Sunderland, Mass. 606 p.

Pickett-Heaps, J. 1976. Cell division in eucaryotic algae. *Bioscience* 26: 445–50.

Scagel, R., G. Bandoni, G. Rouse, W. Schofield, J. Stein, and T. Taylor. 1966. *An Evolutionary Survey of the Plant Kingdom.* Wadsworth Publ. Co., Belmont, Calif. 658 p.

Stein, J. 1971. Cytology, genetics and evolution. In J. Rosowski and B. Parker (Eds.) *Selected Papers in Phycology.* Univ. Nebraska Press, Lincoln. pp. 458–62.

Swanson, C. and P. Webster 1977. *The Cell.* Prentice-Hall, Inc., Englewood Cliffs, N. J. 304 p.

Cyanophyceae — blue-green algae

THE DROUET CLASSIFICATION

Unicells and Colonies
Endospore Formers and Exospore-
    Producing Genera
Filaments

**CLASSIFICATION WITH BACTERIA:
THE PROKARYOTIC FORMS**

CYANOPHYTE MOVEMENT

BLUE-GREEN ECOLOGY

NITROGEN FIXATION

BLUE-GREEN SYMBIONTS

Origin of Cell Organelles

**EVOLUTION OF BLUE-GREEN PLANT
BODY TYPES**

BLUE-GREEN FOSSILS

REFERENCES

Blue-green algae are found in almost all environments including soil, fresh water, and the marine environments, including open ocean and coastal habitats. They possess characteristics that set them apart from other algae. They have many features in common with the bacteria and are at times classified in a separate kingdom established for these primitive or prokaryotic organisms. Cyanophyceae have sometimes been placed together in the Division Schizophyta and Kingdom Monera, but some researchers prefer the name Kingdom Prokaryota.

BLUE-GREEN ALGAL THALLUS TYPES

Plant body types in the Cyanophyceae are quite simple and limited. There are unicells, colonies, and branched and unbranched filaments (Fig. 3-1). The branched filaments, some of which can be multiseriate, are the most complex plant bodies found in this group. Most blue-green

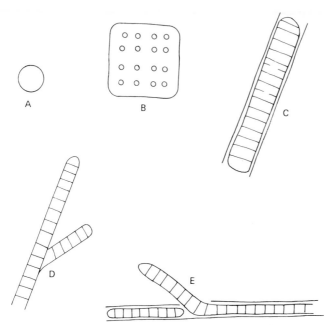

Figure 3-1.   Plant body types found in the blue-green algae. They may exist as unicells (A), colonies (B), unbranched filaments (C), true branched (D), and false branched (E) filaments.

algae are enclosed in an outer layer of wall material, the sheath. The aggregation of cells or filaments within sheaths provides additional plant body types. In some filamentous forms a portion of a sheath ruptures, allowing one or both ends of a broken chain of cells to protrude (Fig. 3-1), resulting in a false branching of filaments.

Cell divisions in filaments may be diffuse, occurring within many cells of the organism, such as an *Oscillatoria* (Fig. 3-16) or intercalary, restricted to a few cells in a basal region, as in *Rivularia* (Fig. 3-20). Sometimes even apical cells divide. Generalized division may occur in some colonies, such as in *Microcystis* (Fig. 3-6), but it can be restricted to a particular plane or planes, resulting in more regular colonies.

In addition, in the cyanophytes there are few clearly differentiated asexual reproductive bodies, and there has been no observed fusion of sex cells. In these microscopic algae, which have few reproductive cells, there are thus a limited number of distinct thallus types.

BLUE-GREEN CHARACTERISTICS
1. Pigmentation
Blue-green algae have chlorophyll *a*, beta carotene, and several xanthophylls. However, the possession of another group of pigments, the

phycobiliproteins—not found in most algae, accounts for their distinct coloration. The considerable amount of phycocyanin, one of the phycobiliproteins, is responsible for the blue-green color of many forms. These algae have allophycocyanin and many possess some phycoerythrin, but only occasionally does the phycoerythrin accumulate sufficiently to color the alga red. The phycobiliproteins are located in distinct structures, the phycobilisomes (or pigment granules), found on the outer surface of thylakoids. The thylakoids are single and not aggregated with one or more thylakoids as one will find in many of the other algal groups. There is no plastid, or any other membrane-bound organelle (for blue-greens are prokaryotes). Phycobilins are tetrapyrroles, which are associated with protein. The pigment, the phycobilin, is not released from the protein easily and thus it is the biliprotein, or pigment protein, which is studied. (Other photosynthetic pigments of algae, the chlorophylls and carotenoids, are released from the associated protein and can be studied separately.) The phycobiliproteins are found only in the blue-green algae, the red algae, a smaller group of flagellates called the Cryptophyceae, and in organisms in which these algae are endophytic. There have been tentative reports of the presence of a phycobilin in some simple green algae. The phycobilins are in phycobilisomes in both the blue-green and red algae, but in a number of investigations the phycobilisomes were not found in the cryptophytes. The pigment granule is approximately 35 nm in diameter. There are a number of structural variations in the pigments, especially in different divisions. The forms C-phycocyanin and C-phycoerythrin are in the blue-greens. Originally the C- designated where the pigment was found, in the Cyanophyceae. We now know that pigments originally given the C-designation are not limited to the blue-green algae. *Porphyra*, a red alga, has C-phycocyanin as well as other phycobilins. There are other similar examples.

In general the lower the light intensity the greater the number of phycobilin pigments produced. Under low light *Anacystis nidulans,* which does not form phycoerythrins, can form up to 25% of its dry weight as phycocyanin. Certain forms that are genetically able to produce both pigments exhibit chromatic adaptation; that is, the color of the alga can be controlled by growth in selected wavelengths of light.

2. Storage Products

The reserve food is a branched carbohydrate, glycogen, an alpha 1:4 linked glucan, with branches connected as a 1:6 linkage (Fig. 3-2). This food reserve, formerly called cyanophycean starch, appears as small granules under light microscopy. It becomes red-brown when stained

Figure 3-2. The chemical structure of the food reserve of the blue-green algae. It is an alpha, 1:4 linked glucan, with branching in the 1:6 position.

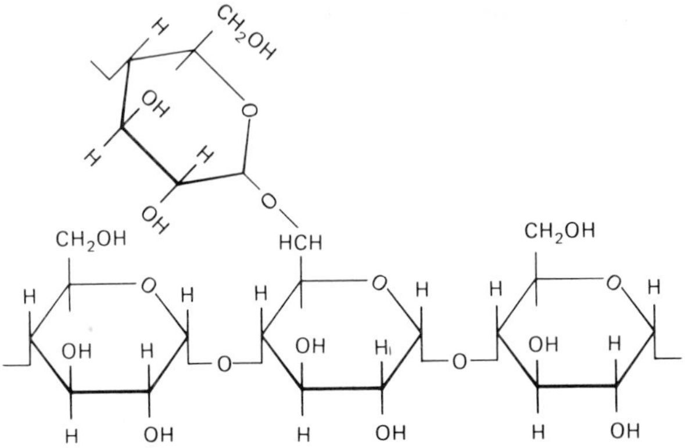

with iodine, as does the branched component of starch, amylopectin. With one strain of *Oscillatoria princeps,* when grown at 25–30°C, a red-brown color was noted with the starch test. But when cultured at 5–10°C, a blue-black color, typical of true starch, was noted. (With true starch we know that the blue-black color of the iodine test for starch comes from the amylose, or unbranched, component.) Thus where the *Oscillatoria* was grown at low temperature, there was little branching of the cyanophycean starch. It remains to be seen just how much variability in formation of food reserves will be seen in this group, especially in the various conditions under which blue-green algae can be found thriving in nature.

3. Motility

Blue-greens possess no flagellated cells, even when reproducing, a characteristic they share with the red algae. Nor do the blue-green algae

Figure 3-3. Diagram of a section of a cell from a blue-green algal filament, showing ultrastructural details. This is a somewhat simplified reconstruction, after Pankratz and Bowen. Courtesy of American Journal of Botany.
Note that the original cell has partially divided, and that the daughter cells have initiated wall formation (a), even before the first cross wall is complete.
(a) cross wall.
(b) thylakoids or photosynthetic membranes.
(c) cylindrical body seen in cross section (upper) and longitudinal section (lower).
(d) polyhedral body, also found in bacteria.
(e) structured granule, also called cyanophycin granule.
(f) DNA strands in the "centroplasm."
(g) glycogen granules.
(h) ribosomes.
(i) lipid globule.
(j) phycobilisomes.
(k) pore.

have flagella resembling the "simpler" bacterial flagellum. However, many blue-green filaments do exhibit both gliding and swaying motions.

4. Walls

The conspicuous wall and sheath of blue-green algae have received a great deal of attention, especially from taxonomists. In the older

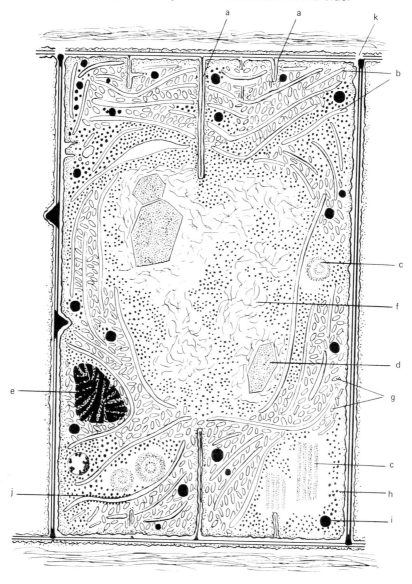

classification of these algae, organisms could be distinguished in part on the basis of the consistency of the sheath material, with watery, firm, and layered types known. These could be demonstrated (in order) in *Anabaena, Lyngbya,* and *Porphyrosiphon.* In addition, the sheath may be colored by various compounds. Typical colors are yellow, brown, red, violet. In some cases the color is sensitive to pH.

The cell wall around individual cells is now known to be quite complex and is composed of four layers, from the inside identified as L I, L II, L III, and L IV. The major components are similar to the wall constituents of gram negative bacteria and consist of some simple sugars, muramic and glutamic acids, diaminopimelic acid, galactosamine, glucosamine and alanine. The L II layer, the mucopolymer, has a globular ultrastructural appearance. Acid hydrolysis of walls and sheaths of several organisms has yielded glucose, galactose, arabinose, mannose, xylose, rhamnose, glucuronic and galacturonic acids (not all of these compounds are found in an individual species). The sugars probably come in part from the sheath material. Electron microscopy has shown 3 nm fibrils throughout an amorphous ground substance. The sheath may be secreted through pores in the wall (Fig. 3-3).

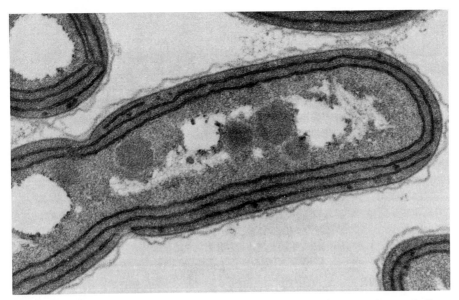

Figure 3-4. Electron micrograph of *Anacystis*-like cell showing cytology of a photosynthetic prokaryote. The outer photosynthetic and central nucleoplasmic regions are clearly visible. 37,000 X. Courtesy of N. Lang.

BLUE-GREEN CYTOLOGY

In an introductory treatment of all algal groups, it is customary to emphasize some cytological information when dealing with microscopic forms, but to rely on gross morphological observations when discussing the macroscopic forms. Because these organisms are the only prokaryotic algae they must be examined carefully in order to see any cytological detail (Figs. 3-3, 3-4, and 3-5).

Before the electron microscope was developed, the blue-greens presented the cytologist many difficulties. But certain details of cell structure described in earlier work were quite accurate. The peripheral cytoplasm, which contained the pigments, was called the chromoplasm and the inner portion, with the nuclear material, was named the

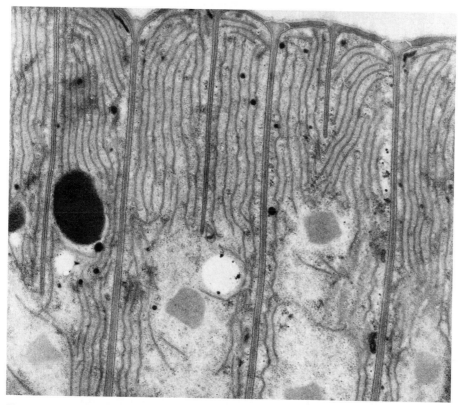

Figure 3-5. Electron micrograph of a filamentous blue-green alga. Thylakoids are not stacked; several cross walls are developing simultaneously. 27,000 X. Courtesy of N. Lang.

centroplasm. Even with the light microscope, it was apparent that there was no membrane around the centroplasm. More recent ultrastructural studies have confirmed that knowledge.

Thylakoids are scattered in the chromoplasm, with pigments in the thylakoids or in granules on them. The thylakoids are typically single, or unstacked, as in the red algae, but in some cases there can be a limited reticulum of membranes. In some organisms, such as *Anabaena* and an *Anacystis,* the thylakoids can be definitely layered, but they are not closely appressed in groups of two or more, as in many of the algal classes.

Glycogen granules, gas vacuoles, ribosomes, and small lipid globules are distributed in the chromoplasm (Fig. 3-3). Gas vacuoles consist of hollow cylindrical vesicles arranged in groups. The membrane around a gas vesicle is banded and composed only of protein. It is much thinner than the typical cell membrane. In some cells gas vacuoles can become abundant within a period of 48 hours and then provide some buoyancy for the cell. If there is maximum photosynthesis and increased turgor, the gas vacuoles collapse. They do not provide a light shield as was once proposed.

Blue-green algae lack a large central vacuole, which is typical of many plant cells. Thus the centroplasm is not vacuolate, but contains the "nuclear" material. DNA can be detected by the Feulgen technique or by modern procedures of isolation and characterization of the compound. In many early reports division figures were presented. However, it appears that in most cases these were due to improper fixation, abnormal material, or confusion with other cell details. Even though typical chromosomes and the mitotic figures common to eukaryotic forms are not seen, replication of DNA does occur and the centroplasm is divided. Genetic information is passed from cell to cell in division, and stability of structure results. Unlike most other organisms, including most algae, DNA is not associated with histones.

Polyhedral bodies (Fig. 3-3) that resemble carboxysomes of some bacteria are involved in photosynthetic carbon dioxide fixation. Heterocysts, which do not fix CO_2 and release oxygen, lack polyhedral bodies.

Cytoplasmic division and wall formation occur in a centripetal direction. They are often continuous processes. Before the cytoplasmic cleavage has completely severed a parent cell into two new progeny, the progeny can show the initial stages of new cytoplasmic division and wall formation (Fig. 3-3) along the outer margins of the cells.

There are no connections between cells in colonial forms, but cytoplasmic connections have been reported in many filaments. The polar nodule of heterocysts is a plug in the vicinity of the pit or hole in adjacent walls.

In filamentous forms the chain of cells is sometimes called a trichome. The trichome, plus the sheath, is the filament. Although it was once believed that some *Oscillatorias* had no outer wall material or sheath, examination in India ink revealed a delicate sheath through which the active *Oscillatoria* cells glided.

There has been no observed sexual reproduction, with fusion of cells or nuclei, but genetic recombination has been reported. For these experiments researchers work with induced mutations such as streptomycin resistance or an amino acid requirement.

REPRODUCTION IN THE BLUE-GREEN ALGAE

Inasmuch as reports of recombination in the blue-green algae are based on data from just a few organisms and on laboratory findings, we will discuss only asexual means of reproduction.

Cell Division

In the unicellular forms cellular division is the sole method of reproduction. However, in many unicells, such as *Chroococcus* (Fig. 3-6), the products of cell division tend to remain within the confines of the parent cell wall. Thus the products of division might not separate for a generation or two.

Cell division can result in a trichome of unlimited size, or an extremely large colony. Thus *fragmentation* or partial grazing of a long filament will result in two individuals.

Exospore and Endospore Formation

Reproduction by spores is restricted to a few forms of blue-green algae. In *Chamaesiphon* (Fig. 3-13) a cell or spore is pinched off at the apex, within the sheath. The spore is surrounded by a portion of the parent wall. Chains of these exospores might be formed in some types of *Chamaesiphon*. When released, the spore can develop into a new individual.

Many endospores are found at one time *within the parent cell* of an organism such as the epiphytic unicellular *Dermocarpa*. Around each spore is a complete new wall.

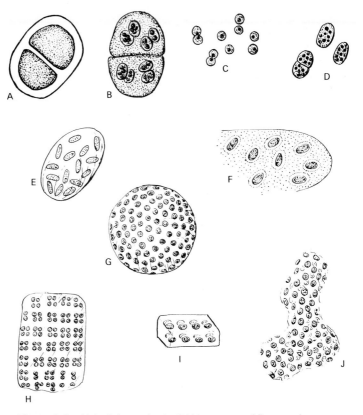

Figure 3-6. Unicellular and colonial blue-greens. Microscopic organisms. The names from the traditional system of classification are used.

A. *Chroococcus*, with a sheath around cells.

B. *Gloeocapsa*, with the sheath stippled; it is frequently pigmented yellow or brown in nature.

C. *Synechocystis*, with two cells dividing and several others recently divided.

D. *Synechococcus*, with one rodlike cell just after division.

E. *Aphanothece*, a colonial form with rodlike cells. Note the common sheath.

F. *Anacystis*, with an individual sheath around each cell.

G. *Coelosphaerium*, with cells found only at the periphery of a spherical colony. Cells can be evenly spaced.

H. *Merismopedia*, with the cells arranged in an orderly fashion in two planes.

I. *Eucapsis* similar to H, but the colony is in three planes. The organism is in the form of a cube.

J. *Microcystis*, a colonial type with random arrangement of cells. This is a common and troublesome form.

Hormogonium Formation

Hormogonium formation begins when one of the intermediate cells of a filament dies; the dead cell is called a separation disc. The wall of this dead cell, without the contact and maintenance provided by a living protoplast, soon weakens and tears. Thus two shorter trichomes, called hormogonia, are soon present within the original sheath. Movement of the hormogonium to a new location, or fragmentation of the original sheath, results in new individuals.

Hormospores or Hormocysts

Reproduction through hormospores or hormocysts is relatively uncommon. In these structures a chain of a few cells develops a thick wall and thickened sheath. The structures may resemble an inflated hormogonium or a multicellular akinete.

Akinetes

Akinetes have a thick envelope outside the original wall, are quite long, and usually have a diameter somewhat wider than other cells (Fig. 3-7). The outer envelope is formed during elongation and cyanophycin granules increase in number. The envelope of the *Cylindrospermum* akinete is spiny. Akinetes may be located in a specific position in a

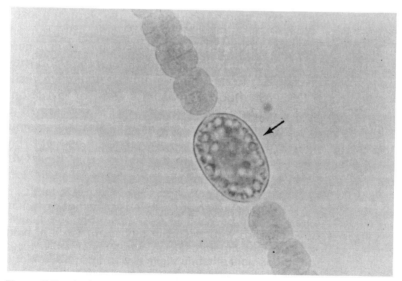

Figure 3-7. *Anabaena* akinete. Note that the heterocyst is larger than the vegetative cell. 648 X.

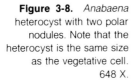

Figure 3-8. *Anabaena* heterocyst with two polar nodules. Note that the heterocyst is the same size as the vegetative cell. 648 X.

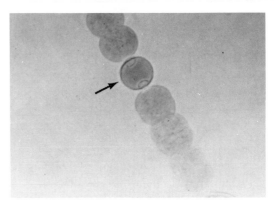

filament; for example, they may have a basal location in some filaments, adjacent to the heterocyst. The precise location has been used in classifying these organisms.

Ultrastructural details of cells change as they become akinetes. The thylakoids may become contorted or arranged in groups of two or three, as in other algal groups. There can be a reduction in glycogen granules and an increase in cyanophycin granules.

Wolk reports that akinetes of *Anabaena cylindrica* are specialized cells for surviving phosphorus deficiency. Akinetes are known to be so resistant that they may survive for 70 years in dry soil. However, akinetes are not essential for blue-green longevity, since some forms without akinetes can live 50 years or more. Akinetes may germinate through a pore or as a result of breakdown of the envelope. A filament eventually forms, but division can occur before or after release from the akinete envelope.

Heterocysts

Until heterocyst ultrastructure was adequately studied, there was some doubt as to whether these reproductive cells were really alive. Only their roles in false branching (see *Tolypothrix* and *Scytonema* in Fig. 3-20) and fragmentation were reported.

Heterocysts are typically smaller than akinetes and not much larger than vegetative cells (Fig. 3-8). They can be distinguished by their thick envelope, which forms nodules at one or both poles, and the homogeneous appearance of the protoplasm, instead of the normal granular appearance of vegetative cells.

The wall thickenings are due to envelope formation outside the original wall. Terminal heterocysts show continuity with just one cell, and thus one polar nodule develops; intercalary heterocysts bear two polar nodules. Heterocysts do not have glycogen and polyhedral bodies. In addition, the thylakoids develop into a latticework and may be concentrated at the poles. Recent developmental studies suggest a complex series of internal changes that accompany envelope deposition and initiation of nitrogen fixation.

Although small percentages of heterocysts can repeatedly be germinated and formation of new filaments has been achieved, perhaps they are not primarily reproductive structures. Evidence now indicates that under certain environmental conditions the heterocyst is the structure specialized to survive nitrogen deficiency. It has been theorized that the heterocyst is a primary location of nitrogen fixation in the blue-green algae, at least when organisms are aerobically grown. Early reports of nitrogen fixation in blue-greens lacking heterocysts can probably be discounted, but in 1969 there was the first confirmed observation of nitrogen fixation in a *Gloeocapsa*. Furthermore, more recent experimentation has shown that there can be nitrogen fixation in nonheterocystous filaments, but only when the level of oxygen has been reduced, that is, under microaerophilic conditions. Laboratory experiments have shown that nitrogenase is inactivated by oxygen. In the center of aggregations of *Oscillatoria* filaments in nature nitrogen fixation was detected, along with reduced pigmentation and photosynthesis.

In laboratory experiments the number of heterocysts increases with gaseous nitrogen as a nitrogen source and is lowest when ammonium is used. In one study, nitrogen fixation was confined to heterocysts when conditions were aerobic; under anaerobic conditions vegetative cells also fixed nitrogen.

The location of heterocysts has also been used in classification. Heterocysts may be located only at the apex, or at both apices of a filament. They may be solitary, in a series, or associated only with akinetes. However, experiments by both Wolk and Stewart have shown that a filament can lack heterocysts when growing well. Then, formation of heterocysts can be stimulated by placing cells in a medium lacking a combined nitrogen source, but with available atmospheric nitrogen. The role of chlorine in heterocyst formation was known for some time, for with *Anabaena* heterocysts developed when chlorides were added.

SOME REPRESENTATIVE GENERA

Among most of the algae it is quite simple to select a few representative genera and present some detail concerning their structure and reproduction. Unfortunately, two quite different systems have been proposed for naming the blue-green algae. The traditional and widely used system of classification relies on cell shape, arrangement of cells in aggregations, and wall morphology. The wall can vary in thickness and is sometimes in distinct layers. One to many trichomes may be in a common sheath. The walls can also vary in rigidity and color. I will introduce this system first and discuss the more recent proposal later.

In a traditional classification, emphasizing cell arrangement and wall-sheath structure, one might divide the cyanophytes into:

1. unicellular and colonial forms,
2. exospore-forming and endospore-producing genera, and
3. filamentous forms.

Because these prokaryotes lack cellular detail, as well as motility and sexual reproduction by way of fusion of egg and sperm, and the unicellular and colonial forms lack specialized asexual reproductive cells, descriptions of genera were quite simple. Distinctions between genera in the group containing unicells and colonies were based on size and shape of unicells, as well as the numbers of cells in colonies, and colony geometry. The amount and consistency of the sheath material were also noted. Using these criteria, numerous genera were described.

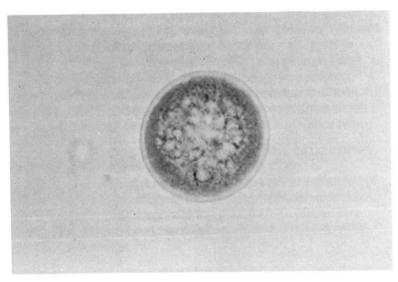

Figure 3-9. Vegetative cell of *Chroococcus*. 3240 X.

Unicellular and Colonial Forms

Chroococcus (Figs. 3-9 and 3-10), *Synechocystis,* and *Synechococcus* (Fig. 3-6) are unicellular forms, although there can be a tendency toward colony formation when products of a recent cell division fail to separate (Fig. 3-10). The *Chroococcus* sheath is always distinct, but it is not easy to distinguish the other two genera because of variability in sheath production. It is immediately clear that this system of classification has a major drawback.

Gloeocapsa, Aphanothece (Fig. 3-6), *Microcystis* (Fig. 3-11), and *Anacystis* (Fig. 3-6) are all forms with colonies of irregular shape. The distinct sheath of *Gloeocapsa* (Fig. 3-6) is often colored in material freshly collected. When any of these colonies fail to form, unicells *resembling* some of the above genera are seen—another problem in classification. As shown in Fig. 3-6 differences in cell size and shape are used in distinguishing some of these forms. Spherical cells of the common *Microcystis* are only a few microns in diameter.

Merismopedia (Fig. 3-12) colonies are flattened (one cell thick) with an orderly cell arrangement.

Eucapsis (Fig. 3-6) colonies are cuboidal, in regular three-dimensional patterns, but one laboratory culture forms just a few colonies among the abundant unicells.

Coelosphaerium (Fig. 3-6) produces regular spherical colonies with a monolayer of cells just at the colony periphery.

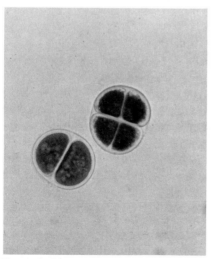

Figure 3-10.   Divided cells of *Chroococcus* in the two-celled and four-celled stages. 1620 X.

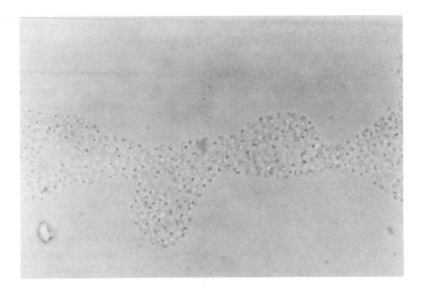

Figure 3-11. *Microcystis* colony. 1620 X.

Many more genera have been described following the pattern presented above.

Endospore-Producing and Exospore-Producing Genera

Dermocarpa, an epiphytic unicell, produces numerous endospores *within* the parent *cell wall.* The spores are released upon lysis of the parent wall.

Chamaesiphon (Fig. 3-13) provides an example of the epiphytic, exospore-forming cyanophytes. Apical spores are produced *within* the parent *cell sheath.* Several spores may be formed in a series (Fig. 3-14) and released easily by rupture of the thin parent sheath. Frequently, aggregations of epiphytes can be found. In some treatments of the blue-green algae the exospores of *Chamaesiphon* are termed endospores.

Filamentous Genera

The filamentous forms, the third group of cyanophytes, have a relatively complex morphology, and thus are easily subdivided. Several genera can be easily recognized. The presence or absence of heterocysts and akinetes, as well as their distribution in the filament, are useful characteristics. In addition, there are a few branched forms, some of which are false branched (Fig. 3-21 and see later). False branching is not found in other groups of algae. Certain forms taper from base to apex, and some of these clearly aggregate into clusters.

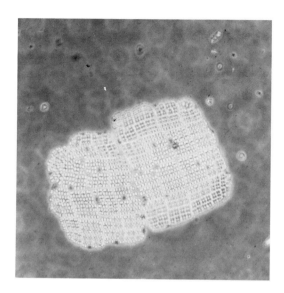

Figure 3-12. *Merismopedia* colony showing organization in two dimensions. 324 X.

Oscillatoria (Fig. 3-15) and *Lyngbya* (Fig. 3-16) are filamentous forms that lack heterocysts and akinetes. The sheath of *Oscillatoria* is frequently difficult to distinguish, whereas it can be seen clearly in *Lyngbya,* especially at the apices of filaments.

Spirulina (Fig. 3-17) and *Arthrospira* (Fig. 3-16) have helical shapes, with the partitioning into cells clearly visible in the latter. *Spirulina* is an elongate helix and can actively move in the plankton. Examination with the light microscope reveals no obvious cross walls. Is the organism then a unicell? Or might it be considered a filament that has lost the ability to form cross walls? Certain species of *"Spirulina"* have been shown to have cross walls when the filaments were examined with an electron

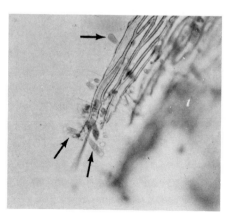

Figure 3-13. *Chamaesiphon,* epiphytic on a moss leaf. 810 X.

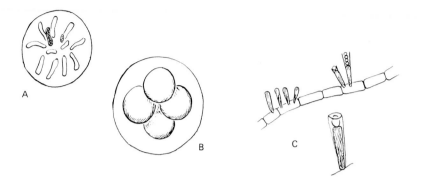

Figure 3-14. Microscopic. A and B. *Glaucocystis,* which is probably a green alga with elongate blue-green-like structures (cyanelles) within. In B the cyanelles are not shown within the four new cells within the parental cell. Some phycologists consider the parental cell to be *Oocystis*-like. C. *Chamaesiphon,* the exospore-forming epiphyte. One cell has formed several exospores at the apex. Individual cells can be precisely lined up on a host cell, such as cells of *Cladophora.*

microscope under high magnification. Perhaps *Spirulina* and *Arthrospira* should be recognized as a single genus.

Anabaenopsis and *Anabaena* (Fig. 3-18) are filamentous forms in which the cell arrangement resembles a chain of beads. In nature, filaments of *Anabaena* with both heterocysts and akinetes can be

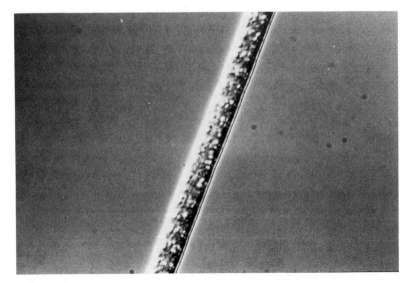

Figure 3-15. *Oscillatoria* filament. Nomarski interference microscopy. 648 X.

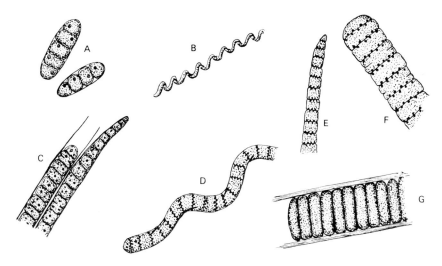

Figure 3-16.   Some filamentous blue-greens. Microscopic.
A. Short filaments called *Borzia*. However, these are similar to, or identical to, filaments of
   one of the other more elongate, unbranched filaments, especially when the latter would
   be young.
B. *Spirulina,* with light microscopy appears to be without cross walls.
C. *Phormidium,* sometimes very similar to *Lyngbya.*
D. *Arthrospira,* a helical form, with obvious cross walls.
E., F. *Oscillatoria,* shown without a sheath, which is difficult to detect, under light
   microscopy.
G. *Lyngbya,* with an obvious sheath.

observed, if one has a population of sufficient size. A heterocyst is found
at each end of the *Anabaenopsis* filament.

Individually, *Nostoc* (Fig. 3-18) filaments resemble those of *Anabaena,*
with heterocysts and akinetes easily found. However, the filaments are all
suspended in common mucilage so that macroscopic colonies can result.
These can be a few millimeters in diameter and appear as dark spherical
balls—*Nostoc* balls. These balls are sold as food in South America (Fig.
3-19). In the Arctic, where soils or rocks are subjected to alternating wet
and dry periods, sheets of *Nostoc* several centimeters in length are often
formed. These sheets are not restricted to cold environments.

Hapalosiphon and *Stigonema* (Fig. 3-20) are branched forms, in which
branching develops by division of a cell in a plane parallel to the long
axis of the main filament. The first cell of this "branch" then develops
polarity and typical cell division then occurs. This cell division is
perpendicular to the long axis of the developing branch. *Hapalosiphon* is
uniseriate whereas the filaments of *Stigonema* are several cells thick.
They are multiseriate.

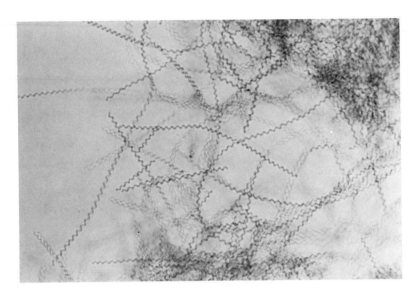

Figure 3-17. *Spirulina.* Aggregation of organisms, some of which are free of entanglement. 810 X.

Scytonema (Fig. 3-20) and *Tolypothrix* (Fig. 3-21) are filaments in which false branching can take place. The trichome does not move within the sheath for it is anchored by the attachment of the heterocyst to the sheath. If there are a few cell divisions in the portion of the trichome between attached heterocysts, elongation of the trichome will occur. But the sheath can rupture if elongation does not keep pace with new cell formation, and one (*Tolypothrix*) or two (*Scytonema*) false branches occur. Continued cell division can produce a "branch" of considerable length. Many years ago the occurrence of single and double false branching in the same population was reported; more recently both have been observed in the same organism in culture. False branching may be valuable in distinguishing some organisms from others, but one cannot rely on a particular organism always having *either* single *or* double false branching.

Gloeotrichia and *Rivularia* (Fig. 3-20) are examples of tapering cyanophyte filaments. *Gloeotrichia* filaments have a terminal heterocyst at the attached portion and the adjacent cell is an akinete. Only basal heterocysts are found in the filaments of the radiating aggregation of *Rivularia* filaments. These aggregations are easily formed by false branching near the heterocyst and then there is a complete separation of the "branch" from the parent filament. Certain tapering forms have narrow, elongate, terminal hairs.

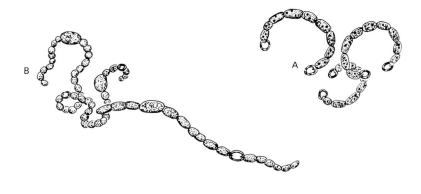

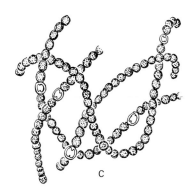

Figure 3-18. Some additional blue-green filaments, with heterocysts. Microscopic organisms.

A. *Anabaenopsis*, with a heterocyst at each end of the filament.
B. *Anabaena*, with several elongate akinetes and four heterocysts. The latter have polar nodules.
C. *Nostoc*. The numerous *Anabaena*-like filaments are all entangled and enclosed in a common mucilage. Macroscopic colonies result.

THE DROUET CLASSIFICATION

For a number of years many people doubted that all of the above criteria used in separating genera were reliable. As pointed out earlier, the production of sheath material was not always uniform and could vary seasonally, in different chemical environments, or when dealing with different communities. Many researchers felt that there had to be

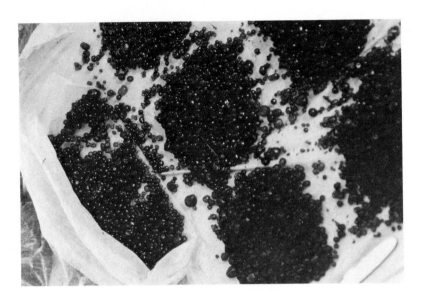

Figure 3-19. *Nostoc* colonies, averaging 5 mm in diameter, for sale in a South American country market. Courtesy of R. Wilce.

extensive morphological variability, especially after an early report of single and double false branching on the same filament! The traditional approach provided no way to deal with the morphological variability. It took the intensive investigations of Drouet to show that an organism such as *Microcoleus vaginatus* could indeed change its form. It is found in fresh water, on the land, and in brackish water, and in remarkable temperature extremes. Drouet's studies showed that organisms in one population could alter the structure of the sheath such that they could resemble many different species in several different genera. Thus he proposed that we rely on different characteristics in order to deal

Figure 3-20. Some additional blue-green filaments. Tapering and branching forms. Microscopic organisms.
A. *Hapalosiphon*, a uniseriate form with true branching.
B. A false branched form, *Tolypothrix*. Note that the single false branch occurs where there is a heterocyst. The false branch resulted when there was a break in the filament. Compare with true branching in A.
C. *Scytonema*, with double false branching. Note heterocysts.
D. *Stigonema*, a multiseriate form with true branching resulting from division in a second plane. Note that the other blue-greens covered on this page are all uniseriate.
E. *Gloeotrichia*, with a basal heterocyst and an adjacent akinete on the filament at the left. Note that the filaments taper from base to apex.
F. *Rivularia*, another tapering form. A radiating aggregation of filaments can be formed.

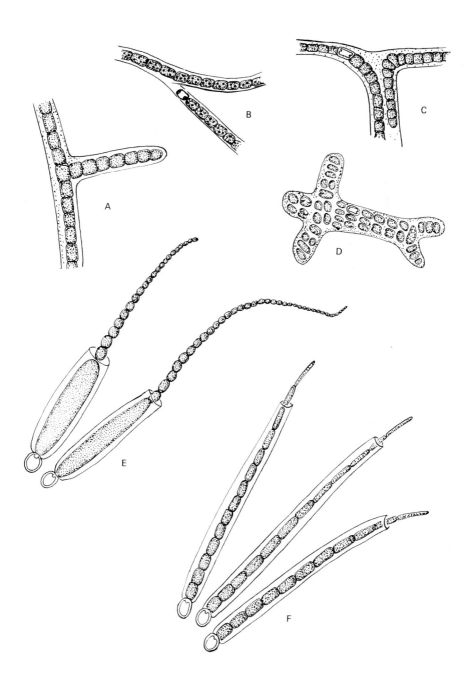

Figure 3-21.
Tolypothrix filament,
with a false branch.
Note heterocyst.
648 X.

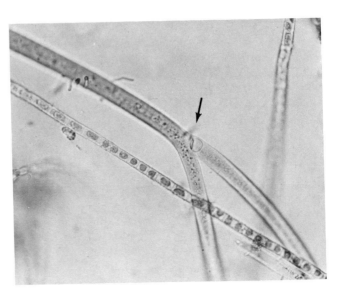

effectively with these genera and to produce a workable means of separating genera.

Unicells and Colonies

With the unicellular and colonial forms Drouet and co-worker, Daily, recognize only six genera. They distinguish them after examining the plane of cell division, cell structure, and arrangement of cells in colonies. The six genera are described below.

Coccochloris is an organism with cylindrical cells that divide perpendicular to the long axis. (Note that there is no use of sheath structure at the genus level and that an unfamiliar name is used for the genus. *Coccochloris* would have been the first valid description of one of these simple unicells.)

Anacystis has spherical cells in a matrix, either irregular or three-dimensional, and possibly in rows.

Johannesbaptistia has discoid cells in a linear series, as well as a matrix enclosing the cells.

Agmenellum cells are spherical, ovoid, or cylindrical, but always in flat or curved plates, with the rows of cells perpendicular to each other.

Microcrocis would be somewhat similar to *Agmenellum,* but the cells are not arranged in a regular manner. Cells are spherical, ovoid, or cylindrical.

Gomphosphaeria cells are spherical and are arranged radially in a single layer, at the periphery of a gelatinous matrix.

Endospore Formers and Exospore-Producing Genera
All exospore- and endospore-producing types, previously classified in at least eight families, are lumped into one genus, *Entophysalis!*

Filaments
Among the relatives of *Oscillatoria,* forms that are uniseriate, unbranched, and without specialized reproductive cells, Drouet recognizes only six genera, *Arthrospira, Spirulina, Schizothrix, Porphyrosiphon, Microcoleus,* and *Oscillatoria.* Note that he still includes *Spirulina* with these filaments, even though it apparently lacks cross walls. Among the relatives of *Nostoc* that have cylindrical filaments he recognizes only *Scytonema, Calothrix,* and *Raphidiopsis.*

   Naturally we can expect similar drastic changes as Drouet continues his work with the more advanced cyanophytes. As one might expect, there are many criticisms of his approach, especially because he places so much emphasis on herbarium specimens and mixed populations collected in nature.

   Much useful information on select groups of blue-greens has been gathered from recent culture studies. Culture studies allow the use of physiological characteristics, such as the ability of the organism to grow in the presence of a specific compound. However, some field investigators do not see the value of such investigations and claim that they are unnatural. In addition, culture studies require laboratory facilities and are time consuming. On the other hand, some laboratory investigators feel that organisms must be examined in culture for precise identification. There is still a long way to go, especially to accommodate both laboratory and field approaches.

CLASSIFICATION WITH BACTERIA: THE PROKARYOTIC FORMS
More than 100 years ago Cohn suggested that the blue-green algae might be somewhat removed from the other algae because of their obvious relationships with bacteria. Since then there have been periods when this relationship was strongly emphasized, and times when we accentuated the differences between the two groups. Since ultrastructural studies indicate that the cytology of both groups is quite similar, there has been a strong tendency to classify these organisms in a division with bacteria, the Schizophyta, or to put both prokaryotic groups in the Kingdom Monera or Kingdom Prokaryota.

   What criteria are evaluated in making such judgments? We know a great deal about the cytology of the cyanophytes, especially at the ultrastructural level. They have some internal membranes (Figs. 3-4 and 3-5), but lack membrane-bound structures typical of eukaryotic algae, higher plants, and animals (i.e., plastids, mitochondria, or a typical nucleus). This fact alone suggests an origin that was early and independent of other algae. The cyanophytes and bacteria are very similar when one notes the presence of the same type of ribosomes, the ability to survive at high temperatures, cell wall structure and chemistry, and details of cell division. In addition, one could cite the usual absence of steroids, the sensitivity to some antibiotics, the ability of some to fix atmospheric nitrogen, the gliding movement seen in certain filaments, as well as the bacterial-type recombination of genetic information. There are considerable similarities, but there are also some clear differences between the two groups.

Photosynthetic bacteria have different pigmentation and chemistry of photosynthesis when compared to blue-green algae. In the Cyanophyceae autotrophy appears to be the rule. Photoassimilation of carbon compounds, and in some cases dark utilization, can take place with *some* of these pigmented prokaryotic organisms. This is in marked contrast to the bacteria, in which there are few examples of autotrophy, in comparison to the broad utilization of organic compounds. One could also emphasize dissimilar plant body types, variations in cell and filament size, differences in asexual reproduction. Certain bacteria are actively motile, propelled by primitive flagella. Flagella of this type are not found in the cyanophytes, or in any eukaryotic form. (Eukaryotic organisms have a much more advanced structure of the flagellum.)

According to present thinking the similarities outweigh the differences and one can consider these organisms as members of the same division.

CYANOPHYTE MOVEMENT

Because the cyanophytes lack the exterior organelles, both flagella and cilia, which we commonly associate with motility, one has to look for other mechanisms in order to understand their movement. Several of the simple filamentous genera exhibit a pronounced gliding motion. *Spirulina* appears to flex and twist slightly as it moves rapidly through the water. Waving and oscillating motions have been described with these or other forms.

Organisms that move usually have a sheath. When the sheath is attached to the substrate, the trichome can move within it, but when the protruding trichome is grasped, the sheath will glide along the group of cells. In some cases it appears that the apical cell is involved, since motion will terminate if it is killed with a probe.

One explanation of blue-green motility relies on secretion of mucilage within the sheath. This mucilage is ejected through pores arranged spirally along the wall, producing motion. Older theories involved charged materials on the cell surface or rhythmic waves of expansion of the trichome surface *or* the mucilage, but supporting evidence was incomplete.

Recent studies with a large *Oscillatoria* emphasize the possible role of protein fibrils in the L III wall layer in motility. Rhythmic waves of contraction of the wall act against the sheath or the substrate and thus accomplish the motion. Right-hand rotation was always seen, and the orientation of fibrils was in a right-handed helix with a pitch of 60 degrees. This theory helps us to understand the observed sudden reversal of motion, and explains why sheath material does not accumulate to a greater degree if mucilage deposition were to provide the propulsion. Pores in the wall are thus probably used to secrete sheath material and are not involved in motility.

Many organisms can glide at a rate of 5 μm sec^{-1}; one *Oscillatoria* can move at 11 μm sec^{-1} (25 mm hr^{-1}). The filaments glide forward and then can reverse the direction, accomplishing the reversal within 1 sec. As the filaments move the trichome rotates within the sheath. More activity occurs in the light and with increased temperature.

BLUE-GREEN ECOLOGY

The blue-green algae are a highly successful, widely distributed group. They are found in and on soils and in both marine and freshwater habitats. They are sometimes important in the marine plankton. *Trichodesmium* is one of the best known of the marine plankters. *Richelia* can occur as an endophyte in the marine diatom *Rhizosolenia*, as an epiphyte on *Chaetoceros*, or it may be a free-living form. In the ocean, other genera can occur in the harsh environment of the upper spray zone and are components of the "black zone." Blue-greens, such as *Microcystis, Anabaena,* and *Aphanizomenon,* can be dominant organisms in the freshwater plankton especially when the CO_2 level is low. They are found in such extremes as the frigid environment of mountain tops and polar regions, as well as on desert soils.

Blue-green algae are frequently cited as examples of organisms that can survive high temperatures, and in some cases even thrive. It is not at all uncommon to see temperature reports of growth in the 50–60°C range, and reports of algae growing at up to 70°C.

From a single hot spring stream investigators have been able to isolate strains of *Synechococcus* (Fig. 3-6) with a variety of temperature optima. These thermal algae may have originated in aquatic environments when average temperatures were considerably above those presently found on the earth. Today they can be pioneer organisms in certain areas, and climax organisms in some eutrophic lakes, where they mat together in troublesome aggregations.

The blue-green algae *Nostoc, Chroococcus,* and *Gloeocapsa* are found in lichens. They can also be endophytes in several types of plants including liverworts, ferns, and cycads. The aquatic fern *Azolla* has an *Anabaena* associated with it.

The toxicity of some genera makes them notorious to those who raise livestock in areas such as the North Central States (Chapter 20).

NITROGEN FIXATION

Some organisms will grow both in nature and in the laboratory without combined nitrogen; they do not require nitrate, ammonium, or an amino acid. These organisms, which can utilize gaseous nitrogen, are called the nitrogen fixers. This phenomenon of fixation among the bacteria was known in classical Greece and Rome, where crops of legumes, with associated nitrogen-fixing bacteria, were rotated with nonleguminous forms. We now know that there are symbiotic relationships of bacteria and legumes such as clover and peas. Other nonleguminous, symbiotic relationships exist, and numerous free-living organisms, including yeasts and both aerobic and anaerobic bacteria, are capable of nitrogen fixation.

Not until 1938 was nitrogen fixation confirmed for blue-green algae, and we now know that many species are capable of nitrogen fixation. An experiment that revealed nitrogen fixation was performed by Beijerinck in 1901. He covered soil with two solutions: water buffered with potassium phosphate and water containing essential nutrients, including nitrate. After incubation for a few weeks, the soil covered with the potassium phosphate solution, developed filaments of *Nostoc, Anabaena,* and *Cylindrosperum,* genera that can fix nitrogen. The soil treated with all essential nutrients had an entirely different flora, including *Oscillatoria* (not known to be capable of nitrogen fixation), diatoms, and some green algae. Perhaps Beijerinck was selecting for nitrogen fixing genera with his nutrient-poor solution.

A number of early papers on nitrogen fixation in blue-greens lacking heterocysts, such as *Gloeocapsa* and *Phormidium,* were open to question because of experimental technique. However, the phenomenon was confirmed in some filamentous heterocyst-bearing genera such as *Anabaena, Nostoc, Tolypothrix, Cylindrospermum, Calothrix, Aulosira,* and *Mastigocladus.* Some free-living forms, and many algae growing with fungi in lichens, in the liverworts *Blasia* and *Anthoceros,* the fern *Azolla,* and cycad *Zamia* are capable of nitrogen fixation. Some of these associations are intracellular, as well as in special pockets within the organism.

As with many areas dealing with cyanophyte physiology, the scarcity of pure cultures retarded progress. It was possible that the nitrogen fixation was taking place in the bacterial associate. Pure cultures are obtained by a variety of procedures, but radiation or treatment with antibiotics are commonly employed. However, in killing the bacteria with UV or antibiotics, might one not be altering the genetic makeup of the now axenic blue-green?

There is now no doubt that nitrogen fixation can occur in cells other than heterocysts. Certain forms of *Gloeocapsa* can grow in the absence of combined nitrogen. Several filamentous genera that lack heterocysts have now been reported to fix nitrogen when grown with reduced levels of oxygen. However, with these organisms it is not always clear just how often they are found under such conditions in the wild, so that it is unclear how often nitrogen fixation actually occurs in nature.

Recently Carpenter and Price used a marine *Oscillatoria* to demonstrate that there was cell "differentiation" in filaments. When *Oscillatoria* filaments aggregate, median cells have reduced pigmentation and apparently do not incorporate carbon dioxide into cells. Thus, with little or no photosynthesis and oxygen evolution, the nitrogenase activity continues.

Nitrogen fixation is accelerated if preceded by starvation, and it is depressed with light intensities above saturation. Boron, calcium, cobalt, iron, and molybdenum are all necessary for nitrogen fixation. In some cases more of the element, such as molybdenum, is needed if atmospheric nitrogen is the source for the growth of the organism.

When a typical nitrogen fixer, which can form heterocysts, is grown in the laboratory, there may be few or no heterocysts in an actively growing culture in a typical complete medium. If the alga is then transferred to a medium that has no combined nitrogen, the number of heterocysts will increase markedly.

There is general agreement that the first product of nitrogen fixation is probably ammonia, which is quickly incorporated into amino acids. Ammonium in the medium will suppress heterocyst formation in *Anabaena cylindrica*. Labeling experiments usually show the labeled nitrogen in a variety of compounds. Nitrate usually does not inhibit heterocyst formation, but in experiments with one *Nostoc* species heterocysts did not form.

When viewed under the electron microscope, heterocysts present during active nitrogen fixation show an accumulation of cyanophycin granules, which contain arginine and aspartic acid. Other experiments clearly show that about 90% of the nitrogen fixed in heterocysts is rapidly transferred to vegetative cells. All nitrogen fixers can be grown autotrophically, even though slowly. An absolute growth requirement has not been demonstrated.

The rate at which nitrogen fixers liberate soluble nitrogenous substances into solution does not always parallel growth of the organism. The liberation can be accelerated by growth in cultures deficient in iron. Small amounts of certain amino acids, for example, aspartic and glutamic acids and alanine, as well as polypeptides, are released. This can be shown in free-living forms, as well as in forms such as *Nostoc* growing with a fungus as a lichen.

Fixation of nitrogen has now been reported in several oceanic filamentous forms, such as an *Oscillatoria* from the Sargasso Sea, *Trichodesmium,* and *Richelia* (in *Rhizosolenia*) in the North Pacific. With the increased nitrogen available to these photosynthetic organisms there is an increase in primary production.

Japanese experiments with mass culture demonstrated that a *Tolypothrix,* under limiting conditions, developed 4 grams of dry weight per square meter per 12 hour light period; 0.24 g of the total was fixed nitrogen. This amounts to 780 lb nitrogen per year. With *Anabaena cylindrica* there were 13 g and 0.9 g per square meter per 12 hr light period, or about a ton and a half of nitrogen per acre per year. *Tolypothrix* has been used in rice fields in attempts to increase yield. The organism is grown on a sand surface and a small inoculum put in the field. In one four-year experiment there was a 128% increase in yield over the nonfertilized control ponds. The blue-green grows better under field conditions when rice is in the paddy, probably because the rice crop releases carbon dioxide. In some experiments with rice crops daphnids caused a sudden crash of the algal population. In California, floating blue-green mats were found to damage emerging rice seedlings.

Experiments with nitrogen fixing blue-green algae in nitrogen-poor tobacco fields were not encouraging.

BLUE-GREEN SYMBIONTS

There are now numerous examples of organisms that have a blue-green alga living within them, as well as other examples of cellular organelles that probably had a blue-green ancestry. Identification of the alga is often difficult. Some researchers even suggest that these are red algae! But first I will cover the symbiotic relationships. *Glaucocystis nostochinearum* (Fig. 3-14) can serve as an example. The symbiosis has been called syncyanosis, in which the infected one is the host, the endophyte is termed the cyanelle, and the two-membered complex is called the cyanome.

   Glaucocystis resembles several unicellular green algae, except that the "plastids" are blue-green. The plastids are rod-shaped, randomly distributed, or have a radial arrangement. They lack a wall (as would be found in a blue-green alga) and divide by centripetal growth of a median constriction. The pigments are associated with lamellae. There are no cyanophycin granules, glycogen granules, cylindrical bodies, or other organelles typical of blue-green algae. But superficially the cyanelle resembles *Aphanothece*.

   The host in the *Glaucocystis* syncyanosis is thought to be a green alga that has lost its plastids, and, instead, the cyanelle serves as the photosynthetic component. The host is often called an *Oocystis*-like organism. The presence of basal bodies in the host has confused the picture because they are found associated with flagella. However, their presence does not necessarily indicate motility by flagella; *Glaucocystis* is not flagellated.

   These organisms are clearly the association of two previously independent organisms. One can see an earlier stage in the development of a syncyanosis in the association of *Richelia* with *Rhizosolenia*. Both genera are clearly recognized when they are together, and the *Richelia* can also live in the plankton, not associated with the diatom. In the *Glaucocystis* syncyanosis the association has become obligatory, with some modification of the blue-green. The first step in the formation of the association could be the loss of the blue-green wall.

Origin of Cell Organelles

Work with endophytic blue-green algae has suggested that plastids in other algae, as well as higher plants, originated when blue-green algae became trapped in eukaryotic cells. Similar evidence exists for the origin of mitochondria from endosymbiotic bacteria. Both mitochondria and plastids are more similar to prokaryotes than are any of the other cell organelles; for example, both are bound by a double membrane, and both can grow and divide somewhat autonomously, with only partial control by the nuclear DNA.

   When photosynthesis evolved some 3 billion years ago, there could have been forms developed with carotenoids, the phycobilins, and, among the chlorophylls, just chlorophyll *a*. These free-living algae would be similar to the blue-greens as we know them today. An organism of this type, invading *another type* of organism lacking plastids, would have been the first syncyanosis (Fig. 3-22). Naturally we would have to account for the origin and survival of the second type of organism, the one that acted as host cell. If the syncyanosis developed successfully, the cyanelles could lose some cell structures and begin to look like plastids as we know them today. This could account for the origin of the red algae or rhodophytes. An additional endosymbiosis, with an animal-like cell as host, a flagellated cell, may have provided the plastid of the Cryptophyceae. In this way one could account for the phycobiliproteins in Cryptophyceae, in the red algae, and in the free-living blue-green algae. In other phylogenetic schemes one has to propose the separate origin of phycobilins.

Certain of the primitive prokaryotic organisms could have evolved chlorophyll *b,* while a separate line could have evolved chlorophyll *c.* Phycobiliproteins would be absent in these two new lines (Fig. 3-22). The group with chlorophyll *b* would have invaded an organism lacking plastids and would give rise to the euglenoids and the green algae. Higher plants would originate from this line. Lewin's recent report of a prokaryote that apparently has chlorophyll *b* and lacks phycobilins would add considerable weight to this proposal. In a similar fashion, a whole line of organisms with chlorophyll *c* evolved once prokaryotes bearing that pigment became endosymbionts. Most *free-living* prokaryotic forms with chlorophylls other than chlorophyll *a* would not have survived until the present. At least we have not yet observed a prokaryote with chlorophylls *a* and *c.*

Note that we are concerned here only with the plastid. Eukaryotic cells that served as hosts for all the "invasions" discussed above must also have evolved. Some of these cells would be flagellated, for example, for the euglenoids and the Cryptophyceae, and perhaps most other groups, while others would not, for example, the host cell for the red algal line.

Clearly a great deal of speculation is involved in this proposal. Although a number of questions remain unanswered, the scheme does give us a new way to look at the origin of the various algal groups. Some investigators who subscribe to the endosymbiont theory feel that perhaps *some* dinoflagellates and euglenoids arose when a *eukaryotic* organism invaded an animal cell. This theory is based on the occurrence of some binucleate dinoflagellates in present time and the number of membranes that surround chloroplasts in these organisms. The details are beyond the scope of this volume. However, endosymbiosis is also discussed in Chapter 16.

EVOLUTION OF BLUE-GREEN PLANT BODY TYPES

The large number and variety of simple plant body types in the cyanophytes encourages speculation regarding evolution of different morphological lines. The information presented here in detail for the blue-green algae can be applied later to organisms in other groups.

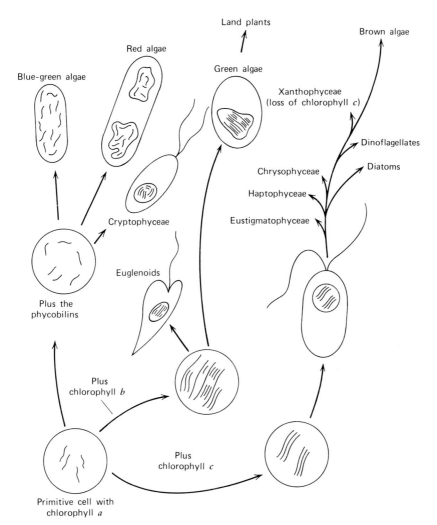

Figure 3-22. The evolution of various groups of algae based on the theory of endosymbiosis. The primitive cell, with chlorophyll *a,* is diagrammed at the bottom left. Pigments are associated with thylakoids, which are designated by random and/or grouped lines. When the primitive photosynthetic cell types invaded other nonphotosynthetic cells, a plastid was formed. The plastid is designated by a circle or irregular shaped body within the host cell. Based on presence of various chlorophylls and the phycobiliproteins in some primitive cell types, and their becoming endosymbionts, several lines of evolution can be presented.

The unicell is considered the primitive cell type, and, in this case, it lacks motility. (In other algae the flagellated unicell is considered primitive.) The unicell can be either spherical or rod-shaped and it may possess a sheath (Fig. 3-23).

Blue-green algae developed along four main lines. First, some of the organisms divided and remained within the old parent cell sheath, forming the first colony. If there was some gelatinization of the sheath and accompanying growth or stretching, large irregular colonies similar to *Microcystis* (Fig. 3-11) would develop. With a little more precision in the arrangement of the cells, a more organized colony such as *Coelosphaerium* (Fig. 3-6), with cells only at the periphery, would be developed.

A second line would include colonies with rectangular or cuboidal shapes (Fig. 3-23). If cell divisions were restricted to two planes, a flattened colony such as *Merismopedia* would develop. Some breaking of the colonies, or an occasional lack of division in a cell or two, can give somewhat irregular colonies. When there are also divisions in a third plane, cuboidal colonies result.

The third and fourth lines, which include most of the well-known blue-green algae, develop initially in the same way. When cell division takes place, the resulting cells remain together, sharing adjacent walls. Continued division in this one plane produces an unbranched filament (Fig. 3-23). An occasional division in a second plane yields branched filaments (Fig. 3-23). Some of these true filaments do not remain uniseriate, that is, as filaments with one row of cells in a series.

With the blue-green algae there is another evolutionary line (Fig. 3-23), and this includes false branching forms. Divisions in a second plane do not occur. Instead as filaments continue to elongate, the length of the chain of cells increases more rapidly than does the length of the filament as a whole. When some of the cells stick to the sheath, as do heterocysts, the cramped intermediate cells burst out at the side. At times only one filament protrudes at the rupture, and this phenomenon is termed single false branching. Double false branching also occurs (Fig. 3-23).

Because of changes in the plane of division, the arrangement of cells within the sheath and the variation in amount and consistency of the sheath, a variety of morphological types have arisen in the blue-green algae. Similar lines have developed in other groups of algae.

BLUE-GREEN FOSSILS

Although almost all cyanophytes are microscopic plants, there is considerable information on them from the fossil record. They are considered an archaic group, with advancement to the present plant body types occurring early in their evolution. Blue-green algae or organisms morphologically quite similar to them, have been known for at least 2 billion years. There are reports from 3 billion years ago, when bacteria also would have been present on the earth.

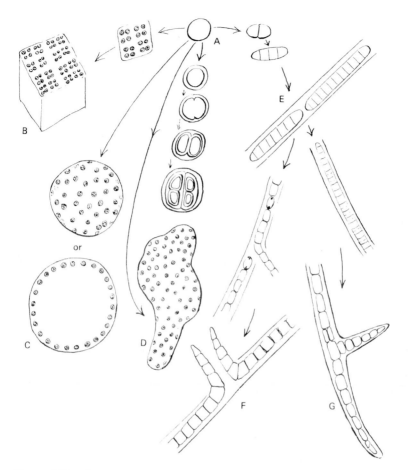

Figure 3-23. The evolution of plant body types in the blue-green algae.
A. A unicellular form, which could have evolved into:
B. a flat plate and then cuboidal type because of divisions in two or three planes.
C. a spherical form with cells only at the periphery, shown in both surface view and optical section (lower).
D. an irregular colonial form such as *Microcystis*, with random divisions.
E. filamentous forms, by the cells remaining joined after division. In this case, there could have been the development toward false branching, as in F, or toward true branching as in G. True branching would involve a division in a second plane.

In the Rhynie Chert, 380 million years of age, there are excellent cell preservations with cells, sheath, and heterocysts. In the Gunflint formation in Canada, approximately 2 billion years old, there are preservations with unicellular blue-greens. In the Bulawayan formation, approximately 3 billion years of age, we have recorded macroscopic calcareous deposits, called stromatolite-like structures. It is felt that they were deposited by ancient blue-green algae. In the Fig Tree series of South Africa, about 3.1 billion years old, there are preservations of cells that are *possibly* the most ancient cyanophytes.

REFERENCES

Barghoorn, E. and S. Typer. 1965. Microorganisms from the Gunflint chert. *Science* 147: 563–77.

Barghoorn, E. and J. Schopf. 1966. Microorganisms three billion years old from the Precambrian of South Africa. *Science* 152: 758–63.

Bourelly, P. 1970. *Les Algues d'eau Douce*. III. Les Algues bleues et rouges. N. Boubee & Cie, Paris. 512 p.

Carpenter, E. and C. Price. 1976. Marine *Oscillatoria* (*Trichodesmium*): Explanation for aerobic nitrogen fixation without heterocysts. *Science* 191: 1278–80.

Carr, N. and B. Whitton. 1973. *The Biology of the Blue-Green Algae*. Univ. Calif. Press, Berkeley, California. 676 p.

Desikachary, T. 1959. *Cyanophyta*. Indian Council of Agricultural Research, New Delhi. 686 p.

Drouet, F. and W. Daily. 1956. Revision of the coccoid Myxophyceae. *Butler Univ. Bot. Studies* 10: 1–218.

Drouet, F. 1968. Revision of the classification of the Oscillatoriaceae. *Monogr. Acad. Nat. Sci.*, Philadelphia 16: 1–341.

Drouet, F. 1973. *Revision of the Nostocaceae with Cylindrical Trichomes* (formerly Scytonemataceae and Rivulariaceae). Hafner Press, New York. 292 p.

Fogg, G., W. Stewart, P. Fay, and A. Walsby. 1973. *The Blue-Green Algae*. Academic Press, London and New York. 469 p.

Fritsch, F. 1945. *The Structure and Reproduction of the Algae*. Volume II. The University Press, Cambridge. 939 p.

Gantt, E. 1975. Phycobilisomes: light-harvesting pigment complexes. *Bioscience* 25: 781–8.

Khoja, T. and B. Whitton. 1975. Heterotrophic growth of filamentous blue-green algae. *Br. Phycol. J.* 10: 139–48.

Lang, N. 1969. The fine structure of blue-green algae. *Ann. Rev. Microbiol.* 22: 15–46.

Lang, N. and J. Waaland. 1971. Bibliography on Cyanophyta. In J. Rosowski and B. Parker (Eds.) *Selected Papers in Phycology*. Univ. Nebraska Press, Lincoln. pp. 754–9.

Lewin, R. 1975. A marine *Synechocystis* (Cyanophyta, Chlorococcales) epizoic on ascidians. *Phycologia* 14: 153–60.

Schopf, J. and D. Oehler. 1976. How old are the eukaryotes? *Science* 193: 47–9.

Singh, R. 1961. *Role of Blue-Green Algae in Nitrogen Economy of Indian Agriculture*. Indian Council of Agricultural Res., New Delhi. 175 p.

Stewart, W. 1974. *Algal Physiology and Biochemistry*. Univ. Calif. Press, Berkeley. 989 p.

Stewart, W. 1977. A botanical ramble among the blue-green algae. *Br. Phycol. J.* 12: 89–115.

Chlorophyceae — green algae

EVOLUTION OF THE GREEN ALGAE   Chlorophytes and Charophytes

REFERENCES

Green algae are common organisms in fresh waters, in oceans (both coastal benthos and planktonic forms), and in and on soils. They can tolerate temperature extremes, but perhaps not temperatures as high as some blue-greens withstand; snow algae are known. Green algae are frequently the algal partners of lichens.

We perhaps know more about the chlorophytes than we do about other types of algae. This is due in part to their wide distribution, their resemblance to higher plants, and the ease with which a number of forms can be brought into culture. Thus you will see a great deal of literature dealing with organisms in the Chlorophyceae and many positive statements about structure, mechanisms, and to some extent, classification.

Since a great deal is known about many of the individuals in this group, one can approach classification with a little more confidence. But as you might expect, the final chapter has not yet been written. Relying on morphological features, 20 years ago one would have thought of two subgroups of green algae. More recently, Round has proposed five categories, the system followed in this book. It presents organisms in categories which can be easily grasped by the student.

However, recent ultrastructural evidence has clarified details of cell division, and it is clear that many phycologists will want to subdivide the green algae into the chlorophytes and charophytes. We discuss details of this exciting development when we talk about filamentous genera.

GREEN ALGAL THALLUS TYPES

There are green algal unicells, colonies, and filaments, and in this case, as contrasted with the cyanophytes, there are motile unicells and motile colonies. Membranous forms, multinucleate types, and advanced green algae with separation of the thallus into nodes and internodes also are noted.

Unicells and colonies, especially those that are nonmotile, are found in various sizes and shapes. Some examples can be seen in Fig. 4-1. Many colonies, including both motile and nonmotile types, have an upper limit

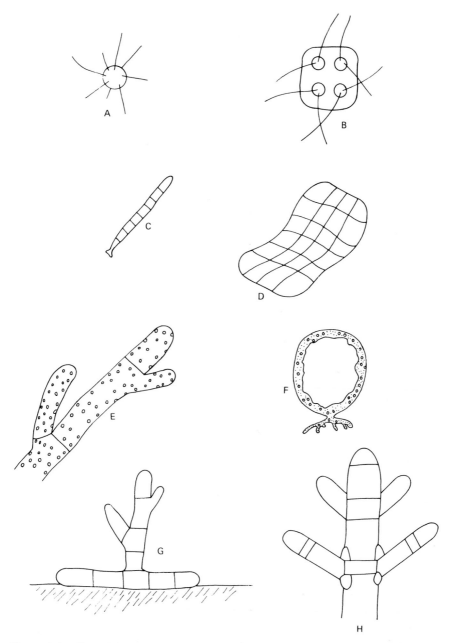

Figure 4-1. Plant body types seen in the green algae. These include unicells (A), colonies (B), filaments (C), parenchymatous forms (D), organisms with multinucleate cells (E), or others with many nuclei and no cell walls around a nucleus or groups of nuclei (F). Also figured are inital stages of the development of a complex plant body (G, H). Some forms are heterotrichous, (G), and others have branches at nodes (H).

on size. These organisms lack vegetative cell division; thus, when they are initially formed, they have all cells that will appear in the adult form. During maturation cell size and thus colony size increase.

Unicells and colonies are usually microscopic, but they can range from just a few microns in diameter (a *Chlorella* cell) to at least a meter (*Hydrodictyon* nets) in length.

Numerous types of filaments can be found. Certain types are unbranched, including *Klebsormidium* (*Hormidium*), *Spirogyra,* and *Oedogonium,* or branched, such as *Stigeoclonium.* Many filaments are composed of cells with only one nucleus, as in all the examples cited. However, other filaments are composed of multinucleate cells; for example, the unbranched *Chaetomorpha* and the branched *Cladophora.* In certain cases, these filaments are attached by a simple modification of the basal cell, as in *Oedogonium.* However, a number of organisms exhibit the heterotrichous pattern of growth (Fig. 4-1). There is a basal prostrate system from which erect filaments develop, for example, in *Stigeoclonium.* This growth pattern is also common in the red and brown algae.

In a few green algae division in two planes produces parenchymatous sheets one cell thick. An example of such growth would be the marine *Monostroma.* Its close relative, *Ulva,* would have divisions in a third plane, and thus is a membrane two cells thick.

The green algae which are considered to be the most advanced morphologically are the relatives of *Chara.* These organisms have apical growth and a differentiation into nodal and internodal cells. The internodal cells elongate considerably. Branches, which form in several directions, are cut off at the nodes and soon develop apical growth. At the nodes, certain cells can be produced that grow around the internodal cell, producing a cortex that is one cell layer thick.

GREEN ALGAL CHARACTERISTICS

1. Pigments

The green algae contain the same pigments as found in higher plants, chlorophylls *a* and *b,* alpha and beta carotene, and many different xanthophylls. The xanthophyll composition is not the same for all species. Some xanthophylls are abundant when the organism is young and healthy, and others appear with aging of the population. Lutein, violaxanthin, and neoxanthin are typically dominant prior to aging. The pigments are always found in a membrane-bound plastid, here accurately called the chloroplast, with the thylakoids in groups of 2 to 6. Even further stacking is obvious when portions of thylakoids arrange themselves in layers called grana. This arrangement is considered typical in higher plants. In some forms the presence or absence of grana depends on the availability of nutrients, or age of cells. The shape of the plastid is important in the classification of many forms at the genus level. A good example would be the genus *Spirogyra,* which has a plastid in the form of a helix.

2. Storage Products
The reserve food is true starch, composed of the unbranched chain of glucose residues, the amylose portion, and the branched or amylopectin portion. It is alpha 1:4 linked with 1:6 linkage for branching. See Fig. 3-2 for an alpha 1:4 linked compound in the cyanophytes. Frequently the starch is formed in granules associated with a proteinaceous body in the plastid called the pyrenoid, but some forms lack a pyrenoid. Reserve food may be stored as oil.

3. Motility
Motility of organisms, or reproductive cells, is quite common in most of the members of this group, but motion by flagella or cilia is absent in *Spirogyra* and its relatives. Typically both whiplash flagella are of equal length, but forms with slightly unequal flagella are not unknown. Furthermore, types with one or many flagella, four as in *Carteria,* or a tuft of flagella as in *Oedogonium,* are known. Organisms such as *Volvox* spend most of their vegetative state actively swimming. A group of flagellates with scales on flagella, as well as body scales, have been removed from the green algae and placed in the class Prasinophyceae. Several investigators dealing with algal ultrastructure feel that the class Prasinophyceae should be abandoned and all organisms classified with other green algae.

4. Walls
Usually there is a cell wall associated with a green alga, but a few forms lack a wall, and some reproductive bodies are also naked. One component of the wall is cellulose, but walled types such as some of the colonial relatives of *Volvox* lack cellulose. In the latter case the wall material is called pectin, material which is said to be found outside the cellulose wall in many other forms. Chitin, sporopollenin, calcium carbonate, and even silica are reported in one or more genera. Many green algae have types of wall ornamentation found useful in

classification. These include scales, a rough texture, thick walls in distinct layers, warts, ridges, and spines. At the ultrastructural level some walls are rather simple and are composed of fibrils. Other algae have very ornate walls, even at the ultrastructural level.

REPRODUCTION

Because of the diversity of reproductive events in the green algae reproduction can be discussed thoroughly only when considering individual organisms. Many highly successful forms reproduce only asexually, relying on cell division or release of motile and/or nonmotile spores. In the most primitive green algae, cell division, and thus reproduction, takes place by division *within* a parent cell wall. None of the parent cell wall is incorporated into the walls of the progeny. The products of division are eventually released from the parent wall. Thus these organisms lack vegetative cell division.

Sexual reproduction is not uncommon, with various genera possessing isogamy, heterogamy, and oogamy. Meiosis can take place in the germination of the zygote or in the formation of spores or gametes. Life histories are sometimes quite complex, with several genera exhibiting alternation of dissimilar generations.

SOME REPRESENTATIVE GENERA: FIVE GROUPS

In order to cover a large number of common green algae in a short time they will be treated in five distinct groups. The *first* will include forms that have motile reproductive cells and cells that are usually uninucleate. In the *second* group there are just a few organisms that have a unique form of cell division and multiflagellated reproductive cells. The *third* group includes the multinucleate forms, both those algae that lack cross walls, except at times of reproduction, and others which have multinucleate cells. In group *four* are unicells and filaments that lack motility by means of flagella as well as motile reproductive cells. In group *five* we shall cover *Chara* and relatives which have nodes and internodes, with branches originating in a whorl at the nodes. These organisms have very complex reproductive structures.

Group One
Motile Organisms Lacking Vegetative Cell Division

The organisms that spend their vegetative state actively swimming are put together. These include unicells, two-dimensional and three-dimensional colonies, with the most advanced form, the common *Volvox*. In addition, all organisms that would be grouped here would have cells with a single nucleus, a single plastid, and flagella at the anterior pole. Typically, there are two flagella, but there is some variety. These organisms lack the ability to divide vegetatively, but must always divide within the parent cell wall. Then the products of division are released as unicells or colonies. Thus a four-celled colony cannot become an eight-celled colony by division of each of the individual cells, nor can one of the cells of the four-celled colony divide so that a five-celled colony results.

Chlamydomonas. A Typical Unicell.   One member of this group is the common *Chlamydomonas* (Fig. 4-2), one of the green algal weeds. Found in soil and in both marine and freshwater habitats, it can tolerate environmental changes such as increased nutrient load in a stream. It grows well in some sewage oxidation ponds. An organism that has such a wide distribution can usually tolerate various laboratory conditions and thus these organisms are quite easily isolated into culture. For this reason a great deal of laboratory experimentation was and is still concerned with these motile unicells.

 Chlamydomonas cells are spherical, elongate, or egg-shaped, with two flagella of equal length at the anterior pole (Fig. 4-2). They are typical of the cells of many of the green algal flagellates and thus some time will be spent with them. Each cell has a single massive plastid, which can be cuplike or parietal. Unless the organism is actively growing and healthy, the shape of the plastid is obscured. Thus it might not always be apparent that a large pyrenoid is present in the plastid. When one cannot determine plastid shape in many of the green algae, it is impossible to identify the organism to genus.

 Within the depression of the cup plastid is a single nucleus. Just below the flagellar insertion one can see two contractile vacuoles. If they are examined for a period of 10 to 30 seconds, it can be determined that they pulse in an alternating rhythm. There is a firm cellulose wall, and when found in environments where the calcium level fluctuates, abundant mucilage can be deposited around the cellulose. In such a case motility would be impossible.

 Asexual reproduction. A *Chlamydomonas* cell actively swims, photosynthesizes, and grows. When it has reached the time for division, the nucleus divides and the protoplasm is cleaved, with new walls deposited around the individual protoplasts (Fig. 4-2). Depending on the

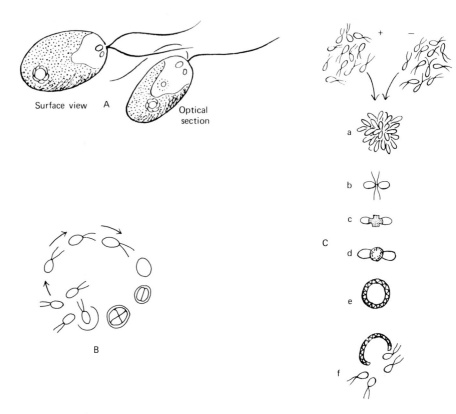

Surface view A Optical section

B

C

a b c d e f

Figure 4-2. A green unicellular flagellate. Microscopic.

A. *Chlamydomonas* is shown in both surface view and in optical section. The two flagella, with two contractile vacuoles below them, are seen at the apex. A pyrenoid with starch plates can be seen in both cells, and the single nucleus in the optical view.

B. The asexual life history of a *Chlamydomonas*, or for any unicellular flagellate in this group. (Modified, if there is no wall on the unicell). The cell loses its flagella and becomes larger and spherical. Then there is division and release of new flagellates from within the parent cell wall.

C. The sexual cycle of a heterothallic *Chlamydomonas*. Plus and minus mating types are mixed in large numbers. Soon one sees: a. A clump of equal numbers of plus and minus, with flagella toward the center of the clump. b. A pair, observed when one plus and one minus swim from the clump. c. Fusion of the cells at their apices. d. Completed fusion of the cytoplasm with the gamete walls apparent at the sides. e. An enlarged, spiny zygote. f. Four flagellated cells, two plus and two minus, formed as a result of meiosis of the zygote.

size of the cell dividing, and environmental conditions, from 2 to 16 or more new protoplasts can be formed within the parent cell wall. Eventually each cell will have a wall and two flagella. The parent cell wall is ruptured or hydrolyzed, and the flagellated individuals swim away. With this type of asexual reproduction (Fig. 4-2), one can soon achieve a large population of *chlamydomonads*. Such cells can be called either zoospores or vegetative cells.

Sexual reproduction. The sexual cycle of *Chlamydomonas* has also been studied extensively. Usually gametes closely resemble zoospores morphologically, and if they are not successful in finding a gamete of the opposite mating type they can eventually behave as zoospores. Examination of pure cultures of several strains of *Chlamydomonas* has shown that mating can be invoked in some cultures that were started from a single cell. Such strains would be homothallic. In other cases populations from two different isolations must be mixed. Thus there are two mating types, and these can only be isolated by randomly picking, culturing, and mating several different strains. The latter are heterothallic (Fig. 4-2).

When the mating types are identical in size and appearance, they are simply designated plus and minus, and the grade of sexuality is termed isogamy, inasmuch as both gametes swim and have an identical morphology. Populations can be made actively gametic in culture, in many cases, by lowering the nitrogen level in the medium, but some strains mate at both high and low nitrogen concentrations in laboratory experiments. However, one species that responds to nitrogen depletion in the laboratory will mate in river water collected just below a sewage treatment plant, or in river water to which additional nitrogen has been added.

When actively swimming plus and minus populations are mixed, there is an aggregation of approximately equal numbers of plus and minus cells in clumps (Fig. 4-2) with the anterior end of each cell toward the center of the clump. Flagella contact between opposite mating types can then be achieved. Substances that attract cells of the opposite mating type, as well as substances responsible for several steps in mating, have been studied.

In some forms a tandem pair (Fig. 4-2) emerges from a clump, and then another and another. The pair may swim for hours, with flagella from the minus strain providing the movement. In other strains fusion of pairs is much more rapid, but the organism still has motility, because the quadriflagellate zygote can swim actively. In the first case, the pair eventually stops swimming and the protoplasts fuse at the anterior ends

Figure 4-3. Freeze-etch preparation of a *Chlamydomonas* gamete pair, showing fusion at the site called a protoplasmic bridge. 34,000 X. Courtesy of Brown and McLean.

(Fig. 4-3), eventually completely slipping out of the gamete walls (Fig. 4-2). A spherical zygote with the attached earlike gamete walls can be found in some cultures. The zygote may develop a thickened wall, and in some cases the wall appears irregularly thickened, with numerous warty projections. Such a diploid zygote is named a zygospore (Fig. 4-2). Some phycologists consider this an unforunate term, since in the plant kingdom most spores are haploid.

Germination of the zygote is always accompanied by meiosis, and the mating types are segregated (Fig. 4-2). Thus one can recover two plus strains and the two minus strains from zygote germination (Fig. 4-2). When the zygote is large the two divisions of meiosis are accompanied by another division (mitosis). Thus four plus and four minus flagellated cells are released from one zygote.

Since a mating reaction can be induced, zygotes formed and germinated, and the progeny isolated into new populations, *Chlamydomonas* has been used in genetic studies. New populations often can be ready for mating within 10 days of the initial pairing. The chlamydomonad system is one of the best systems for examining genetic processes in a photosynthetic organism if one wishes short generation times.

As one might expect, there is also abundant literature concerned with control of sexuality in these haploid organisms. Investigators have not only examined the ultrastructure of vegetative cells and gametes, but have also followed the sexual process at the ultrastructural level. The sexual process has been investigated in just a few of the more than 500 species described for this genus. Perhaps some organisms lack sexuality, but if one examines or cultures just one strain of a heterothallic species, sexual reproduction will not be observed.

Other Unicellular Forms.   Emphasis has been placed on *Chlamydomonas,* not only because of the numerous studies that have been conducted, but also because it can serve as a type specimen for most of the unicellular volvocalean genera. Brief descriptions of several unicellular relatives follow.

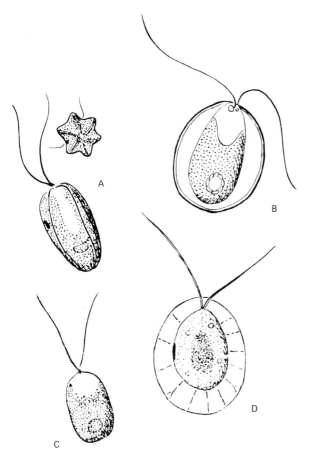

Figure 4-4. Some unicellular green algae. Microscopic. (*A*) *Stephanoptera,* shown also in view from the posterior end (above). Note the star-like shape in this view. (*B*) *Dysmorphococcus.* A flagellate with a lorica. (*C*) *Dunaliella,* a marine flagellate lacking a wall. (*D*) *Haematococcus,* with the protoplasmic strands connecting the protoplast to the wall. Note the stigma or eyespot at the left. In the center of the cell large accumulations of a xanthophyll may give it a red color.

Dysmorphococcus (Fig. 4-4) cells lack a wall, but this is not apparent at first glance. There is a firm outer covering with numerous small pores. Since the protoplast is not in contact with this outer covering, it is termed a lorica, rather than wall. The flagella pass through two of the small pores.

Carteria (Fig. 4-5) can be considered as a quadriflagellate *Chlamydomonas*. Marine, freshwater, and soil representatives are known in both genera. Over 500 species of *Chlamydomonas* have been described. There are far fewer *Carteria* species, and there is not the same wealth of information available for *Carteria* that has been published on *Chlamydomonas.*

Polytoma is a walled volvocalean form that is quite similar to *Chlamydomonas,* but it lacks a plastid. It could be a mutant *Chlamydomonas* or a case of parallel evolution among nonphotosynthetic types.

Chlorogonium cells are unlike other unicellular flagellates. They are fusiform, that is, long and thin, and tapering at both ends. The flagella are about half the body length.

Haematococcus (Fig. 4-4) appears at first glance to have a lorica, but there is a clear difference. Most of the protoplast is not in contact with the wall, but there are several to numerous cytoplasmic strands connecting the body of the central protoplast to the wall (Fig. 4-6). This genus is also of interest in that it can remain dormant in temporary pools of small size, or in small urns, and develop when there is sufficient water. When not growing at their maximum rate, the cells accumulate abundant xanthophyll bodies. The water in which they are growing can thus appear red.

Several unicellular types lack a wall or lorica.

Stephanoptera (Fig. 4-4) is found in both marine and freshwater habitats. The anterior portion of the cell appears to be pinched in so that in cross section there are four to six lobes. Flagella are quite long, longer than the cell. Forms are adapted to high salinity, so that they appear in salty lakes or in tide pools that become concentrated by evaporation.

Dunaliella (Fig. 4-4) cells have a variable shape. Without the irregular margin at the apex, they appear simpler than *Stephanoptera.* They, too, can live with high salinity; forms in tide pools can withstand three to four times the concentration of sea water.

Pyramimonas cells also lack a wall and resemble *Carteria* in that they too have four flagella.

Colonial Forms. The Gonium-Volvox Series.   *Gonium* is the simplest colonial representative in the *Gonium-Volvox* series. The cell

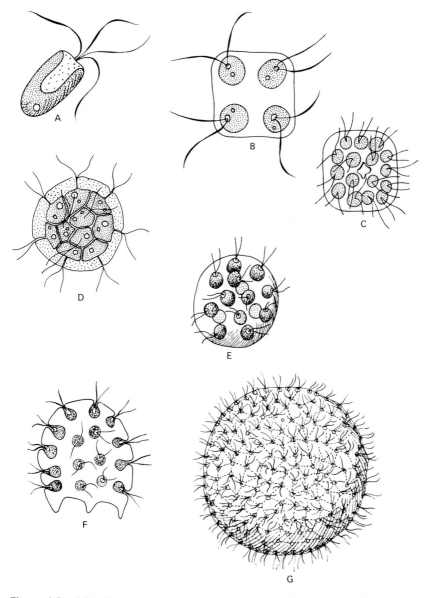

Figure 4-5. A flagellated unicell, and several colonial forms. Microscopic. A. The quadriflagellate, *Carteria*. B, C. *Gonium,* two-dimensional colonies. B has four cells, and a 16-celled form is seen in C. Note the hole in the common mucilage in the center of the colony in C. D. *Pandorina,* with the three-dimensional colony, and cells adjacent to each other. E. *Eudorina,* with a small colony of a few cells shown. F. *Platydorina,* with tail-like appendages. Two-dimensional. G. The largest colony of the group, *Volvox.* Colonies may be of sufficient size to be seen with the naked eye.

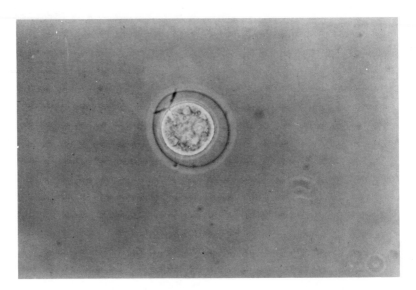

Figure 4-6. Vegetative cell of *Haematococcus*. Cytoplasmic strands are visible, along with the lower portions of two flagella. 1000 X.

number may be as few as four, and these organisms are found in two-dimensional colonies, typically in a flat plate. *Pandorina, Eudorina,* and *Volvox* are spherical colonies; the cells of *Volvox* are arranged only at the periphery of the large colony (Fig. 4-7).

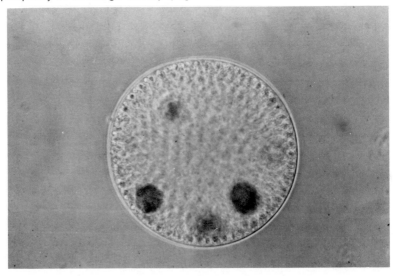

Figure 4-7. *Volvox* colony with several new colonies forming within. 300 X.

Each of these genera has been studied extensively both in the laboratory and in the field. The available literature is extensive. Numerous cultures of each are available in collections, and there have been at least minimal studies on their distribution, mating reactions, morphology and life histories, nutritional responses, ultrastructure, and genetics.

The basic cell type of these genera is that of *Chlamydomonas,* with two flagella, two contractile vacuoles, a single haploid nucleus, a stigma in a cuplike plastid, and a pyrenoid in the chloroplast. Cellulose is lacking; however, the protoplasts may be merely suspended in a common mucilaginous envelope. In the *Gonium-Volvox* series one can see an evolutionary pattern when cell number, reproduction, or presence of stigmata are examined.

Cell number. This varies from 4 to over 500, and even to 5000 in *Volvox* (Fig. 4-5).

Colony shape. Most of the colonies are three-dimensional, but *Gonium* (Fig. 4-5) forms flat, two-dimensional colonies with flagella at one side or around the edge of the colony. *Platydorina* (Fig. 4-5) is another flat colony type. *Pandorina* (Fig. 4-5) colonies are ellipsoidal, with the cells adjacent to each other, whereas *Volvox* colonies are spherical with cells separated from each other. *Volvox* may or may not be connected by cytoplasmic strands through the mucilage. In *Eudorina* (Fig. 4-5) the view from the side shows cells arranged in tiers.

Asexual reproduction. Frequently, but not always, cells capable of division divide simultaneously. In forms such as *Gonium* and *Pandorina* all cells may divide and form new colonies, but in the higher forms there is usually some specialization, with a reduction in number of reproductive rows. Some of the cells, such as the anterior row or rows in *Eudorina,* may be sterile, whereas in *Volvox* there is specialization in formation of reproductive and vegetative cells early in the construction of the parent colony. Because of unequal division of cells during early cleavage, as many as 50 larger reproductive cells, the gonidia, are formed. Only these are reproductive cells.

Colonies with the same number of cells as the parent are formed by repeated cleavage of the reproductive cell. Thus there are many divisions of one cell to form the few thousand cells of a *Volvox* colony. Only in *Gonium* are the anterior faces of cells exposed toward the outside. In all other cases division and the orientation of cells result in flagella being formed toward the center of the new colony. Thus there must be inversion during the latter stages of colony formation. The opening through which the inversion took place is then sealed and the newly formed colony may move within the parent colony. Eventually the parent wall is ruptured, and the new young colony can then swim away.

Sexual reproduction. In the simple colonial form *Gonium* initial stages of gamete production are exactly like asexual colony formation, but the cells do not join to form a colony. Flagellated eggs and sperm are released from the parent colonies. In *Eudorina, Pleodorina,* and *Volvox* sperm are formed in cuplike packets, released as such, and swim as a colony to the egg cell. They then break apart and individual sperm may penetrate the colony near the egg.

In *Volvox* sperm packets in a colony may be few to many in number and are formed by division of enlarged cells. The female cell or oogonium resembles a gonidium. There are usually a limited number of eggs and sperm packets formed by each colony.

Gametes of *Gonium* are isogamous. The intermediate volvocalean forms are heterogamous and oogamy is the rule with *Volvox.* In *Volvox* and *Eudorina* proteinaceous substances, acting as hormones, are produced by the male strain. In culture these substances stimulate the formation of sexual colonies.

Specialization of cells in colonies. In addition to specialization of a few cells for reproduction, one can also see some changes in cell size or distribution of cell organelles. This is most apparent with eyespot size. *Gonium* cells all have a similar sized eyespot, but in *Volvox* the anterior cells have a larger stigma, perhaps only in the four most anterior cells.

Zygote germination. In *Gonium* there may be four single zoospores or a simple four-celled colony resulting from meiosis, but in all other forms there is usually only one cell, a flagellated one, left from the divisions. That haploid cell eventually forms a new colony, although reproduction may proceed through a few intermediate, smaller colonies before the large parent-type colony is formed.

Conclusion. Just a few of the many evolutionary changes that have occurred in these algae have been discussed. Others will occur to you as you examine the algae from nature or in culture. The organisms are quite interesting for many reasons, but provide an evolutionary dead end because they lack vegetative cell division, a feature they share with the next group of organisms presented.

Other motile colonial forms.   In the following genera colonies are not inverted during development. When cell division occurs the flagella are correctly positioned for movement of the colony.

Pyrobotrys colonies are arranged in small numbers of tiers of four cells each, with the flagellar end of all cells pointing toward the anterior end of the colony. Individual cells have two flagella each. There is no common sheath surrounding all cells of the colony.

Spondylomorum colonies have 8 or 16 cells in the same arrangement, but the cells bear four flagella each. Both genera are freshwater forms, but are known where there is abundant organic matter.

Stephanosphaera has cells of the colony surrounded by a common sheath. It is a rare, but striking, colonial form. The colonies are usually composed of eight naked cells with obvious protoplast projections. All are arranged in a ring within a gelatinous envelope, with flagella (two) from each cell all pointing toward the anterior end of the colony. At Mountain Lake, Virginia, it is found in temporary pools, only about a liter in volume, in limestone cliffs at a mountain top. How did this flagellate get there, and how does it survive desiccation?

Nonmotile Forms Lacking Vegetative Cell Division

This group of algae contains some of the most abundantly distributed microalgae in the world. It consists of unicells and colonies that are *not motile* and always lack vegetative cell division. When there is a cell division or series of divisions, they are involved in reproduction. These organisms release the products of division as individuals, or young colonies, from within the parent cell wall. Along with the previously studied flagellated types, they are among the easiest of algae to obtain in axenic culture, that is, unialgal, bacteria-free culture. In nature bacteria would be in the water surrounding the organism, or attached to the wall. By dilution one can eliminate free living bacteria, and the escape from the parent cell wall enables us to culture the recently released products without bacteria. Members of this group such as *Chlorella, Scenedesmus,* and *Chlorococcum* (Fig. 4-8) have been cultured in the laboratory for many years. Is it possible that the choice of these organisms for laboratory studies has overemphasized their importance in nature?

In several representatives of this group it is possible to see how the environment can affect the type of reproductive cell. Flagellated and nonflagellated cells can be observed. If an organism can form both motile and nonmotile reproductive cells, the former are called planospores, and usually not the more common term, zoospore. Then the nonmotile type is termed an aplanospore. In actively growing *Chlorococcum* cultures planospores are formed, because sufficient nutrients and water are present. But in older cultures, and presumably in a somewhat dry soil, aplanospores develop.

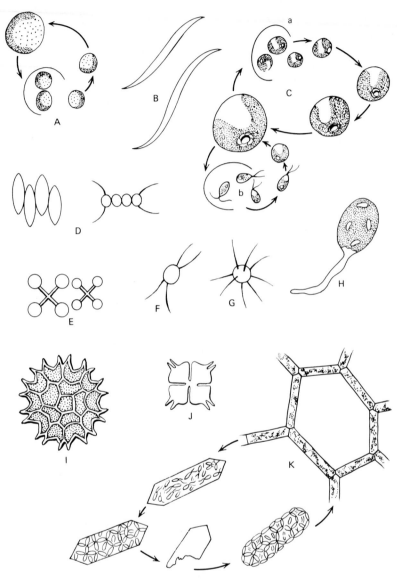

Individual cells and cells of colonies may be either uninucleate or multinucleate. Cell division can be repeated by bipartition or by progressive cleavage of a multinucleate protoplasm. With progressive cleavage only the young cells are uninucleate. As they increase in size cells become multinucleate. When cytoplasmic division is triggered, there are multiple cleavages of the protoplasm. Each nucleus, along with other

Figure 4-8. Forms that lack both flagella on the vegetative cell, as well as vegetative cell division. All but *Hydrodictyon* are microscopic.
A. *Chlorella*, with the simple life cycle, i.e., division and release of cells from within the parent cell wall.
B. *Ankistrodesmus*. In some forms the cells are not twisted, while in others the filaments are twisted together resulting in a colonial form.
C. *Chlorococcum*. When it reproduces by aplanospores (cycle a), it resembles *Chlorella*. But it may form planospores (cycle b).
D. *Scenedesmus* with both a nonspiny and a spiny colony.
E. Four-celled colonies of *Dictyosphaerium*. Note the common wall material joining the cells.
F. *Chodatella*, with spines only at the apices.
G. *Golenkinia*, with spines over the surface.
H. *Protosiphon*, with a basal rhizoid.
I. The flattened colony of *Pediastrum*. Note the horn-like processes.
J. A four-celled *Pediastrum*.
K. The water net, *Hydrodictyon*. Each cell forms flagellated cells within, and those form a miniature colony, eventually released from the mother cell wall. The net may be as large as 0.5m.

cell organelles, becomes surrounded by a wall and reproductive cells are formed. In organisms with repeated bipartition there is one nucleus, or two just before division of the protoplasm. Further nuclear divisions are followed immediately by cytoplasmic divisions.

When the organisms divide and produce planospores, there are a number of possibilities.

1. If the planospores are released singly through a pore or breakdown of the wall, the organism is a unicell. *Chlorococcum* (Fig. 4-8) may reproduce in this way.
2. If the planospores are released as a group within a flexible membrane, the vesicle, the resultant organism may be either a unicell or a colony. Unicells are formed when there is a break in the vesicle and the individual cells become scattered, as in *Hormotilopsis*. Colonies develop if the products of division remain in the vesicle and become attached, as in *Pediastrum*.
3. If the planospores are retained within the parent cell and attach to each other, the organism will be colonial. *Hydrodictyon* nets are formed in this way.

However, if aplanospores are formed, or if the organism merely produces new cells by division, there are other possibilities.

4. If the cells are released, a unicellular organism results. *Chlorococcum* may reproduce this way, as do *Ankistrodesmus* (Fig. 4-9) and *Chlorella* (Fig. 4-8).

Figure 4-9. Population of *Ankistrodesmus* cells. 1600 X. Courtesy of C. Thompson.

5. If the division products are retained and attach to each other a colony results, as with *Coelastrum* and *Scenedesmus* (Fig. 4-8).

A new insight into the study of the morphology and classification of these organisms was provided by Starr during the middle 1950s. Large spherical green unicells, or related packets of such cells, had given both field and laboratory workers considerable problems when attempting to identify them. In a comparative culture study of many of the spherical unicells that were known to reproduce by planospores, Starr found that these organisms could be distinguished if one determined the type of plastid, the type of planospore, and the presence or absence of a pyrenoid. The following plastid types were noted: cups, axial forms with irregular processes, nets, separate discs or asteroid types. Planospores may be naked or possess walls; the planospores may have two flagella of equal or unequal length. A walled planospore with flagella of equal length was termed the *Chlamydomonas* type. If it was the reproductive cell of an organism with a cup plastid, the organism would be a *Chlorococcum*. On the other hand, a naked planospore from a unicell with a netlike plastid would be a *Spongiochloris*. In a similar way other genera could be conveniently delimited.

After studying some of these organisms from natural collections, we learned that at times it is difficult to determine the plastid type, if it is obscured by food reserve, and we might not always be fortunate enough to see the flagellated cells. One has to be patient enough to look at living field material for long periods, or else to rely on laboratory cultures. For those who want to make generic and specific determinations based on observations of a few cells, or based on examination of material fixed in the field, Starr's system is not apt to work. Field and laboratory workers see this problem of identifying these organisms differently. Field investigators would be unfairly hampered if they had to culture much of their material, while laboratory investigators, with a workable system, *might* be unnecessarily complicating the "natural" classification of green unicells. Two compromises appear possible. Either field and laboratory investigators work jointly on projects so that each can contribute his or her expertise, or we suggest that most unicells observed in the field be lumped into a few categories. All would be aware of the fact that these categories would not mean the same thing to laboratory and field investigators.

Sexual reproduction is quite common in many of these unicells and colonies, with all phases of the life history haploid. Meiosis takes place in the germinating zygote. On occasion more complex cycles are detected, with both haploid and diploid vegetative cells.

Some Representative Types.   *Chlorella* (Fig. 4-8) is the best known representative, and certainly is among the smallest of the members. Reproduction in *Chlorella* is uncomplicated, for it can only produce daughter cells (sometimes called autospores) within the mother cell wall, and these are never flagellated. From 2 to 32 are formed at each division. *Chlorella* divides about as rapidly as any green plant, in somewhat under 2 hours at maximum growth rates. Thus the organism is widely used in physiological studies, particularly those investigations dealing with growth of algae as a source of food or as a gas exchanger, that is, supplying oxygen while utilizing the carbon dioxide that humans exhale.

Chlorella cells have a single, delicate, shallow cuplike plastid which may or may not have a pyrenoid. The wall is usually quite thin, with no ornamentation. Resistant spores are unknown.

Chlorococcum (Fig. 4-8), the type genus, is unlike *Chlorella* in that larger cells (10 to 40 μm) and flagellated cells are known. The latter have a wall and flagella of equal length. The asexual cycle is shown in Fig. 4-8; some forms have a sexual phase. Flagellated cells are *Chlamydomonas*-like with a wall and with flagella of equal length. Over two dozen species of *Chlorococcum* are known and their morphological and physiological differences have been documented.

Scenedesmus (Fig. 4-8) is the best known of the colonial forms within the group. We usually think of this genus as a four-celled colony, but 2-, 4-, 8-, and 16-celled colonies are now known, as well as unicellular stages. As with *Chlorella,* it was thought until recently that sexual reproduction was lacking, but we now know that it can be induced in many non-spine-forming types by growth at low temperature with low nitrogen levels. This genus has representatives that grow quite rapidly and thus there are a great deal of physiological data available.

Morphological variability has now been documented with *Scenedesmus.* Thus a spiny type that resembles *Scenedesmus quadricauda* (Fig. 4-8) can be induced to form a population of unicells. In such an environment *S. quadricauda* types would be completely missing and the unicells would resemble *Chodatella* (Fig. 4-8).

There are many other genera in this group which at times can be dominant in a freshwater environment (Figs. 4-10 and 4-11). Some are diagrammed in Fig. 4-8, but one should consult references to freshwater green algae for information on individual species and genera.

Some Filamentous, Tubular and Sheetlike Forms

In this group, as in the remainder of the green algae, there is vegetative cell division. If a cell divides vegetatively, and the products of division remain attached, a juvenile filament is formed. Continued divisions would result in a filament of indeterminate length. If there is an *occasional* division in a second plane, followed by growth at that point, a branched

Figure 4-10. *Dictyosphaerium* 16-celled colony. 1600 X.

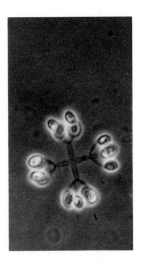

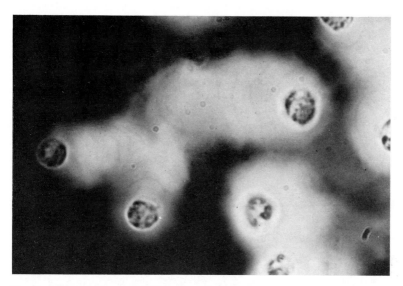

Figure 4-11. Mucilage production. Several primitive green algae can produce abundant mucilage, as figured here. 1600 X. Courtesy of R. Hilton.

filament is formed. *Equal* numbers of divisions in two planes result in a membranous sheet one cell layer thick. Division in a third plane would add thickness to the sheet or membrane.

Filaments. Filamentous forms of the above type, with uninucleate cells and reproduction by biflagellated or quadriflagellated cells, could all be considered together.

Stichococcus (Fig. 4-12) would be the simplest member of this group, with unbranched filaments that have a tendency to fragment. An incomplete band or a bandlike plastid is noted and pyrenoids are lacking. Recently, a *Stichococcus*-like organism was shown to develop a helical plastid when cultured under very favorable conditions. With slower growth the plastids were typical. There are no flagellated cells, asexual or sexual, in this primitive genus.

Klebsormidium (Fig. 4-13), until recently called *Hormidium,* is a little more advanced, for filaments do not fragment as easily as do those of *Stichococcus.* There is one pyrenoid in each parietal bandlike plastid. Reproduction occurs by means of biflagellated zoospores.

Ulothrix (Fig. 4-12) filaments are also unbranched, but with several pyrenoids per bandlike plastid. The margins of the band can be quite irregular. Zoospores are quadriflagellate and gametes are biflagellate. When a zoospore settles on a substrate, a distinct holdfast is formed,

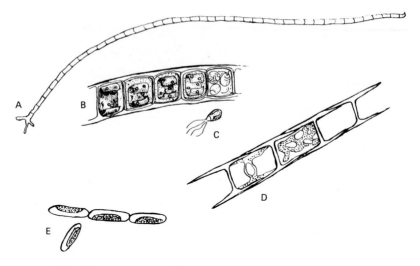

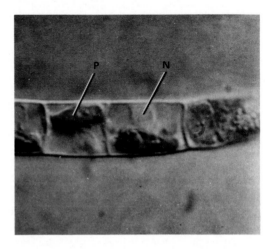

Figure 4-12. Filamentous green algae.

  A. and B. *Ulothrix*, with an elongate filament, and the holdfast (shown in A). Detail of some cells is shown in B. Note the bandlike plastid, and the flagellated cells formed in the cell to the right. One quadriflagellated cell (C) has released from the parent cell. Filaments may be several millimeters long.

  C. *Ulothrix* quadriflagellated zoospore. Microscopic.

  D. *Microspora*, with a cell shown in optical section to the left and one in surface view to the right. Note the interconnecting "H" pieces of the wall.

  E. *Stichococcus*, with a parietal plastid. The organism can be in short filaments. There has been a recent report of a helical as well as a parietal plastid in the same population of a strain in culture. Microscopic.

Figure 4-13. *Klebsormidium* filament showing plastid and nucleus. Courtesy of J. Cain. 1000 X.

and *Ulothrix* filaments can be seen attached by this structure. However, filaments of some length often become detached.

Microspora (Fig. 4-12) filaments are also unbranched; each cell has a reticulate plastid. The "H-shape" arrangement of cell wall components is quite distinctive for this genus. Filament walls are composed of segments that in outline resemble the letter H. The letter H is composed of the upper and lower halves of walls of adjacent cells, including the cross wall. *Microspora* resembles *Tribonema,* a yellow-green alga (Xanthophyceae). The two can be distinguished in several ways. *Tribonema* cells are usually several times longer than wide, at times a little bulbous, with many plastids per cell giving a distinct yellowish color to the filament. But *Microspora* cells are usually a little broader, not more than two times longer than broad with one reticulate (frequently obscured) plastid that is quite green. In addition, the wall is thicker in *Microspora* and sometimes laminated.

Stigeoclonium (Fig. 4-14) and *Chaetophora* are two somewhat similar branching filamentous forms, with bandlike plastids and branches that taper toward the tip. In both genera the branches can end with hairs at the tips, but these are usually more pronounced in *Chaetophora,* hence the name. The two are distinguished by the mucilage that surrounds *Chaetophora* filaments, thus creating a macroscopic colony.

Draparnaldia (Fig. 4-14) filaments are a little more complex. They are composed of a larger central filament with smaller treelike branches occurring periodically along the axis.

The last three genera are found in freshwater streams attached to rocks. They can form both zoospores and gametes. Meiosis is zygotic.

There are numerous other related genera, but for these one should consult monographs or texts on freshwater or marine algae. Certain genera might be of interest in the evolution of forms toward the land habit and toward the vascular plants.

Recent excellent cytological and ultrastructural studies point out that in several species of *Ulothrix, Schizomeris, Stigeoclonium,* and *Fritschiella* new walls in dividing cells develop between dividing nuclei and grow toward the lateral walls. However, in *Stichococcus, Klebsormidium,* and one *Ulothrix* species, the cell wall between newly divided cells is formed centripetally, that is, inward from the lateral walls. When attempting to put these organisms into orders and families, we will have to take these facts into consideration. In addition, we also realize that we should not have relied solely on the presence or absence of branching to delimit families and orders. Mattox and Stewart suggest that unbranched filaments can be primitive members of several groups.

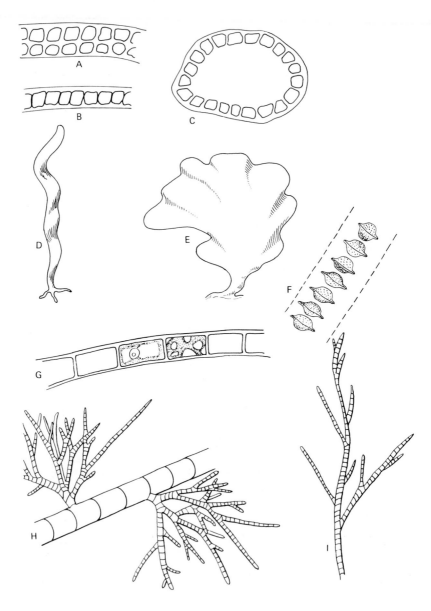

The Tubular and Membranous Forms in Group One.

Enteromorpha (Fig. 4-14) and *Ulva* (Fig. 4-15) represent not only the marine forms, but also the tubular and sheetlike green algae. The two develop at first like a young *Ulothrix* filament. Later divisions in a second

Figure 4-14. Other green algae, some marine. Macroscopic, except for cross sections.

Ulva. This genus is shown in both A and E. The form the plant body takes is diagrammed in E, while a cross section, showing that it is two cells thick, is shown in A.

Enteromorpha is diagrammed in both C and D. D shows the plant body type, with the tubular form, and a cross section is shown in C.

Monostroma could resemble E, but the organism would be thinner than *Ulva.* This would be evident when one examines the cross section, B. The organism is one cell layer thick.

Microspora is shown in G. This is a filamentous form.

Radiofilum, with a wide mucilage sheath and distinct cell shape, shown in F.

H is a diagram of *Draparnaldia,* with a large central filament and smaller branches in tufts. Filaments measured in mm.

Stigeoclonium is shown diagrammatically in I. Note that branches of *Draparnaldia* resemble the plant body of this genus. Filaments measured in mm.

plane produce a sheet, and the first division in a third plane makes the membrane two cells thick. *Enteromorpha* first develops as a long and narrow membrane two cells thick. When the layers of cells separate, a hollow tube that is one cell thick results. Separation might not always be complete. In fact, at least one organism is tubular at the base at maturity and membranous at the apex. It has been called both *Ulva linza* and *Enteromorpha linza,* with the latter name preferred on this continent.

Figure 4-15. Intertidal zone along a rocky coast, with numerous *Ulva* plants exposed. Courtesy of N. Proctor.

When the young membrane is two cells thick (Fig. 4-14), if the layers do not split, but rather the total number of cells increases, a large sheet called *Ulva,* the sea lettuce (Fig. 4-15), is formed. The organism is widely distributed in marine coastal waters and is attached to rocks, wood, rope, or other algae. Cross wall formation is in a centripetal direction.

Reproduction in *Ulva* (Fig. 4-16) can be considered as representative of the group, but there are many variations on this single pattern. In order to study the life history completely, one must gather three different plants, and they are all almost identical in appearance, with only the color of the margin as an aid in distinguishing types of membranous forms. In all cases, at the time of reproduction there are internal divisions of marginal cells, forming flagellated cells.

One plant (Fig. 4-16) would form quadriflagellated zoospores within the parent cells. These would be released and could be followed to a

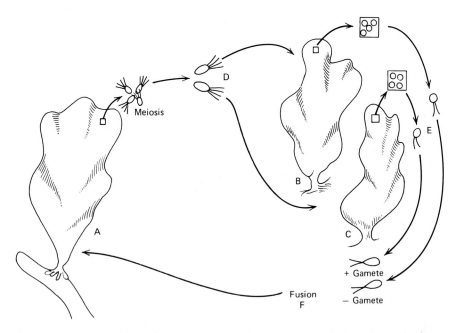

Figure 4-16. *Ulva* and its life history. Plant A is the sporophyte and in a small portion of the thallus, designated by the small square on the upper part of the plant body, quadriflagellated zoospores are formed *after* reduction division takes place. These cells (D) give rise to gametophytes (plants B and C). The latter are identical in form to each other *and* to the sporophyte (A). They produce biflagellated gametes (E), which are designated + and −. These pair and fuse (F) and form a plant similar to the one with which this life history was initiated. Macroscopic plants 0.2 m in length are common.

similar substrate where they would develop into *Ulva* plants. But if the flagellated cells from a plant derived from a quadriflagellated zoospore are examined, one would see that only biflagellated swarmers would develop (Fig. 4-16). If the investigator were fortunate enough to select two such plants of the correct mating types, he (she) could determine that the biflagellated cells were gametes. Both isogamy and anisogamy have been reported. Fusion occurs and the zygote develops into the type of membranous plant first mentioned in this paragraph. This plant (Fig. 4-16A), which is diploid and produces zoospores after reduction division in marginal cells, is called a sporophyte, and the other two membranous forms, which develop biflagellated gametes in marginal cells, are called gametophytes (Figs. 4-16B and C). The gametophyte generation and sporophyte generation are identical in appearance (isomorphic), with sporic meiosis occurring in this alternation of generations. We will see that this type of life history is not uncommon in the algae.

Monostroma (Fig. 4-14) membranous plants develop much like *Enteromorpha,* but at maturity there is a splitting of the tube, and the sheet that develops is just one cell layer thick.

Ulva, Enteromorpha, and *Monostroma* are found in brackish water. The latter two genera have been reported inland, in saline environments. *Enteromorpha* grows in rivers in England.

All three genera have been grown in culture, but it is not always possible to obtain typical thalli in axenic culture (without bacteria). In some cases either the bacteria themselves have to be physically present, or the extract from a culture of the bacteria must be present. For example, *Monostroma* forms only single cells or simple filaments in axenic culture, but with a pink marine bacterium, or a filtrate of the bacterium, typical sheets can be formed.

With one *Ulva* strain single cells or cells with rhizoids developed in axenic culture, and with the addition of some bacteria only *Enteromorpha*-like forms developed. The bacterium must be physically present, for an extract of the bacterium will not allow development of the *Enteromorpha*-like stage. Provasoli has initiated much of this work. Now Bonneau has shown that other axenic strains of *Ulva* produce flattened thalli, which look like *Ulva*. But when one looks at the culture of many strains of *Ulva* by both of these investigators, a great deal of variability in strain behavior is apparent. No longer can we think of *Ulva* as having a simple life history. In certain cases there are many forms possible in any pure population so that a single species may develop stages resembling previously described species either of *Ulva,* or of *Enteromorpha!*

Group Two

Oedogonium and Allies

All the green algae treated thus far can be considered as belonging to one group. The cells are typically uninucleate, a simple plastid is the rule, meiosis is generally zygotic, and, in general, there are other

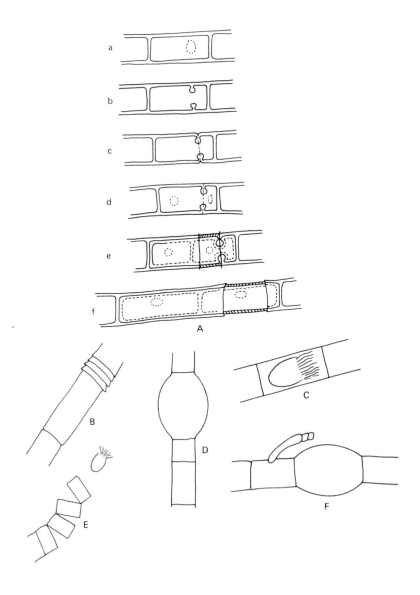

similarities among many of the unicells, colonies, and filaments. (There are always exceptions to the rule, including the different plastids in many of the unicellular, zoospore forming types, as well as the sporic meiosis in *Ulva*.)

But *Oedogonium* (Fig. 4-17) and relatives have features that immediately set them aside. These organisms are simple filaments that may or may not be branched. *Oedogonium* is unbranched. It is found in fresh waters and is sessile, at least when the filaments are young. But as with other elongate filaments, there can be fragmentation and thus free-floating types. The cells are elongate, are uninucleate (with the nucleus quite visible even in unstained material), and have a reticulate plastid (Fig. 4-18). The orientation of the net plastid is along the axis, so initially the plastid appears to be composed of a series of elongate bands.

The method of cell division outlined in Fig. 4-17 is the first characteristic that sets these organisms apart from those examined thus far. New lateral wall material for the dividing cell is deposited in a ring near the parent end wall. Nuclear and cytoplasmic divisions then take place. Next the parent wall separates just outside the point where the ring is positioned. The ring of new wall material is then "stretched" out to cover the elongating protoplasts. One result is the formation of a polar cap (Fig. 4-19) near the cross wall of a dividing cell. Since the division of cells in the filament is at least for a time restricted to select cells in the filament, several polar caps are seen on certain cells (Figs. 4-17 and 19). The number of caps indicates the divisions that have already taken place.

A second distinctive characteristic is the flagellated cell. Whether it is a zoospore, a sperm, or an androspore, it is multiflagellated (Fig. 4-17).

Figure 4-17. *Oedogonium,* cell division and form. Microscopic. The complete filament is measured in mm.

A. The manner in which new walls are deposited in *Oedogonium.* Depicted are (a) a cell with a single nucleus, (b) the ring of wall material, (c) the ring again, also apparent from the exterior, (d) the cell with two nuclei, (e) the protoplasts, after division and a limited amount of growth, (f) and finally the two new cells, with the new wall material coming from the ring seen in the first diagram of this series. Note that the polar cap is formed as a result of the division.

B. A cell with several polar caps of wall material.

C. A cell which has formed one large, multiflagellated zoospore.

D. The enlarged cell that will form an egg, with a smaller cell below.

E. The shortened cells (antheridia) that break apart and release two multiflagellated sperm each.

F. A dwarf male plant immediately below the egg. See text for details.

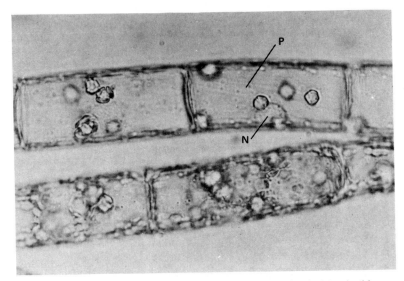

Figure 4-18. *Oedogonium,* showing the single nucleus and reticulate plastid. 1000 X.

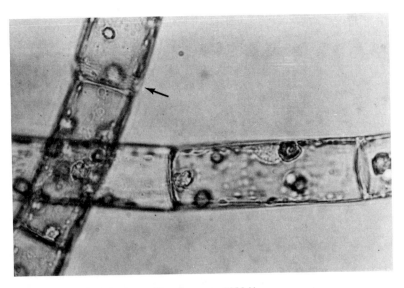

Figure 4-19. *Oedogonium,* with polar caps. 1000 X.

The smaller sperm and androspore have about 30 flagella each, while as many as 120 flagella are found in a ring at the apex of a zoospore. Since the flagella are paired, it is felt that they are really the flagella of an evolved synzoospore. An ancient *Oedogonium* cell, which *would have produced* 60 biflagellated zoospores, now *does not cleave* the protoplasm, but forms a large, multiflagellated reproductive cell.

When the single zoospore is released, it can swim for a while and then settle on another alga, a rock, or even in culture to the side of a culture flask. A large brownish holdfast is formed and the cell elongates and divides, forming a large unbranched filament.

Oedogonium species may be homothallic or heterothallic. The egg is formed by the equal division of a vegetative cell, with one cell becoming the oogonium and the other the supporting cell (Fig. 4-17). A pore is formed in the wall and the egg is now obvious because of a slight autoplasmolysis. Certain elongate cells divide several times into numerous cells which are shorter than they are wide (Fig. 4-17) and which release sperm in pairs. The sperm swim toward the egg, through the pore, and fertilization occurs. Meiosis is zygotic. All of these stages, especially the details of fertilization, have recently been described by Hoffman.

When the male and female filaments are identical in size, and reproduction occurs as described above, and when fertilization occurs directly in a homothallic strain, the organism is called a macrandrous *Oedogonium.* But the organism takes different forms, and these provide a third distinctive characteristic for this group.

In the nannandrous forms the flagellated cell, which we would believe would fertilize the egg, approaches the developing egg, and then settles on the adjacent supporting cell. It develops into a miniature *Oedogonium* (Fig. 4-17) which soon forms flagellated sperm cells. The first flagellated cell in the series is then called an androspore and the second the sperm.

If one can isolate one filament, which then can reproduce sexually, without the formation of a miniature male stage, the strain is homothallic.

If one can isolate one filament, which can reproduce sexually, but only after the formation of a miniature filament (produced when the androspore germinates), the strain is homothallic, with a dwarf stage. Sperm cells, the second flagellated stage, would be produced by the miniature filament.

If one must isolate two filaments in order to complete a sexual cycle, with eggs formed by one strain and sperm by the other, then the strain is heterothallic.

If one must isolate two filaments in order to complete a sexual cycle, with androspores formed by one strain, and eggs (as well as sperm on the miniature filaments) are formed on the second filament, the strain is also heterothallic, but with a dwarf phase. The positioning of the dwarf male filament could insure fertilization in an evolving organism.

More than 200 species of *Oedogonium* have been described, and one must see the organism while reproductive in order to determine the species. Naturally, great emphasis is placed on size of cells, shape of the egg and zygote, as well as its ornamentation. One wonders about the validity of so many species in an organism that has such a limited morphology. Morphology is used almost exclusively in classification, although it is now possible to initiate studies with species in culture, species in which reproduction can be controlled.

Bulbochaete is another common representative of this group. The organism is distinct because it is a branched filament, with bulbous hairs comprising the branches. It is thus appropriately named. The organism is attached by a basal cell.

Oedocladium is a terrestrial or aquatic representative of the group. There are no hairs in this freely branched organism. It demonstrates the heterotrichous pattern of growth, with colorless rhizoids of terrestrial forms penetrating the soil. Once thought to be an uncommon form, this genus can be collected in colder months in the southeastern United States.

Group Three
Multinucleate Types
A number of the sac-like or filamentous algae either lack a cross wall or have fewer cross walls than in organisms covered thus far. The cross

Figure 4-20. Multinucleate or coenocytic forms. Macroscopic, measured in centimeters, except the cell detail.
A. *Cladophora* is a branching filamentous type. Note the position of branches.
B. *Rhizoclonium,* which is infrequently branched, or even unbranched.
C. *Chaetomorpha,* with enlarged, barrel-like cells.
D. *Valonia* diagram, showing the very young coenocyte at the right and a typical segmented plant body at the left.
E. *Acetabularia,* with a rhizoidal system, a stalk, and a cap.
F. *Halimeda,* a calcareous form.
G. and H. *Codium,* shown 1/3 life size in G, with a portion seen under magnification in H. The plant is multinucleate and composed of strands intertwined; utricles project from the strands. Three utricles are seen in H.
I. *Caulerpa.* A great deal of variability is seen among the various species. Length up to 1 meter.

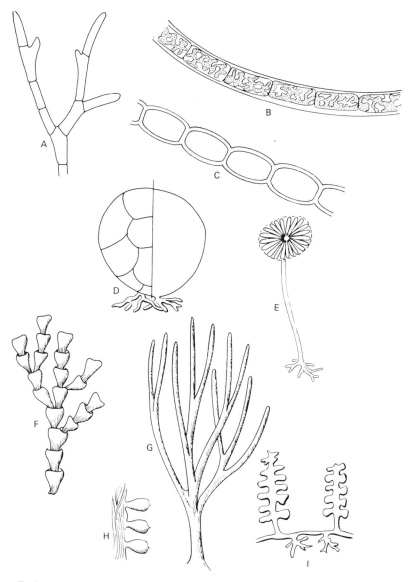

walls in some sacs are absent only in vegetative portions. Thus we see multinucleate cells or filaments called coenocytes. These are plant bodies with many nuclei and with cross walls formed only at the time of reproduction.

Cladophora (Fig. 4-20) is a branching filamentous type, with multinucleate cells, that is found in both marine and freshwater habitats.

In some lakes one species aggregates in "balls," probably as a result of bottom current action. *Cladophora* can be a nuisance alga on the coast of the Great Lakes where it is increasing in numbers. Filaments are frequently epiphytized, presumably because of the lack of an outer mucilage wall. Gametes are biflagellated and zoospores are quadriflagellated. Both are formed directly from divisions of protoplasm in terminal cells. At least some freshwater and marine species reproduce by an alternation of isomorphic generations.

Spongomorpha is a branched type, once considered by some to be a *Cladophora*. The distinguishing feature is the twisting of branches and the joining of axes by means of short hooklike or descending branches. Sometimes the joining is by means of rhizoidal branches. At least one *Spongomorpha* has a heteromorphic alternation of generations with the sporophyte phase a smaller form, identical to *Codiolum*. (*Codiolum*-like stages might be the alternate generations of genera other than *Spongomorpha*.)

Rhizoclonium (Fig. 4-20) is an infrequently branched (Fig. 4-21), or even unbranched, form which is both marine and freshwater. Branches may be small and colorless, like rhizoids.

Figure 4-21. *Rhizoclonium*, showing a small branch. Many times branching is not observed on *Rhizoclonium* filaments. 800 X. Courtesy of D. Sheak.

Basicladia has infrequent branches, and they are basal. The habitat of this genus is the back of a turtle, especially the snapping turtle. These turtles are called moss backs.

Chaetomorpha (Fig. 4-20) is an unbranched, mostly marine species, which possess barrel-like cells in very coarse and sometimes large filaments. The entangled filaments remind one of coarse steel wool (Figs. 4-22 and 4-23).

Other green algae, usually marine and tropical forms, are siphonous, and at one time were considered by some in one order, the Siphonales. One group of coenocytes has the plant body divided into a few multinucleate segments.

Valonia (Fig. 4-20) is a sac-like marine form, which may be the size of a chicken egg, with attachment by means of rhizoids. There is a reticulate plastid in each multinucleate segment of the plant body. The organism was considered diploid, with gametic meiosis, but recently there is an indication that at least some forms have an alternate phase.

Anadyomene is basically filamentous, but with aggregations of filaments, giving it a fan-shaped appearance.

A second group of siphonous forms has a radial symmetry, and although the organisms can be 3 cm or more in height, they have but

Figure 4-22. *Chaetomorpha*, entangled in other seaweed, and attached. Courtesy of R. Wilce.

Figure 4-23. *Chaetomorpha* in a tide pool. Courtesy of R. Wilce.

one large nucleus, until reproduction. Flagellated gametes are formed in cysts, which have a cap or operculum.

Acetabularia (Fig. 4-20) species resemble an umbrella (Fig. 4-24), both with or without the webbing of cloth. There is radial symmetry of the upper portion or cap. The genus is the best known of the group, because of the morphogenetic studies of Hämmerling. Just a few years ago it was a very useful organism in determining the role of the nucleus in development, as well as in detecting messenger RNA, once called the morphogenetic substance. There is one free-living generation, with gametic meiosis. A complete cycle occurs yearly. Many forms are covered with lime.

The last group of multinucleate greens to be discussed is what most phycologists now call the true Siphonales. The organisms are coenocytic with numerous disclike plastids; many are calcified. Many are presumed to be diploid, with gametic meiosis, but at least one has a heteromorphic alternation of generations with sporic meiosis.

Derbesia (Fig. 4-25) was described as a branching tubelike plant with a prostrate basal portion growing in or on soft calcareous substrates. The erect portion is tubular and slightly branched. Septations are found on the latter at the time of reproduction. Although we do not have all the cytological data for confirmation, it appears that many times this form is

Figure 4-24. Population of *Acetabularia* under water. ¼ X. Courtesy of N. Proctor.

but one stage of a more complex heteromorphic life history. (See *Pedobesia* for those cases in which the life history is not heteromorphic.) The prostrate tubular form is the sporophyte and the small sac-like branches are zoosporangia. Zoospores are multiflagellated, with the appendages in a whorl. The life history is presented in Figure 4-26.

What does the gametophyte look like? In some cases it is a sac-like plant, not unlike a *Valonia* lacking septations. This gametophyte phase was initially described as a *Halicystis*. This is but another example of early failure to realize that two dissimilar plants were in fact stages in the same life history. Each stage thus received its own generic name. The sac has basal rhizoids attaching the coenocyte to the substrate. At reproduction we can distinguish grass-green female gametophytes from the dull olive-green males. For some time it has been known that these plants were sexual, but it was assumed that meiosis was gametic. This timing of meiosis is impossible when one considers that the sac and tubular forms are alternates in the same life history (Fig. 4-26). The sacs are haploid, and the zygote develops into the tubular portion, which originally was called *Derbesia*. This has been confirmed both in culture studies and in nature. One would then assume that the tubular portion was diploid, with meiosis occurring in the development of the zoospores.

Figure 4-25. *Derbesia* cultured in the laboratory in a Petri dish. Scale = 1 mm. Courtesy of J. Page.

Several points should be emphasized:

1. The *Derbesia-Halicystis* combination should be called *Derbesia,* because the former has priority according to the rules of nomenclature. Should the name *Halicystis* be abandoned? Perhaps we should first be certain that most, if not all, alternate with tubular forms. Some *Halicystis* species might not have a *Derbesia* stage.

2. Not all *Derbesia* forms are connected with *Halicystis.* Several species have *Bryopsis* species as alternates in heteromorphic life cycles!

3. Some *Derbesia* forms, the tubular stages, have abundant growth in nature where no sac-like coenocytes are present. In culture studies one "*Derbesia*" had a direct life cycle; that is, the zoospores produced another "*Derbesia*" stage. In such cases sexuality would have been lost. Because of the rules of nomenclature dealing with the individual species in question, this organism should be known as *Pedobesia* (Fig. 4-26). However, the *Pedobesia* culture study might not be readily accepted by some phycologists.

As with other plants we shall examine, the length of day, and thus season of the year, must be considered when attempting to understand these life histories. In one *Derbesia* studied, the multiflagellated zoospores were developed when days were long. In another, zoospores formed at all day lengths, but in long day conditions a typical *Halicystis* was formed; under short days a slender, rhizoidal form was produced.

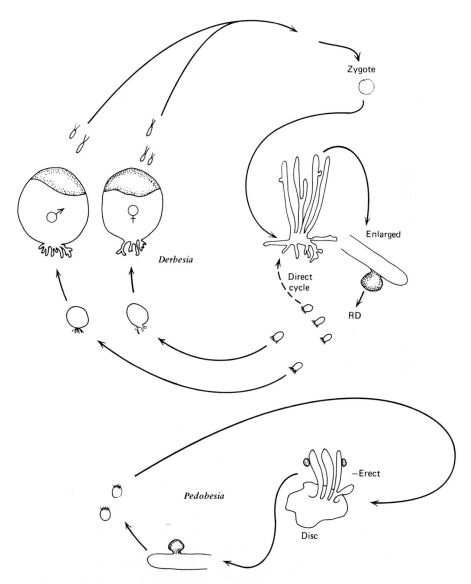

Figure 4-26.   Upper. The heteromorphic life history of *Derbesia,* formerly described as certain *Derbesia* and *Halicystis* species. Lower, the direct life history of *Pedobesia,* formerly a few *Derbesia* species. It has been reported that at times *Derbesia* can also have a direct life history, as indicated by the broken small arrows. If meiosis (RD) occurred, this direct cycle would produce a haploid plant!

Bryopsis plants at first appear to be pinnately branched filaments, but on closer examination one sees no cross walls on vegetative portions. The plants are tufted and bushy and grow attached by rhizoids to a rocky substrate. Gametes are formed in branch apices after the formation of a cross wall. The gametes are biflagellated, with the female larger. Generally, the life history is said to have one free-living form, with gametic meiosis in a diploid plant, but in several cases *Derbesia*-like alternate phases are known.

Caulerpa (Fig. 4-20) is found in some temperate waters, but mostly in the tropics. The plant grows along the substrate by a tubular process, with rhizoids in groups below and erect photosynthetic portions above. Although cross walls are absent, there is some support by means of internal thickenings of the walls. Forms up to a meter in length can be found.

Codium (Figs. 4-20 and 4-27), which resembles a green spaghetti mop, does not at first appear to be a coenocytic form. It is composed of numerous branching and interwoven coenocytic filaments (Fig. 4-20). A cross section of the plant body reveals that the outer portion is composed of bulblike growths developing from the central tubular strands. These bulbs or utricles (Fig. 4-20) are photosynthetic. Sacs that bear the gametes develop laterally from the utricles.

Figure 4-27. *Codium* growing in 7 m water. Note many hairs except at the growing tip. ¼ X. Courtesy of R. Wilce.

Codium is known to reproduce by fragmentation, with growth of filaments at the free end. These filaments provide contact with a new substrate. Laboratory developmental studies show that, unless there are sheer forces from agitation, there is an undifferentiated mass of tubular filaments. Even dedifferentiation of mature forms bearing utricles occurs without culture agitation. Distribution of the weedy organism is first limited by available substrate (Fig. 4-28). *Zostera,* eel grass, excludes *Codium* by increasing siltation. In estuaries distribution is limited when salinity is lower than 22 ppt or temperature is above 16°C.

Penicillus (Fig. 4-29) looks like an old-fashioned shaving brush, with numerous rhizoidal growths penetrating the softer, sandy substrate in which it grows. It is calcified.

Halimeda (Figs. 4-20 and 4-30) with obvious segmentation, is a calcareous form, once thought to be an animal. The life history details are obscure. The hardened wall contains aragonite.

Figure 4-28. *Codium fragile* on *Aequipecten* irradians. Courtesy of J. Ramus.

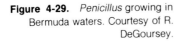

Figure 4-29.   *Penicillus* growing in
Bermuda waters. Courtesy of R.
DeGoursey.

Group Four
Spirogyra and Allies, Including Desmids

These plants can be distinguished from the other green algae because
they lack motility in any phase of the life history, at least motility by
means of flagella or cilia. It will be seen that their gametes move by an
amoeboid motion and that some desmids secrete mucilage. Thus the
desmids can be propelled for short distances. In this group are found
only single-celled forms or unbranched filaments, and all cells are
uninucleate and haploid.

   Spirogyra (Fig. 4-31) is the best-known genus. It is commonly found in
fresh waters throughout the world, especially in small ponds or
temporary pools in wet areas. Unbranched filaments are composed of
mostly elongate cells joined end to end in long strands. There is an

Figure 4-30. *Halimeda* thallus. Courtesy of R. Wilce.

abundance of mucilage around the filament so that aggregations feel slick to the touch. One or more elongate, bandlike, helically twisted plastids are located in each cell (Fig. 4-32). Close examination of the plastid will reveal that the band does not lie flat against the cell wall, that the margin is serrate, and that there are numerous pyrenoids per plastid. The plastid has a very elaborate structure. Cross wall formation proceeds first in a centripetal direction, but then a cell plate is initiated.

Reproduction is either by fragmentation, made easier in some forms because of the large indentation in the end wall of certain species, or by sexual reproduction. The latter may be scalariform (Fig. 4-31), involving adjacent filaments, or lateral (Fig. 4-31), with a conjugation tube from one cell joining with the next cell in the same filament. Meiosis is zygotic.

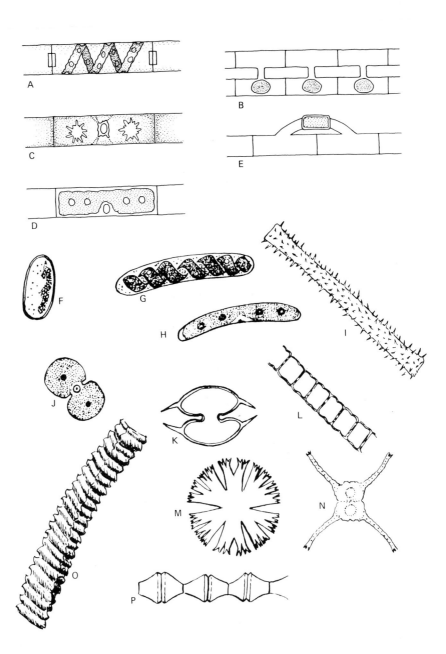

Figure 4-31. *Spirogyra* and its allies. Microscopic. Masses of filaments are macroscopic.

A. *Spirogyra*. Note replicate end walls, helical plastid, and the single central nucleus.

B. Scalariform conjugation, found in several of the filamentous forms. Note the zygotes in lower filament, and the three conjugation tubes.

C. *Zygnema*, with two stellate plastids, and a central nucleus in each cell.

D. *Mougeotia*, with a flat, sheetlike plastid.

E. Lateral conjugation, with a zygote in the conjugation tube between two cells.

F. *Mesotaenium*, a saccoderm desmid. Note the simple plastid, and contrast its form with the plastids of other organisms on this page.

G. *Spirotaenia*, another saccoderm desmid, with a helical plastid.

H. *Roya*, with four pyrenoids. Saccoderm type.

I. *Gonatozygon*.

J. *Cosmarium*.

K. through P. All are placoderm types.

K. *Arthrodesmus*. Do not confuse with *Xanthidium*.

L. *Hyalotheca*, a filamentous desmid. Mucilage boundary is not shown.

M. *Micrasterias*.

N. *Staurastrum*, with horns.

O. *Desmidium*, with a twist to the filament.

P. *Bambusina*, earlier called *Gymnozyga*.

Zygnema (Figs. 4-31 and 4-33) is about as widely distributed as *Spirogyra*, and it can be distinguished by the two large axial chloroplasts per cell, with the central nucleus between. As in *Spirogyra*, the nucleus can be easily detected suspended by cytoplasmic strands from the adjacent walls.

Mougeotia (Fig. 4-31) has a flat bandlike plastid suspended in the center of the cell, extending the entire length. When lateral sexual reproduction takes place, one can find the zygote formed in the conjugation tube (Fig. 4-31). Thus there must have been equal movement of each gamete, or protoplast, during the initial stages of reproduction.

The desmids are also included in this group. They are unicells and unbranched filaments. Cells of many desmids typically have a median constriction which sets apart two semicells. Note that the constriction is lacking in *Spirotaenia* (Fig. 4-31). When cells divide, semicells separate, and the new semicells are gradually formed between the old. When the new semicells are complete, unicells separate and two individuals result. The identical process merely adds length to filaments. New semicells are not precisely in line, and thus a twist soon forms in a filament (Fig. 4-31o).

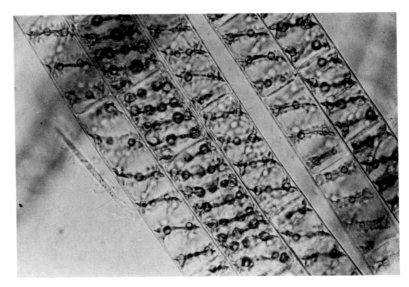

Figure 4-32. *Spirogyra* filaments. 300 X.

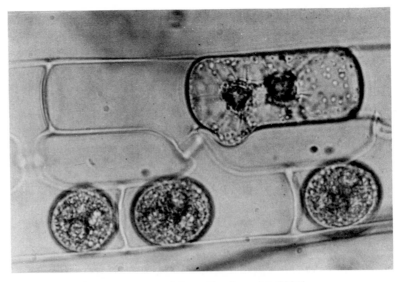

Figure 4-33. *Zygnema,* one vegetative cell and zygotes. 600 X.

The first group of desmids, the placoderm type, has segmented walls and pores in the walls. In sexual reproduction the conjugation tube, when present, is not as well defined.

Cosmarium (Figs. 4-31 and 4-34) is one of the most common of the desmids. It is a single cell with a very obvious constriction in the center of the cell. In the central area the nucleus can be seen; it is frequently apparent in living material. The cells are flattened, as can be seen in end view. There is considerable variation in shape and in texture of the surface, including even a warty appearance. Some species are not biradiate (the two arms of the flattened cell), but triradiate, and thus might resemble some forms of the *Staurastrum* (below). Sexual reproduction results in a very spiny zygote.

Closterium (Figs. 4-35 and 4-36) cells are elongate and sickle-shaped. The median constriction may be *very* slight. Clusters of gypsum crystals, moving about by Brownian movement, can be seen at the two apices. Sexual reproduction, involving homothallic or heterothallic populations, has been observed in both the field and laboratory (Fig. 4-37). Two cells adjacent to each other (a chance phenomenon in a nonmotile genus) each break at the juncture of the semicells. The mating types are surrounded by a gelatinous matrix. The protoplasts move toward each other and fuse, along with nuclear fusion (Fig. 4-37). The zygote has numerous small bumps on the wall.

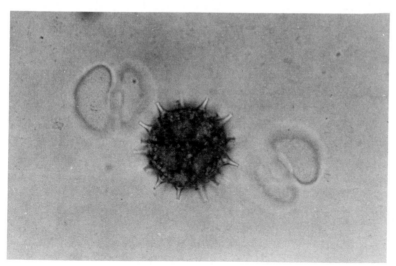

Figure 4-34. *Cosmarium* zygote, with empty walls of two gametes. 600 X.

Figure 4-35. *Closterium* vegetative cell. 300 X.

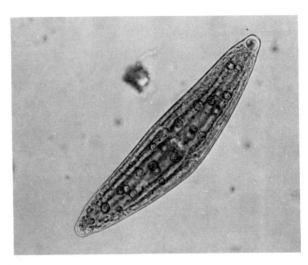

  Staurastrum (Fig. 4-31) cells are quite variable, usually with ornate horns. But the smaller, less ornate forms can be confused with *Cosmarium.*

  Arthrodesmus (Fig. 4-38) cells have two spines on each semicell and on first glance look somewhat like a two-celled *Scenedesmus.* However, the median constriction is not complete and thus one is looking at semicells.

  Micrasterias (Fig. 4-31) is a very much flattened genus, with repeated dichotomous branching of the short arms of each semicell.

  The filamentous desmids might be represented by *Hyalotheca* (Figs. 4-31 and 4-39), in which there is a hint of a median constriction,

Figure 4-36. A zygote of *Closterium.* Note that each of the parent cells has split at the mid-region. The contents of the cells have fused, forming the zygote.

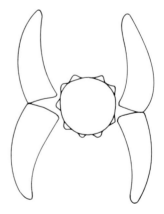

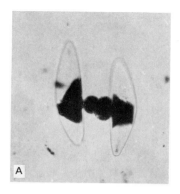
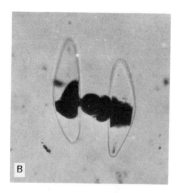
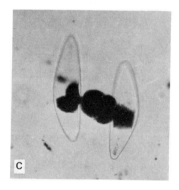
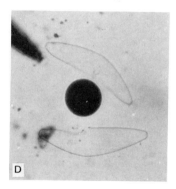

Figure 4-37. *Closterium.* Four stages in sexual reproduction beginning with two cells initiating fusion (A) and ending with a zygote (D). Courtesy of P. Cook. 200 X.

Desmidium (Fig. 4-31) in which a definite spiral twist is apparent, and *Bambusina* (Fig. 4-31) which has ornate, barrel-shaped cells.

Other desmids, the saccoderm type, have unsegmented walls, no wall pores, and have a conjugation tube. There is no median constriction evident in the elliptical *Mesotaenium* (Fig. 4-31), the elongate, spiny *Gonatozygon* (Fig. 4-31), and the elongate *Spirotaenia* (Fig. 4-31). The latter has a spiral plastid.

Group Five
Chara and Relatives
Charas have quite complex plant bodies, when compared to those previously covered, perhaps with a parallel only in the siphonous types. Growth is by divisions of an apical cell, and there are nodes, where "branches" and "leaves" originate, and internodes (Figs. 4-40 and

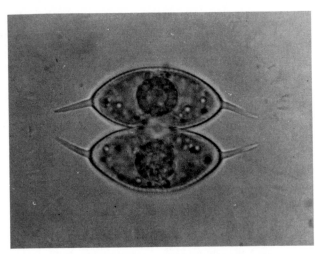

4-41). These are not true branches and leaves, as found in vascular
plants, for the algae do not have the conducting tissues, xylem and
phloem. Some prefer to use the terms branchlike and leaflike. Every
other cell of a filament elongates considerably. Branches occur from the
shorter cells, the nodal area.

A cortex can develop by growth of nodal cell derivatives up or down
along the internodal cell (Fig. 4-40). The main point of interest, however,

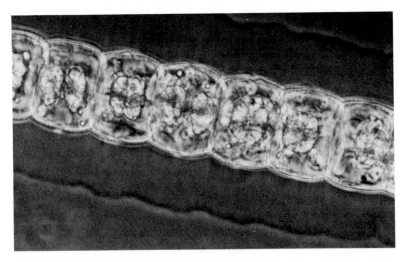

Figure 4-39. *Hyalotheca*, a filamentous desmid, with a broad expanse of mucilage.
600 X.

would be the sex organs and the interpretation of the origin of sterile cells associated with them. Could they be just a little more primitive than the antheridia and archegonia of land plants?

A longitudinal section of the plant apex will show a very precise development of nodes, internodes, "branches" or "leaves" (Fig. 4-40). The latter structures develop in whorls (Fig. 4-40), giving the plant an appearance similar to an *Equisetum*. Elongation of the internode cells, increasing their length many times, results in plants up to 25 cm or more in length. Attachment to a muddy substrate in a pond is by means of a well-developed rhizoidal system.

Individual cells are uninucleate and have many discoid chloroplasts. The organisms are frequently used for demonstrations of cytoplasmic streaming in internodal cells. The entire plant body, including reproductive structures, may be calcified; fossil *Charas* are known. Fossilized female reproductive structures are frequently illustrated in textbooks.

The three genera commonly found are *Chara* (Fig. 4-40), *Nitella,* and *Tolypella*. All are found in fresh water, but a few brackish water species have been recorded. None are marine.

There has been considerable discussion concerning the sexual reproductive bodies. The male structure is called the globule (Fig. 4-42) and the female the nucule (Fig. 4-42). Both structures are precisely formed by divisions of the apical cell, with the sexual structure on a short stalk or pedicel.

The female structure bears one egg, which is surrounded by elongate and twisted tube cells and capped by one or more sets of dome cells (Fig. 4-40). *Chara* has one tier of dome cells on the female reproductive body, while *Nitella* has two tiers. The male structure bears numerous antheridia in filaments, which develop from central cells in the globule. In addition, there are spoke cells and shield cells (Fig. 4-40) in the complex organ.

As one examines the form and the development of both the male and female structures, it appears that a precise relationship to higher plant sex organs, as well as similar structures in other algae, is not clear. Perhaps those found on *Chara* are a little more advanced than on most algae, because of the sterile cells, but they are not in a direct line toward the evolution of sex organs on land plants. Based on how one interprets the development of the globule and nucule, and what weight one gives this development and the sterile cells, *Chara* and *Nitella* can be classified in several ways. In some treatments of the algae they are even put in

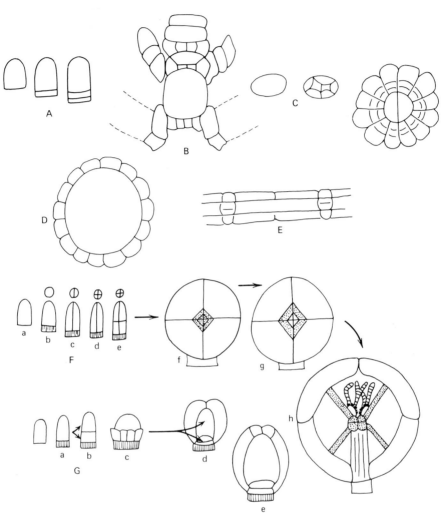

their own division, indicating independent origin, but not with close affinity to land plants. Here I classify them as one of five separate groups in the green algae.

EVOLUTION OF THE GREEN ALGAE

In the past, the green algae have been classified in a variety of ways, and in the present treatment they are put into five separate groups. Thus

Figure 4-40. *Chara,* structure and development. All detail microscopic.
A. The apical cell and the division at the base to form nodal cells.
B. A longitudinal section through the apex of the plant, showing the alternate nodal and internodal areas, and how these are close to each other until there is elongation of the internodal cell. Note the branches forming.
C. Cross sections. First at the apical cell. Middle, at the node when there have been several divisions to give cells that can form branches. Finally, the cross section near the base of branches.
D. Cross section at an internode, showing the large central cell and the cells of the cortex.
E. Surface view, showing how the cortications arise from the nodes and grow up and down the surface of the internodal (central) cell.
F. Formation of the male structure. In the upper series (a), the cells are in cross section, while in the lower series (b-g) the view is lateral. The series shows the divisions that take place to form the male structure, which is mature in h.
G. Formation of the female structure. A series of divisions (a-e).

emphasis is placed on motility, vegetative cell division, the type of sex organs or complications of the sexual cycle, or the multinucleate condition. In addition to showing possible relationships, this division places organisms conveniently in five distinct categories.

The subject of cytological investigations of green algae which deal with their evolution was introduced in Chapter 2. Much of the early work was done with *Ulothrix* and its relatives. Now that the ultrastructure of a number of the green algae has been examined in some detail, some researchers feel that there are probably just two main evolutionary lines in the green algae.

Figure 4-41. *Chara* plant body, fertile. Life size.

Figure 4-42. *Chara,* with both male and female reproductive structures. 80 X. Courtesy of L. Shubert.

Chlorophytes and Charophytes

After examining the data on nuclear division and cytokinetic phenomena, Pickett-Heaps, as well as Stewart and Mattox, has separated the green algae into the chlorophytes and the charophytes. Several decades ago the green algae were put into two classes with these same names, but only *Chara* and a few other organisms were in the second class. That classification was based on gross morphological features and the morphological complexity of the sex organs.

The charophytes as now defined have a persistent, open mitotic spindle. An open spindle is one that is not surrounded by a nuclear envelope. Rather, the nuclear envelope fragments. Because of the persistent spindle, the two nuclei produced by division are separated by some distance. The microtubules involved in cytokinesis are perpendicular to the eventual plane of cytokinesis. A cell plate and phragmoplast such as found in land plants are typically formed (see Chapter 2).

The chlorophytes, which include many of the organisms covered up to this point, have closed spindles, that is, a spindle formed within the nuclear envelope. The spindle apparatus collapses early, soon after the nuclei have divided, so that the new nuclei remain very close. They may even flatten against each other. The microtubules involved in cytokinesis are between the newly produced nuclei and are parallel to the plane of cytokinesis. In many cases this arrangement of cytokinetic tubules is called a phycoplast (see Chapter 2).

With the recent interest in microtubules some attention has been paid to green algal zoospores. The battery of microtubules associated with the flagellar apparatus has been called the cytoskeleton.

Ultrastructural studies demonstrate differences in the zoospore cytoskeleton of chlorophycean and charophycean algae (Fig. 4-43). The former type of cytoskeleton, found in many flagellated genera and

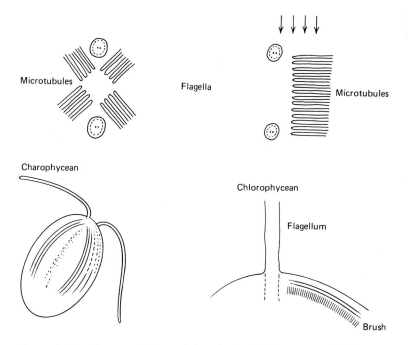

Figure 4-43.   Diagrammatic presentation of microtubule arrangement in the chlorophycean and charophycean types of zoospores.

Left. Chlorophycean cytoskeleton. *Upper,* view from the apex, showing the flagella in cross section, as well as the cruciate arrangement of groups of microtubules (in four's). *Lower,* diagram of a chlorophycean zoospore cytoskeleton.

Right. Charophycean cytoskeleton. *Upper,* view from the apex, showing the flagella in cross section, and a broad band of microtubules. If viewed from the side (arrows), there is the view seen in the lower diagram. *Lower,* the brushlike structure below the band of microtubules.

zoospores of both filamentous forms and certain multinucleate algae, has four cruciately arranged flagellar roots. The roots may have the same number of microtubules. Thus there would be four groups of four microtubules each (4-4-4-4). But other arrangements such as 5-2-5-2 are also found.

Other filamentous algae, such as *Klebsormidium* and *Chara* have the charophycean type of zoospore, which has a cytoskeleton quite similar to those of flagellated cells in higher plants (liverworts, mosses, *Equisetum,* and *Zamia*). The flagella in the charophycean type are associated with a broad band of microtubules, composed of 16 to 30 tubules (Fig. 4-43). This ultrastructural feature is clearly unrelated to the nuclear and cytokinetic phenomena, but again points to two separate groups of green algae.

How much weight should be placed on cytological, especially ultrastructural, detail? Certainly phenomena associated with nuclear division are most basic and are considered conservative features or characteristics. As additional genera are examined and more data are available, it appears that these ultrastructural characteristics should be considered fundamental when we are classifying the green algae.

Some recent research with enzymes associated with peroxisomes has provided additional data to support the chlorophyte-charophyte separation. In higher plants, glycolate oxidase, along with catalase, are found in peroxisomes or microbodies. These enzymes are involved in the oxidation of glycolate; catalase is involved in the removal of hydrogen peroxide released in this process.

Some research reports have stated that algae lack peroxisomes, but quite similar bodies *as well as* glycolate oxidase and catalase have been found in several green algae. All of these algae belong to the charophyte line!

Green algae in the chlorophyte line use glycolate dehydrogenase in their glycolate metabolism. It is not known where this activity is located. (However, in some algae glycolate is oxidized in mitochondria.) With these organisms there is no hydrogen peroxide production.

These separate lines of research support the separation of this group into two evolutionary lines. The charophytes, ancestors to higher plants, possess glycolate oxidase, have a cytoskeleton with a broad band of microtubules, and have a persistent, open mitotic spindle. The chlorophytes possess glycolate dehydrogenase, have a cytoskeleton with a cruciately arranged group of microtubules, and have a closed mitotic spindle, with early collapse of the apparatus.

REFERENCES

Bonneau, E. 1977. Polymorphic behavior of *Ulva lactuca* (Chlorophyceae) in axenic culture. I. Occurrence of *Enteromorpha*-like plants in haploid clones. J. Phycol. 13: 133–140.

Bourrelly, P. 1966. *Les algues d'eau Douce.* Initiation a la systematique. Tome I. Les algues vertes. N. Boubee and Co., Paris. 511 p.

Chapman, V. and D. Chapman. 1973. *The algae.* MacMillan and Co., London. 497 p.

Cook, P. and L. Hoffman. 1971. Bibliography on Chlorophyta. In J. Rosowski and B. Parker (Eds.) *Selected Papers in Phycology.* Dept. Botany, Univ. Nebraska, Lincoln. pp. 768–76.

Fritsch, F. 1935. *The Structure and Reproduction of the Algae.* Vol. I. The University Press, Cambridge. 791 p.

Gayral, P. 1975. *Les algues: Morphologie—Cytologie—Reproduction—Ecologie.* Doin, Paris. 166 p.

Pickett-Heaps, J. 1975. Green Algae: Structure, Reproduction and Evolution in Selected Genera. Sinauer Assoc., Inc., Sunderland, Mass. 606 p.

Proctor, V. 1975. The nature of charophyte species. *Phycologia* 14: 97–113.

Round, F. 1973. *The Biology of the Algae.* St. Martin's Press, New York. 278 p.

Sager, R. and S. Granick. 1954. Nutritional control of sexuality in *Chlamydomonas reinhardi. J. Gen. Physiol.* 37: 729–42.

Smith, G. 1950. *The Fresh-Water Algae of the United States.* McGraw-Hill, New York. 719 p.

Starr, R. 1955. *A Comparative Study of* Chlorococcum *Meneghini and Other Spherical, Zoospore-Producing Genera of the Chlorococcales.* Indiana Univ. Press, Bloomington. 111 p.

Starr, R. 1964. The culture collection of algae at Indiana University. *Amer. J. Botany* 51: 1013–44.

Stewart, K. and K. Mattox. 1975. Comparative cytology, evolution and classification of the green algae with some consideration of the origin of other organisms with chlorophylls *a* and *b. Bot. Rev.* 41: 104–35.

Stewart, W. 1974. *Algal Physiology and Biochemistry.* Univ. Calif. Press, Berkeley. 989 p.

Trainor, F., J. Cain, and L. Shubert. 1976. Morphology and nutrition of the colonial green alga *Scenedesmus:* 80 years later. *Bot. Rev.* 42: 5–25.

Wood, R. and D. Harrington. 1971. Bibliography on the Charophyta. In J. Rosowski and B. Parker (Eds.) *Selected Papers in Phycology.* Dept. Botany, Univ. Nebraska, Lincoln. pp. 811–5.

Euglenophyceae — the euglenoids

When we think of organisms that exhibit both plant-like and animal-like characteristics, we usually think of *Euglena* and its relatives (Fig. 5-2). These organisms are almost exclusively freshwater forms and are quite interesting to study, not only because of their unusual features, but also because they are large algae, at least for unicells. They are found in both moving and stationary water, and in the soil and brackish muds. They are especially common, and can achieve bloom proportions, when the organic level is increased, as in farm ponds. Euglenoids commonly require at least one vitamin, and strains of *E. gracilis* are used in vitamin B_{12} assays.

EUGLENOID THALLUS TYPES

The euglenoids occur mostly as flagellated unicells. The size and shape of the cells vary considerably, even in a single organism. Many representatives are capable of considerable change in shape. When they are not moving by flagellar activity, limited motion is accomplished by changing cell shape (Fig. 5-2). During some stages of the life history, the organism can assume a spherical shape, or it might encyst. Unicells may be enclosed in a common gelatinous matrix, forming a primitive colony. However, few members are colonial, and the most advanced plant body type is a sessile, dendroid colony.

EUGLENOID CHARACTERISTICS

1. Pigments

This is one of the few algal groups that has both chlorophyll *a* and chlorophyll *b*. Other pigments are beta carotene and a series of xanthophylls. Sometimes the xanthophyll, astaxanthin, is called euglenorhodone or haematochrome. *Euglena sanguinea* can accumulate sufficient astaxanthin to give a red color to the small pond in which it is growing. The pigments are in plastids which may be lenticular, bandlike, ribbonlike, reticulate, or stellate. The thylakoids may be in groups of from two to six, but are usually in threes. There are no girdling lamellae. In the individual cells a red-orange eye spot is apparent.

2. Storage Products

The food reserve is paramylon or paramylum, an insoluble crystalline carbohydrate. It is composed only of linear aggregations of glucose

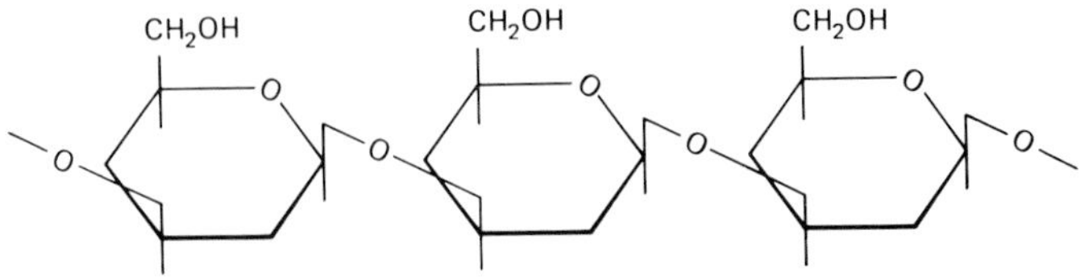

Figure 5-1. Chemical structure of paramylon, a beta, 1:3 linked glucan.

residues, a beta 1:3 linked glucan (Fig. 5-1). Chrysolaminarin (discussed with diatoms in Chapter 8) is also a beta 1:3 linked glucan, but it is branched. Since the food reserve of brown algae is also a 1:3 linked glucan, such compounds are among the most abundant carbohydrates on earth. The food reserve can be seen with the light microscope, for rather large paramylon bodies are formed, sometimes just a few per cell. They are quite symmetrical. In some species a "pyrenoid" has been reported, a center around which the food reserve is formed. Paramylon can be either inside or outside plastids, depending on the genus observed.

3. Motility

Euglenoids are typically flagellated. One flagellum is commonly reported in some forms, but probably a second, shorter flagellum is not seen, especially when it does not emerge from the reservoir, or "gullet" as in *Euglena*. The emergent flagellum of *Euglena*, with one row of hairs, is the tinsel type. Some euglenoid flagella have additional smaller hairs. When two flagella are present, they may be either of equal or unequal length. Three flagella have been reported in one species, and *Hegneria* is said to have seven.

Figure 5-2. Euglenoids. Microscopic.
A. Typical *Euglena* cell with the emergent flagellum showing, the anterior reservoir, numerous plastids, a single nucleus, and two paramylum bodies. The striations, the bands of the periplast, are also evident on some of the larger forms, even under light microscopy.
B. Various shapes that one *Euglena* can assume, as it moves its periplast.
C. A cross section of one of the bands of the periplast, very much enlarged when compared to the scale of the previous diagrams. Observed with the electron microscope.
D. An elongate *Euglena* cell.
E. *Phacus*, a flattened relative of *Euglena*, with a rigid periplast.
F–H. *Trachelomonas*. F, G, and H show various forms of the lorica, with a collar and various kinds and numbers of spines.
I. *Trachelomonas* as it would appear in section. There is an outside lorica, numerous plastids (shaded) adjacent to the plasmalemma, and a nucleus.

4. Cell Covering

Euglenoids have no wall. However, they do possess a cell covering, the pellicle or periplast, located just inside the plasmalemma (Fig. 5-2). If one were inclined to think of a wall as a necessary plant characteristic, then

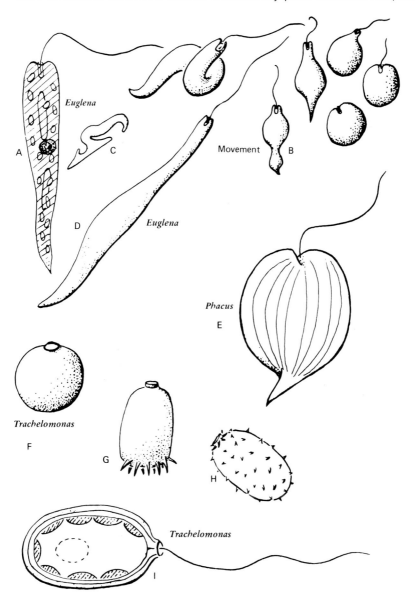

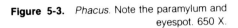

Figure 5-3. *Phacus.* Note the paramylum and eyespot. 650 X.

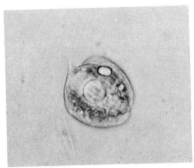

euglenoids are not plants. But do animals photosynthesize? It is difficult to make distinctions among organisms in different kingdoms if one relies on a single characteristic, such as presence or absence of a wall or ability to photosynthesize. It is best to examine many characteristics.

With *Euglena spirogyra* it is easy to show that the pellicle is actually composed of interlocking elongate bands, which are wrapped around the cell in a spiral manner. A cross section of a pellicle band, observed under the high magnification of EM, shows considerable structure. The "hook" of one band fits into the "groove" of the next band (Fig. 5-2). Thus the bands can slide along each other, *along* their long axis. When this sliding occurs, there is motion of the whole euglenoid (Fig. 5-2). This movement has been called metaboly.

The pellicle may be quite ornate on the exterior, with a series of warts apparent along one of the ridges of the pellicle band. Some ornamentation is exterior to the membrane. These warts can also be seen with the light microscope, and with *E. spirogyra,* the ornamentation of one band can be seen moving in one direction. Beneath the pellicle there are microtubules and mucilagenous vesicles. The vesicles could be involved in production of the pellicle components.

Figure 5-4. *Trachelomonas* cell with the opening in the lorica toward the right. Note spines, even on the collar of the opening. 650 X.

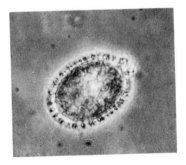

The structure of the pellicle of *Phacus* (Fig. 5-3) is similar to that of *E. spirogyra,* but there is no movement of one band against another, and thus no metaboly. *Trachelomonas* cells (Fig. 5-4) also possess a pellicle, but in vegetative cells this is surrounded by an elaborate lorica. With the light microscope the lorica appears to be a solid cell covering with an opening through which the flagellum emerges. In SEM studies some forms still appear to be solid, while others have numerous small holes. The number, size, and distribution of spines on the lorica have been found useful in classifying species.

One might wish to question the distinction made between wall and pellicle, inasmuch as both are really cell coverings. However, the fact that the plasma membrane is outside the pellicle, and inside a typical plant wall, should make the distinction clear.

The mucilage produced typically by some forms, and formed only at specific points in the life history, must be also mentioned when discussing wall components. This material is clearly outside the plasmalemma.

REPRODUCTION

Euglenoids can increase their numbers by cell division, even while they are in motion. After nuclear division and duplication of cell organelles, one will find two series of organelles associated with locomotion. There is a cleavage between the two canals leading into the reservoir. This cytokinesis then apparently follows the line of a pellicle band in a helical path toward the distal end. In a similar fashion a *Trachelomonas* cell may divide and soon one of the products of division emerges from the lorica, but just to the rim of the opening. Once this one protoplast is outside, and while still attached, the pair can swim in tandem while one new lorica is formed.

Division can also occur while there is abundant mucilage around a group of cells; thus there is an increase in numbers in nonmotile forms. In addition, these flagellates can assume spherical shapes and encyst. There can be numerous concentric layers of wall material around the cyst. Under favorable conditions, the protoplast within the cyst wall enlarges, the cyst is ruptured, and a new individual emerges. Perhaps euglenoids survive some types of adverse conditions while in such a morphological state.

Sexual reproduction has been reported for *Euglena,* as well as for other forms. However, inasmuch as these observations were made over

60 years ago, and no one has recently demonstrated sexual reproduction, even in numerous recent culture studies, there must remain some doubt. The older reports show fusion of sister nuclei, forming a zygote nucleus. Meiosis occurs and four daughter cells result, once cytokinesis takes place.

REPRESENTATIVE GENERA

A number of the organisms classified with the euglenoids and often studied with them are colorless forms that have evolved from the same or similar stock. Here we examine only the photosynthetic members of this group.

Euglena

This organism (Fig. 5-2) is a unicellular, free-swimming, photosynthetic representative in which the cell is never completely rigid. The euglenoid motion may be at times quite pronounced. There are many plastids per cell; they may be discs or ribbonlike and each may or may not have a pyrenoid.

Eyespot, or Stigma.   The pigments of the eyespot are not located in a plastid or chromatophore. However, they are located in several membrane-bound vesicles that are adjacent to the border of the reservoir. Lutein, beta carotene, and cryptoxanthin are pigments found in stigma vesicles. Microtubules are associated with the *Euglena* stigma.

Flagellum.   In a "typical" *Euglena* (Fig. 5-2) only the long flagellum emerges from the reservoir. It bears a single row of elongate hairs. The long flagellum has a conspicuous thickening at one point at the proximal end. The thickening is seen on that portion of the flagellum still within the reservoir, near the stigma and it could be the photoreceptor. Thus when the organism rotates while swimming, the eyespot filters the light that reaches the flagellum swelling or photoreceptor. Apparently the stimulus of alternately receiving filtered and nonfiltered light enables the organism to respond in its motion to the direction of the light. On the other hand, some researchers feel that the photoreceptor could be the stigma or structures associated with the stigma.

A shorter flagellum, once thought to be a portion of a forked base of the long flagellum, remains within the reservoir. When there are two or more emerging flagella, the short type is lacking.

Nucleus.   This organelle is large and division figures are quite interesting. However, some of the associated phenomena are unlike what we typically think of as almost universal in nuclear division. Chromosomes divide and sister chromatids are formed. However, the chromosomes lack centromeres and the nucleolus does not "disappear," but it also divides. In addition, those microtubules involved in division are formed only within the nucleus, the nuclear membrane remains intact, and there is no typical spindle.

Vacuole system.   The anterior invagination is called a gullet or reservoir. Since it is not normally involved with food gathering in the plastid-bearing euglenoids, it would be best to use the latter name, reservoir. Not only are the flagella attached at the base of the reservoir, but the stigma is also located here. Smaller vacuoles discharge into a larger vacuole, which is adjacent to the reservoir. The large vacuole discharges into the reservoir. Recently it was observed that protein macromolecules in the reservoir can be recovered by vesicles. There is thus uptake by pinocytosis.

Astasia
Astasia is a colorless member of the group. Although I will not discuss the remaining colorless forms, this organism is of interest because some forms, without plastids and eyespots, are otherwise identical to species of *Euglena.*

Phacus
This organism has a very conspicuous pellicle. Before we thoroughly understood the structure of the pellicle we could see the striations on the surface of *Phacus.* These are the edges of individual components of the *Phacus* pellicle. The bands of the pellicle are rigid, and thus there is no metaboly with this organism. Motility is strictly by action of the flagellum.

It is while these cells are in motion that one can see the flattened appearance. Many species are quite flattened, with a disc shape (Fig. 5-3). From the posterior end a rigid, curved tail emerges.

LOSS OF PLASTIDS
If euglenoids are placed in the dark, they become colorless. It might take several divisions of the population, but eventually only colorless proplastids are present in cells. (Green algae do not typically lose their color or plastids in darkness.) When placed back in the light, the plastid structure is re-formed and the green color returns. In the 1940s Provasoli et al. reported that colorless mutants could be produced by exposure to

streptomycin. Now we know that both heat treatment and UV can also produce these mutants. These colorless euglenoids closely resemble the colorless protozoan genus, *Astasia.*

The loss of both chlorophylls *a* and *b* tells us something about the synthesis of these pigments, and that the dark block is early in the development of the chlorophyll molecule. Some have suggested that *Euglena* is now evolving toward *Astasia.*

NUTRITION

Some euglenoids, as *E. gracilis,* strain Z, are widely used as assay organisms for vitamin B_{12}. Because they have an absolute requirement for the vitamin, the level of growth in a medium with an unknown quantity of the vitamin will indicate how much vitamin B_{12} is present.

CLASSIFICATION

Some feel that the euglenoids are really not true algae, and thus they are not treated in all phycology books. This decision would be based on their animal-like characteristics. But can we ignore a group of photosynthetic organisms that appear to be intermediate between the plant and animal kingdoms? It might be sufficient to indicate that they probably did not arise from any other algal group, but do have some algal features. On the other hand, they could have evolved, by way of endosymbiosis, with algae containing similar pigmentation. Later they developed their distinguishing characteristics. For a modern treatment of the classification of these forms, consult recent works by Leedale.

REFERENCES

Dodge, J. 1973. *The Fine Structure of Algal Cells.* Academic Press, New York. 261 p.
Gojdics, M. 1953. *The Genus Euglena.* Univ. Wisconsin Press, Madison. 268 p.
Gomez, M., J. Harris and P. Walne. 1974. Studies on *Euglena gracilis in aging culture.* I. Light microscopy and cytochemistry. II. Ultrastructure. *Br. Phycol. J.* 9: 163–93.
Kivic, P. and M. Vesk. 1974. Pinocytotic uptake of protein from the reservoir in *Euglena. Arch. Microbiol.* 96: 155–9.
Leedale, G. 1967. Euglenida/Euglenophyta. *Ann. Rev. Microbiol.* 21: 31–48.
Leedale, F. 1967. Euglenoid flagellates. Prentice-Hall, Inc., New Jersey. 242 p.
Leedale, G. and P. Walne. 1971. Bibliography on Euglenophyta. In J. Rosowski and B. Parker (Eds.) *Selected Papers in Phycology.* Dept. Botany, Univ. Nebraska, Lincoln. pp. 797–802.

6

Rhodophyceae — red algae

**THE COMPLEXITY OF RED ALGAL
LIFE HISTORIES**

CALCAREOUS RED ALGAE

REFERENCES

The red algal group contains organisms that are found in fresh waters, the ocean, and both on and in soil. However, they are somewhat restricted in distribution, for only a few unicellular forms inhabit the soil. There are a few true freshwater forms (Fig. 6-12), while the marine species are found mostly along the coasts. In addition, the marine red algae, which outnumber all of the other macroalgae in the ocean, are most abundant in warmer waters. Many are calcareous.

The red algae are not among the largest of the seaweeds; their length or diameter can usually be measured in centimeters, with sizes between 5 and 25 cm quite common. This is in contrast to the brown algae, many of which can be measured in meters. There are few of the unicellular, colonial, and simple filamentous types, so common in other groups of algae (e.g., the diatoms and green algae). As we shall see later, the brown algae also lack such simple plant body types.

Using knowledge of red algal morphology, reproduction, and life histories, it is convenient to divide these organisms into the more primitive forms, the bangiophytes, and the more advanced florideophytes. We shall discuss the criteria used in making this separation later in the chapter.

RED ALGAL THALLUS TYPES

Primitive Red Algal Thallus Types

Some red algae, such as *Porphyridium*, *Rhodella*, and *Rhodosorus*, are unicells (Fig. 6-1). Cell division is their only means of reproduction. At times unicells aggregate, mainly because the wall material sticks to that of adjacent cells. Uniseriate or multiseriate, branched or unbranched filaments are found (Fig. 6-1). There may be heterotrichy, that is, both a prostrate and erect system, as in *Erythrotrichia*. This is in contrast to the

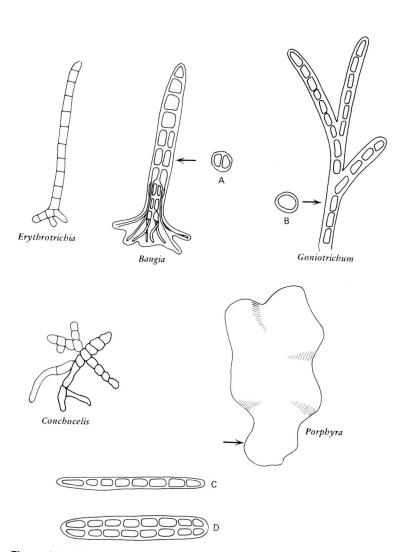

Figure 6-1. Morphological types that can be found in the simpler red algae. These include unicells, unbranched and branched filaments, and a blade (*Porphyra*). The latter may be one or two cells thick (C,D). Some filaments are uniseriate (B) and others are multiseriate (A). The cross sections, A and B, are taken at the points on the thalli marked by arrows.

morphology of *Bangia* (Fig. 6-6), in which the lowermost cells of the erect filament produce rhizoids. The aggregation of rhizoids then firmly anchors the plant to the substrate. *Bangia* is also multiseriate.

Porphyra has the form of a sheet or membrane, firmly attached by rhizoidal growth (Fig. 6-11). Growth of such a plant body, as well as of most of the other organisms in the primitive group, is by division of almost any cell of the thallus, sometimes called diffuse growth. But apical cell division, at least during the initial stages of development from a spore, is possible.

Thallus Types in the Advanced Red Algae

These algae are all branching filamentous forms. In most cases the wall material of adjacent filaments adheres so that a pseudoparenchyma is formed. There is typically apical growth, but in a few red algae in this advanced group, such as some coralline algae, growth is intercalary. All advanced red algae, or florideophytes can be divided into encrusting forms, which adhere closely to the substrate, and those which are erect, and thus frequently more conspicuous.

In the crustose forms, as the spore germinates, filaments radiate from the initial point of spore contact (Fig. 6-2). Then short upright filaments develop from this basal portion; the erect filaments may or may not branch. Filaments are so densely packed together that branching is limited. A great variety of form is possible from this simple growth pattern.

Most advanced erect red algae exhibit the heterotrichous pattern of growth. There is a prostrate system from which the large erect filaments

Figure 6-2. Morphological types in the more advanced red algae. Many of the red algae exhibit a heterotrichous pattern of growth (a-f). A spore (a) on a suitable substrate (f) develops a prostrate system (c, e). In addition, erect filaments can develop (b, d).

Crustose red algae develop by a radiating growth from the germination of a spore (g-i). In lateral view (j-l) it becomes apparent that short erect filaments (arrows in j) can develop from the radiating base. These branched or unbranched erect filaments, which give the crust a limited thickness, may be so densely packed that cell lineage is not immediately apparent (k). In other cases the filaments remain somewhat distinct, but joined in a common matrix (l).

Erect red algae are either uniaxial (m,n) or multiaxial (o). In these filamentous forms the central filaments (p) are typically filaments of unlimited growth, whereas the lateral filaments (q) are capable of only limited growth. Branching may be alternate (m) or opposite (n), or branches may form in a whorl. A cross section of a multiaxial thallus in shown (r).

Heterotrichy

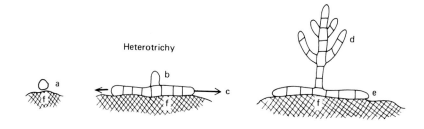

Crustose forms

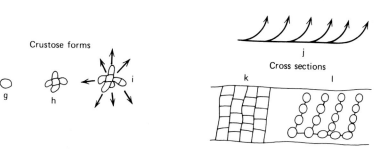

Cross sections

Erect forms

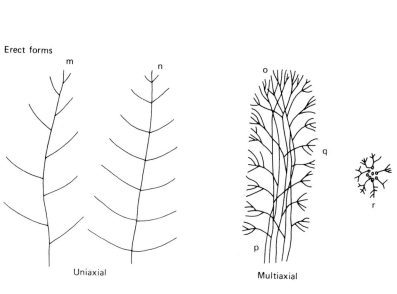

Uniaxial

Multiaxial

develop. The prostrate portion serves as an anchoring base, and it can give rise to new growth in another season, or after damage to the remainder of the plant. The erect system may have one central filament (uniaxial) or several filaments combine to form the axis (multiaxial) (Fig. 6-2). The axial filaments, filaments of unlimited growth, not only provide the "skeleton" framework, but also are mainly responsible for the overall size of the plant. The shorter exterior filaments, which are branches of the axial system, are photosynthetic and are called filaments of limited growth. In a cross section of a thallus with several axial filaments one gets the impression that they make up the medulla where there is little or no photosynthesis, and that the filaments of limited growth are the cortex (Fig. 6-2).

With these erect red algae, variation in form is achieved in a variety of ways previously mentioned, including the duration of growth of filaments, pattern and extent of branching, and the formation of a pseudoparenchyma (Fig. 6-3). If there is elongation and/or enlargement of certain axial cells (Fig. 6-2), quite different forms result. Cell enlargement up to 40,000 times has been noted! Often near the point of contact between elongated central cells, numerous filaments of smaller diameter begin to develop adjacent to the cell surface. Subsequently they grow both up and down the central filament surface, forming a cortex.

Some branches are located essentially in only one plane, resulting in a flat plant body. In other forms, branching is in two planes, giving thickness to the thallus. One type of branch is formed at nodes in a whorl, but branching can also be opposite or alternate. With these possibilities it is not surprising that there are so many different forms of red algae.

The most highly evolved red algae, *Ceramium, Polysiphonia,* and their relatives, do not exhibit heterotrichy. Spore development is bipolar, with the lower portion developing rhizoids, and the upper eventually into the thallus we recognize. Continued development of certain rhizoidal filaments toward the substrate results in a most secure form of attachment. These plants are all uniaxial, and they develop from the activity of an apical cell.

Secondary meristems, dividing tissues, may develop in a number of red algae. These naturally occur in *Rhodymenia,* so that juvenile plants appear on the surface of the thallus. But in other cases they develop after loss or damage to a portion of the erect thallus. Thus regeneration can take place.

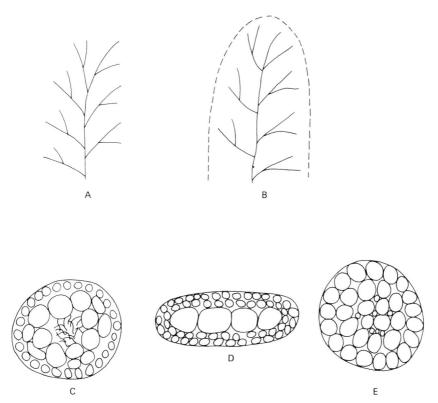

Figure 6-3. Form in the advanced red algae. Some branching filamentous types (A)
may remain as such, whereas others become pseudoparenchymatous (B). The dotted
boundary represents the outer thallus surface. In cross sections of the thallus (C–E) it is
apparent that enlargement of certain cells can alter the form achieved. Diagramatic
presentations are made for *Agardhiella* (C) *Rhodymenia* (D) and *Gelidium* (E).

Pit Connections

Connections from cell to cell can easily be observed in many of the red
algae. These are the pit connections, which EM has shown to be
incomplete septa (cross walls) filled by a dense plug (Fig. 6-4). In cross
section the plug may have the shape of a pulley wheel. However, there
are no pit connections in *Erythrotrichia, Boldia,* and *Compsopogon,* and
they are found only in the *Conchocelis* stages of *Bangia* and *Porphyra.*
Pit connections are most useful in interpreting the sequence of
development of a complex plant body. One can determine the origin of a
cell in a group by locating the pit connection between it and an adjacent
cell.

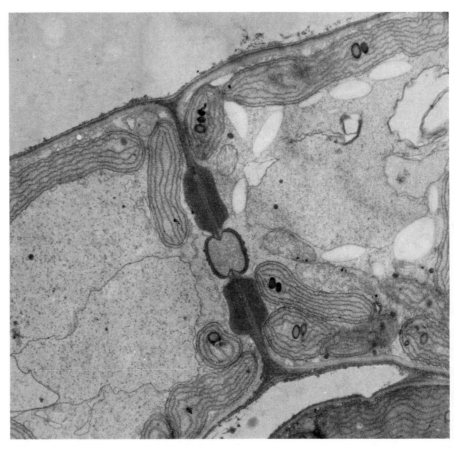

Figure 6-4.   Section of a filamentous red alga showing plug, of the typical red algal pit connection. Courtesy of J. Ramus and the Journal of Cell Biology. 15,000 X.

RED ALGAL CHARACTERISTICS

1. Pigmentation

The pigments of the red algae include chlorophyll *a* and chlorophyll *d,* alpha and beta carotene, a few xanthophylls, *and* the phycobiliproteins. Chlorophyll *d* is present in just a few advanced red algae. All but the phycobilins are in plastid thylakoids, which occur singly. There is, in advanced rhodophytes, one encircling lamella or thylakoid at the plastid boundary. On the thylakoids there are 35 nm structures, the phycobilisomes, which contain the R-phycocyanin and R-phycoerythrin. When the phycobilisomes contain an abundance of phycoerythrin, they are spherical and in orderly arrays on the thylakoid surface. They are disc-like, and might stack to form cylinders, when there is a dominance of phycocyanin. It is the latter pigment, very similar to the phycobilin in the blue-green algae, which gives some red algae a blue-green color. But because the dominant pigment, phycoerythrin, masks other pigments, most red algae are some shade of red. Allophycocyanin is also found in red algae.

2. Storage Products

The reserve foods are floridean starch, found *outside* the plastid, as well as numerous simpler organic compounds. Floridean starch is similar to amylopectin, the branched portion of higher plant starch (Fig. 6-5). With an iodine test a brownish color is formed. It is an alpha 1:4 linked glucan, with branches linked in the 1:6 position. In this respect floridean starch is then similar to cyanophycean starch. In some forms the food reserve is even more branched than floridean starch and is similar to glycogen of animals. Pyrenoids are formed only in the simpler red algae, including some relatives of *Nemalion*.

3. Motility

There is no motility by means of flagellated or ciliated cells; however, there have been reports of restricted amoeboid motion of some spores. Sexual reproduction is well documented for many organisms, especially the more advanced red algae, but the sperm or spermatia are carried passively to the female cell.

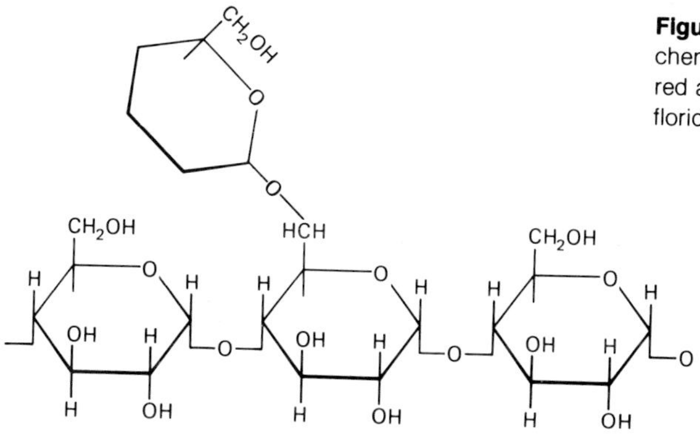

Figure 6-5. The chemical structure of the red algal food reserve, floridean starch.

4. Wall Components

Red algae are important commercially not only as a source of food in some countries, but also because of certain wall components. The walls may contain cellulose, or have a fibrillar component composed of mannans or xylans. But it is the nonfibrillar (outer) component that attracts the most attention. The student first becomes aware of it in assembling herbarium specimens, for red algae adhere exceptionally well to herbarium paper. This property is really due to several chemically different compounds but all are sulphated galactans. They include gelling agents or stabilizers such as carrageenan, obtained from *Chondrus, Gigartina,* and *Eucheuma* and agar, obtained from *Gelidium* and *Gracilaria* (Chapter 20). Other red algae, some of which are important in reef building, for example, *Porolithon,* have a calcareous wall exterior to the type previously mentioned.

REPRODUCTION

Primitive red algae can reproduce by cell division or by the production, release, and germination of simple spores.

Satisfactory information on sexual reproduction is not available for at least some of the primitive red algae. In the more advanced forms, the egg remains attached to the female gametophyte and fertilization takes place. There are numerous postfertilization complications, cell divisions, and cell fusions while the developing diploid structure is still attached to the female gametophyte. These are followed by the growth of numerous short filaments. The eventual result is the formation of carpospores. The diploid tissue producing carpospores represents a distinct stage (the carposporophyte) in many of the life histories. Thus there are gametophytes, carposporophytes, and sporophytes (tetrasporophytes). The latter are independent plants that may be smaller than the gametophyte, or as in many genera, essentially identical in size and shape. Reduction division occurs in the formation of tetraspores. Life histories become as complex as any in the plant kingdom. Several are presented in the following pages.

BANGIOPHYTES AND FLORIDEOPHYTES

In the bangiophytes there are organisms existing as unicells, filaments, and parenchymatous blades. Cell division is usually not restricted to a particular portion of the thallus, but it may be apical. There is minimal cell enlargement. Clear evidence for sexual reproduction is frequently absent; thus we do not completely know many life histories.

The florideophytes are filamentous or pseudoparenchymatous forms that grow by means of the activity of an apical cell, enlargement of products of division (sometimes considerable enlargement), and differentiation. Plants are oogamous, the zygote is retained on the female plant and a "parasitic" phase called the carposporophyte develops. Several types of life histories are known; two types are most commonly observed (pages 156 and 161).

In distinguishing bangiophytes and florideophytes investigators sometimes rely on the number and type of plastids, as well as the presence or absence of both central vacuoles and pit connections. Recent ultrastructural studies of the microscopic phase of several bangiophytes show many similarities to the more advanced florideophytes.

In classifying organisms within these two groups, reproductive structures have been used for the definition of orders. The morphology of the female structure or carpogonium, with associated cells, and postfertilization events are emphasized.

SOME REPRESENTATIVE GENERA AMONG THE BANGIOPHYTES

Porphyridium
Porphyridium (Fig. 6-6) usually serves as an example of a unicellular red alga, although other unicellular forms are known. *Porphyridium* can be found growing on the surface of damp soil, especially in greenhouses, and is also known from marine waters. It has a single axial star-shaped plastid and the nucleus is found between the arms of the plastid (Fig. 6-7). The accumulation of wall material results in the cells adhering (Fig. 6-8), and in aggregation in a layer on soil surfaces. The wall material, if it accumulates at one point, can form a thick stalk. Reproduction is by cell division; no sexual reproduction has been noted.

Erythrotrichia
Erythrotrichia (Fig. 6-6) is an unbranched, filamentous form, usually found epiphytic on other algae in the marine environment. The growth form is an example of heterotrichy, seldom found in the primitive red algae. Reproduction is by means of spores, and nothing is known about its sexual reproduction.

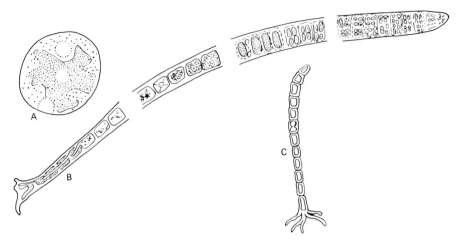

Figure 6-6.   Some primitive red algae.
A. *Porphyridium,* a unicell with a stellate plastid. Microscopic.
B. *Bangia,* in four portions. Rhizoids develop from numerous basal cells. Note that the organism becomes multiseriate, and that divided cells (reproductive) are shown in the segments to the right. Length measured in mm.
C. The epiphytic *Erythrotrichia.* Microscopic.

Bangia

Bangia (Fig. 6-6) is an unbranched, multiseriate, filamentous alga found on rocks in the intertidal zone. It attaches to the rocky substrate (Fig. 6-9) by means of rhizoids arising from the lower cells of the filament (Fig. 6-6). Some distance from the apex one notes that the *Bangia* filament is biseriate, and further toward the base additional divisions can result in a multiseriate filament 15 or more cells in cross section. At the basal portion, with rhizoidal growth possible from several tiers of cells, attachment is quite secure. The organism can survive over the winter by means of this basal portion.

Bangia reproduction.   This life history is the first we will encounter in the red algae in which there are complications. A small filamentous, *branching* red alga, often called *Conchocelis,* has been found to be a stage in the life history of some species of *Bangia.* Later we will see that there are *Conchocelis*-like stages in *Porphyra* species.

The further complication in the life history is that, although there is a *Bangia* stage and a *Conchocelis* stage, the alternation is not obligate. In addition, both stages may have the same chromosome number! For some Atlantic coast (of North America) forms which have been examined

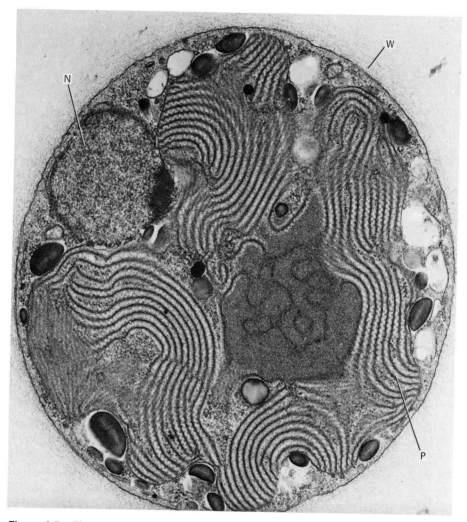

Figure 6-7. Electron micrograph of *Porphyridium* cell. Note wall (W), lobed plastid (P) with phycobilisomes attached to thylakoids, and the nucleus (N). 25,500 X. Courtesy of Gantt, Conti, and Journal of Cell Biology.

in culture, the chromosome number of the *Bangia* and *Conchocelis* portions is 3, while a number of 10 has been reported for Pacific forms.

In culture there has been no evidence for a sexual cycle, with reproduction taking place by asexual spores, often called monospores and carpospores. The term carpospore has been used in an unfortunate

Figure 6-8. *Porphyridium* culture with thousands of cells stuck to the walls of the Pyrex flask. 1½ X.

way in some of these studies. We will see this term used with advanced red algae to describe the spores that result soon after sexual reproduction. With *Bangia,* as well as *Porphyra,* which we discuss next, Conway has suggested that all larger spores be called alpha spores, and

Figure 6-9. *Bangia,* intertidal, firmly attached to the rocks.

smaller, colorless forms be called beta spores. With little solid evidence in a few species for some reproductive phases, these terms now are used by some phycologists.

In experiments with various light-dark cycles, used to simulate both short and long day conditions, any vegetative cell of the *Bangia* stage could produce reproductive cells. If these spores were germinated:

1. In a 24 hour cycle with *less than* 12 hours of light, the germination was bipolar and the organism grew into the *Bangia* stage.
2. In a 24 hour cycle with *more than* 12 hours of light, the germination was unipolar, and the organism grew into the *Conchocelis* stage.

The *Conchocelis* stage, as well as the *Bangia* stage, could be maintained indefinitely by manipulation of L/D cycles. In a 16 L/8 D cycle the *Conchocelis* stage was maintained for 3 years. This response to photoperiod is similar to that of higher plants. Phytochrome is the pigment involved in responses to photoperiod with *Porphyra, Bangia,* and higher plants.

There was no sexual reproduction observed in these studies. According to the older ideas of the life history of *Bangia,* there was zygotic meiosis. This zygote would have resulted from fusion of a free-floating, nonmotile cell and a cell of a *Bangia* filament. If this is found to be true in some cases, then one could still have both *Bangia* and *Conchocelis* stages with the same chromosome numbers.

Other *Bangia* species may respond differently:

1. Some do not react at all to photoperiod.
2. One species has a thallus that is diploid below and haploid closer to the tip!
3. Another species responds to temperature, and completes its life history only with temperature changes. At 9°C, but not at 22°C, there is a *Conchocelis* stage.
4. Sexual reproduction could occur rarely, *if at all.*

Porphyra

Porphyra (Figs. 6-10 and 6-11) is a bladelike marine form that resembles *Ulva.* It occurs in the littoral and upper sublittoral zones attached to rocks and other algae. The blade, a true parenchyma, may be either one or two cells thick (Fig. 6-10). This stage is conspicuous and large, with sheets up to a meter in length. There is also a *Conchocelis* state, and either this stage or the attached portion of *Porphyra* can be responsible for the organism surviving from year to year. Spores could also be involved. Photoperiod changes are also necessary in the *Porphyra-Conchocelis* life history.

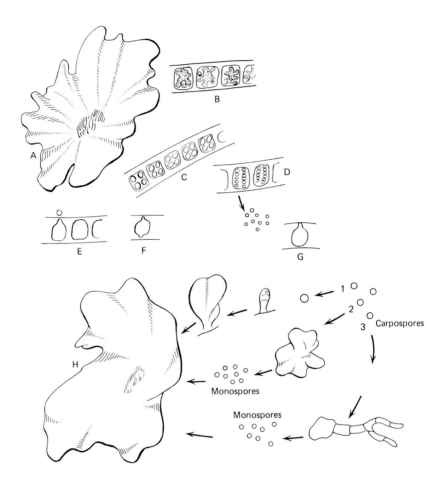

Figure 6-10. Structure and reproduction of *Porphyra*.
A. The *Porphyra* plant. Size: measured in cm.
B. Cross section showing the stellate plastid and the fact that the plant is one cell thick.
C. Generalized diagram to show formation of monospores.
D. Diagram showing formation of spermatia.
E. Showing a carpogonium with one trichogyne.
F. Two trichogynes on a carpogonium.
G. Spermatia moving by currents toward the carpogonium.
H. Three of the interpretations of the life history, including
   1. direct development from the carpospores.
   2. development through a small *Porphyra*, which forms the larger type from monospores.
   3. development through a *Conchocelis* stage, which forms monospores.

Figure 6-11. *Porphyra,* pressed on a herbarium sheet.

As is the case with *Bangia,* the organism is attached by rhizoidal projections from many cells and the individual vegetative cells have a star-shaped plastid with a pyrenoid (Fig. 6-10) as well as a single nucleus. Cell division can occur in all portions of the thallus.

Porphyra reproduction.   Asexual reproduction takes place by means of spores, either monospores or alpha spores. There are reports of amoeboid motion of some spores in some species, but movement is restricted to a distance of a few cell diameters. With both vegetative growth and spore production the *Porphyra* and *Conchocelis* plants could be maintained indefinitely, and apparently some cultures of these organisms and plants in nature reproduce this way.

When we attempt to unravel the complete story of reproduction, involving sexual reproduction and meiosis, the discussion becomes more complex. *Some researchers even suggest that there is no convincing evidence for fusion of sperm* (spermatia in the red algae) *and egg* (carpogonium) in a few species of *Porphyra.* On the other hand, Suto and others in Japan repeatedly make crosses between different populations in attempts to improve the strains used for nori culture (Chapter 20).

First, in order to present certain red algal terminology, one possible *Porphyra* life history involving sexual reproduction is given. Other possibilities will follow and finally we shall see how all observations can be explained, especially in strains where there is little evidence yet for reproduction.

The male cells are nonmotile and colorless; they may be formed in groups of 32, with eight stacks of four each, but the number is variable, depending on the species. The female cell, or carpogonium, has one or two projections (Fig. 6-10) called trichogynes. The latter break through the outer wall material of the *Porphyra* blade and are in contact with the environment. After contact of spermatia and the trichogyne, the protoplasts and nuclei fuse. When the zygote germinates, carpospores are formed.

These carpospores would be diploid in a life history in which reduction division was in the *Conchocelis* stage (sporic meiosis). The carpospores would germinate and develop into a *Conchocelis* stage, which resembles that phase we have previously seen with *Bangia.* Spores would form in a region called a fertile cell row. With sporic meiosis, the haploid number is restored, and spore germination produces another *Porphyra* plant. This type of life cycle would be similar to those we shall see in the higher red algae. The *Porphyra* phase is the gametophyte and *Conchocelis* is the sporophyte (Fig. 6-10).

Other life histories have been reported, for at least certain species of *Porphyra.* In some plants spermatia and carpogonia are formed in the usual way, a zygote is formed, but there is meiosis during the germination of the zygote. With germination of the resulting *haploid carpospores,* there appear to be a number of possibilities in the completion of the life history.

1. Forms in which there is no *Conchocelis* stage.
   a. The carpospores would develop directly into a typical *Porphyra* plant.
   b. The carpospores would develop into a *Porphyra* plant that did not achieve a large size. Spores would be formed and disseminated. They develop into a mature *Porphyra* plant.

2. Forms in which there is a *Conchocelis* stage.
   a. Carpospores would develop into a *Conchocelis* stage which eventually forms monospores. The monospores would germinate and form a *Porphyra*.
   b. Carpospores would develop into a *Conchocelis* stage which then directly produces a *Porphyra* plant. This *Conchocelis* stage does not form spores.

With at least one species of *Porphyra* there has been a report of external fertilization, that is, fusion of dissimilar, nonmotile, reproductive cells, external to the plant.

Why is there so much uncertainty about *Porphyra* life histories? First, it is often difficult to observe fertilization in organisms such as these. Collection of material has to be precisely timed, and, even then, the cytological evidence can be interpreted in different ways. Material in culture could be a great help; certain *Porphyra-Conchocelis* life histories have been completed several times in culture. But when there is a report of external fertilization in cultured material, some will doubt that this really happens in nature. Cultured material might not always show what is typical for an organism. In nature and in culture there have been some suggestions that evidence for carpogonia and trichogynes is really based on observations of portions of fungal components. It would not be

Figure 6-12. *Boldia,* a freshwater red alga discovered in the mid 1950s in cool, rapid mountain streams of the eastern United States.

surprising to have fungal parasites on an alga. Thus some species might lack sexual reproduction.

But can we not confirm the location of meiosis? Good cytological evidence for reduction division is difficult to obtain. With simple plants such as the *Porphyra* and *Conchocelis* stages, it is difficult to know when meiosis is about to occur, so that material can be fixed at precisely the right time.

Perhaps not all species have similar life histories, and we have been attempting to generalize for the genus. It would be unlikely that meiosis would be *either* sporic *or* zygotic in the same species. But an organism might easily bypass spore production, and go directly to formation of the blade from the *Conchocelis* base.

Our information is still incomplete, but we will soon know *where* meiosis occurs in these organisms, *if at all* in some species!

REPRESENTATIVE GENERA AMONG THE FLORIDEOPHYTES

In this group of organisms we will discuss two types of life histories. Either *Bonnemaisonia* or *Nemalion* could be the type example for one, whereas both *Ceramium* and *Polysiphonia* can serve as good examples for the second. Because the haploid and diploid phases in the *Bonnemaisonia*-type are dissimilar, the life history can be called heteromorphic. With the *Polysiphonia* type there is an alternation of isomorphic generations.

Nemalion and Allies

The *Nemalion* plant body (Figs. 6-13 and 6-14) is composed of intertwined, branching filaments. It is found in the littoral zone attached to rocks, is soft and delicate to the touch, and is often said to be wormlike. Plants up to 30 cm have been found, but they are typically 10 cm or so in length. The plant body is sparingly branched (Fig. 6-13).

The center or medulla of the plant body contains many longitudinal, colorless filaments, called the axial filaments, or filaments of unlimited growth (Figs. 6-2 and 6-14). From these, at their margins, tufts of lateral, branching filaments arise in very dense whorls (Figs. 6-2 and 6-14). These filaments of limited growth have beadlike photosynthetic cells that contain one large plastid. A cross section of the thallus, showing the axial system surrounded by the photosynthetic system, is presented in Fig. 6-14.

Reproduction in *Nemalion,* in contrast to that in red algae considered so far, is well understood. The reproductive cells are found within the

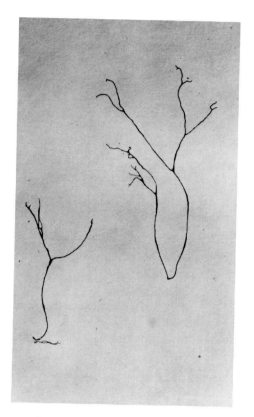

Figure 6-13. *Nemalion* thalli. ½ X.

photosynthetic tuft, but are smaller and lighter in color than vegetative or photosynthetic cells. The female cell, or carpogonium, is the terminal cell in a four-celled carpogonial branch (Fig. 6-14) which is found on a lateral branch of a photosynthetic tuft. Although the plants are homothallic, the early development of spermatia ensures cross fertilization.

Many spermatia may attach to the elongate trichogyne (Fig. 6-14), but there is fusion of just one sperm nucleus with the carpogonium nucleus (Fig. 6-14). Division of the zygote is mitotic. Short filaments, the gonimoblast filaments, then begin to develop from the zygote surface in several directions. With the formation of more filaments in a tuft or cluster, and limited branching, a dense cystocarp develops (Fig. 6-14). The apical cells of gonimoblast filaments act as carposporangia and release carpospores. Upon germination, the latter develop into a sporophyte that is not at all similar to *Nemalion*.

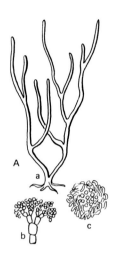

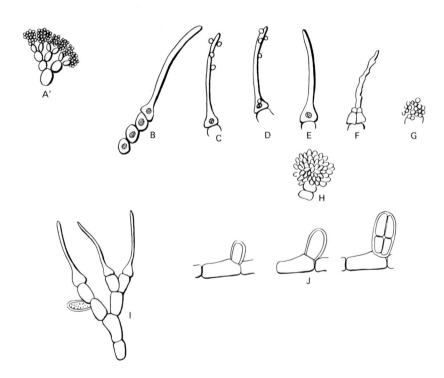

Figure 6-14. *Nemalion* structure and reproduction.
A. Vegetative portions. Plant (a), life size, a photosynthetic tuft (b) and a cross section of the plant (c). In c note central strands, with photosynthetic tufts toward the outside. Length measured in mm.
  A.' A tuft with spermatia at the tip.
B. The four-celled carpogonial branch, with an elongate trichogyne on the carpogonium.
C. Spermatia on the trichogyne.
D. Male (spermatial) and female (carpogonial) nuclei adjacent.
E. The diploid nucleus.
F. There are mitotic divisions, resulting in the formation of young gonimoblast filaments, which are seen at the base of the carpogonium. The trichogyne is withering.
G. The cystocarp, which is a cluster of gonimoblast filaments, with carposporangia at the tips.
H. The mature cystocarp, further developed than in G.
I. An *Acrochaetium*-like plant. This is an alternate in at least some *Nemalion* life cycles, when reduction division does not take place between E and F. Note the hair tips.
J. Development of the tetraspores on an *Acrochaetium*-type plant.

The Nemalion sporphyte.   Umezaki and Fries independently have recently suggested that for at least some *Nemalion* species the sporophyte phase could be similar to the tetrasporophyte of other higher red algae. They have reported germination of carpospores of *Nemalion* into small, branched filaments which resembled *Rhodochorton* or *Acrochaetium* (Fig. 6-14). These filaments gave rise to groups of spores in fours, the tetraspores (Fig. 6-14). It would thus appear that alternation of generations in *Nemalion* is heteromorphic. (Not too many years ago many of us learned that, based on the work of Kylin, meiosis was zygotic. In that case, one would not have looked for a tetrasporophyte.)

Bonnemaisonia.   This organism is also one of a series of algae, in several different classes, in which we will have to describe two plant body types, for the haploid and diploid plants are quite dissimilar in macroscopic appearance, and each formerly had a different generic name (Fig. 6-15).

Bonnemaisonia, the haploid phase, is a bushy, frequently entangled plant about 10 cm in height. The plant body is irregularly branched, with conspicuous main axes and brushlike ultimate branches. Internally it is composed of a multiseriate system, with photosynthetic branches to the exterior. There are numerous, enlarged, fleshy hooklike branches randomly positioned on the thallus. In one species, the hooks resemble what one would find at the end of a tow chain, but there can be different

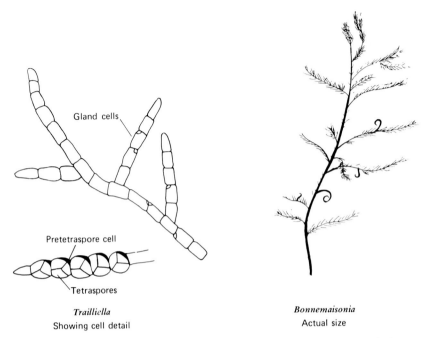

Gland cells

Pretetraspore cell

Tetraspores

Trailliella
Showing cell detail

Bonnemaisonia
Actual size

Figure 6-15. The *Trailliella* and *Bonnemaisonia* stages in the life history of a single organism.

amounts of curvature, with even a complete coil around the small branch of another alga.

The male and female reproductive structures are very similar to those of *Nemalion*. However, during the formation of the cystocarp, sterile filaments, which soon become pseudoparenchymatous, surround the developing cystocarp (Fig. 6-14). An urn, or pericarp, results, with the carpospores within.

The diploid phase of *Bonnemaisonia hamifera,* not too long ago called *Asparagopsis hamifera,* is a smaller, 2 to 3 cm, sparingly branched, uniseriate, filamentous form, which had not at first been recognized as connected with this genus. It had been called *Trailliella intricata.* *Bonnemaisonia* sporophytes appear macroscopically as small tufts attached to other algae, or free-floating. Individual cells are elongate and contain many plastids. Much smaller refractory gland cells are frequently found alternating with photosynthetic cells. Tetraspores, frequently found in a series, are formed as a result of reduction division. A tetraspore upon germination forms a *Bonnemaisonia* gametophyte.

This complete life history has been examined both in the laboratory and in nature. It is complicated by the fact that only female plants are recorded in some areas, at least in any abundance. Carpospores are sometimes formed under such conditions, but they are not viable.

A Life History with Alternation of Isomorphic Generations

Polysiphonia and Ceramium.   These two genera are commonly found in marine waters. Both genera contain a number of species. There is a tetrasporic stage and the life histories contain isomorphic haploid and diploid plants. As we shall soon see, some scholars interpret the life history differently because there is an attached diminutive carposporophyte. One can easily distinguish most forms of these two genera, at least with the aid of a hand lens, for *Ceramium* has clawlike tips, and one can detect the enlarged *Polysiphonia* cells, or siphons.

   Polysiphonia (Fig. 6-17) is an erect plant, growing attached to other vegetation and other algae as well as shells, stones, and woodwork. It is found in the littoral and sublittoral zones.

   With *Polysiphonia* and all members of the order, germination of the spore and subsequent cell divisions do not result in a prostrate system from which erect filaments develop. Instead, with the first division of the spore, a basal rhizoid forms. The upper cell resulting from that division gives rise to the plant body we recognize. There is apical growth (Fig. 6-18) from activity of a single, dome-shaped apical cell. First there is a uniseriate row of cells and then one sees small branches of two types. The first might persist and become true branches (Fig. 6-16), while the second are the trichoblasts (Fig. 6-16).

   Trichoblasts are uniseriate branches that have a limited amount of growth and are usually deciduous. They are either lighter in color than photosynthetic cells, or are colorless, with leucoplasts. They are important in two ways. First, branching occurs in many species by way of trichoblasts, with the branch eventually resulting from development of the basal cell. That is, once the trichoblast falls from the plant, the basal cell, still attached, can grow and a polysiphonous branch develops. Second, the reproductive structures are commonly formed on trichoblasts.

   A short way from the tip of the uniseriate axis one can see the development of the siphons of *Polysiphonia* (Fig. 6-16). By longitudinal divisions of a cell, the pericentral cells (Fig. 6-16) are soon cut off in regular fashion. In some cases, there are as few as four, with a central or axial cell in the middle. There can be more than 20 pericentral cells,

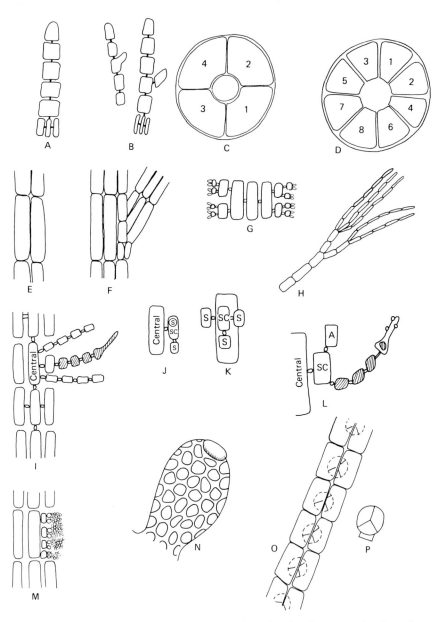

with the number fairly constant in a species. Cortications can be found.
These develop from an orderly division of the pericentral cells
(Fig. 6-16), and one can easily follow the origin by tracing the
arrangement of the pit connections. It is sometimes difficult to see the
pericentral cells, except at the tips or near the tips, in some highly

Figure 6-16. *Polysiphonia* structure and reproduction. Detail of cells and filaments is microscopic.

A. Longitudinal section through an apical portion.
B. Similar section, but at a point where there was a branch. Note multiseriate portion below.
C. Cross section after there were divisions producing a multiseriate region (as in B). Note the central cell and four pericentral cells.
D. Similar to C, but there are eight pericentral cells. In this section, as well as in C, the numbers indicate the order in which the pericentral cells were formed.
E. Surface view of an organism with four pericentral cells.
F. Surface view of a form with more pericentral cells and a branch.
G. Longitudinal section showing central cell, and pericentral cells adjacent to it. From the pericentral cells, on both sides, a cortex has developed. One can follow the development of the cortex by examining the arrangement of pits between cells.
H. A trichoblast.
I. Longitudinal section showing a carpogonial branch on a pericentral cell (now called a supporting cell). The cells above and below (similar cells surrounding the carpogonial branch were not diagrammed) eventually develop into an urnlike container.
J. Detail showing that there are sterile cells associated with the supporting cell.
K. Same as J, but examining the supporting cell face on.
L. The position of the auxiliary cell, formed after fertilization.
M. Spermatia.
N. This is the pericarp. This is around all structures from I–L, especially when the latter are mature. The pericarp with contents is the cystocarp.
O. A *Polysiphonia* plant with tetraspores.
P. A group of the four spores, on a short stalk.

corticated species (Fig. 6-16).

Reproduction may take place by growth from the basal portions after the erect portion has been broken away, or by the sexual cycle.

The female reproductive structure (initially the entire reproductive assemblage is called the procarp) is found in many species on a trichoblast, near the base. The cells of the trichoblast in that region become polysiphonous; the four-celled carpogonial branch (Fig. 6-16) develops from a modified pericentral cell, called the supporting cell (Fig. 6-16). A number of cells develop from the supporting cell.

1. The four-celled carpogonial branch, containing the terminal carpogonium with a trichogyne.
2. Some sterile cells (Fig. 6-16). There may be one below the carpogonial branch and two at the sides of the supporting cell.
3. An auxiliary cell, which is cut off *after* fertilization (Fig. 6-16).

Spermatangia are also formed on trichoblasts. After repeated divisions of pericentral cells of the latter, numerous spermatangia are produced.

Figure 6-17.
Polysiphonia plant
body.

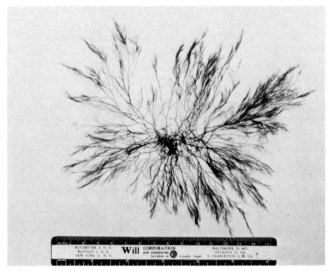

Spermatia are released individually at the tips of very small branches (Fig. 6-16).

Fertilization occurs, and then mitotic division of the zygote nucleus takes place. The auxiliary cell is formed and because of the curvature of the carpogonial branch it is adjacent to the dividing zygote (Fig. 6-16).

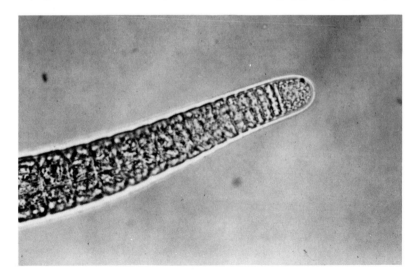

Figure 6-18. *Polysiphonia.* Apex of filament with enlarged apical cell and pericentral cells formed below. 500 X. Courtesy of C. Yarish.

There is fusion of the zygote and the auxiliary cell protoplasts, but *not nuclear fusions*. The diploid nucleus may divide by mitosis.

Next, the supporting cell and the axial cell fuse with the zygote-auxiliary cell; the resulting cell is called the placental cell. There are now a number of mitoses of the diploid nucleus, and gonimoblast filaments grow from the multinucleate placental cell. The terminal cells of these filaments act as carposporangia. After the carpospore is released from the carposporangium, the subterminal cells can grow and form a two-celled branch; the apex is another carposporangium. Successive two-celled branches can be formed. This mode of development can result in numerous carpospores, all of which are diploid, from one fertilization.

While all of this is taking place, vegetative cells of the gametophyte, around the carpogonium and gonimoblast filaments, produce an urn or pericarp. The entire structure including the pericarp is the cystocarp (Fig. 6-16).

Spores can be released through an opening in the pericarp, or inasmuch as the entire pericarp is borne on a trichoblast, the entire pericarp might be carried to another location when the trichoblast falls off. Germination of a carpospore results in formation of a tetrasporic plant.

The tetrasporophytes resemble both male and female gametophytes in *macroscopic* appearance, but when any stage is fruiting they can be easily distinguished. At maturity the tetraspore parent cell is formed on a short branch coming from the central cell, just inside the pericentral cells (Fig. 6-16). The parent cell enlarges and reduction division results in a tetrad visible from the external surface (Fig. 6-19). Segregation results in two spores that will give rise to male plants and two that develop into female plants.

In a *very general* way the life history of *Polysiphonia* is similar to those found with *Ectocarpus* and *Ulva*. In other words the organisms have identical free-living gametophyte and sporophyte phases. In comparing the three life histories the differences are apparent when one looks at the reproductive cells and organs, as well as mechanisms for asexual reproduction. In addition, *Polysiphonia* and its allies have the attached carposporophyte. The latter is often called a third generation by some who would *not* describe the life history as an alternative of isomorphic generations.

Ceramium, with the species *C. rubrum* in mind, is a coarse, freely branched alga with pseudodichotomous branching very apparent at the tips. It is attached to wood, rocks, or other algae. The two main characteristics, the pincerlike tips, and the banded appearance of

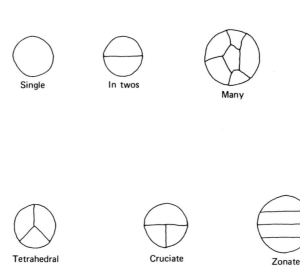

Figure 6-19. Types of tetraspores found in the red algae. In a few cases the spores may be single, in twos, or in groups larger than four. Many of the typical patterns that can be found in the red algae are figured.

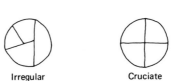

filaments, can be seen with the naked eye, if one has good vision! The organisms are quite polymorphic; they are variable in size and color, with many shades of red, as well as in thickness of the axes. However, the organism still retains its characteristic features, enabling one to recognize it as *Ceramium*. The clawlike tips are the result of pseudodichotomous branching and a slight, inwardly curved pattern of growth, at least during the initial stages of branch development. There is a single axial cell, which eventually increases in size many thousands of times. At the juncture of two such cells, many considerably smaller cells develop around and up the distal cell as well as around and down the proximal cell, providing a cortex. If the cortications from each direction do not meet at the middle of the axial cell, the cortications are incomplete. Incomplete corticiations, or those that are more dense at the nodes, give the characteristic banded appearance.

An Alga in which Meiosis Is not Gametic, Sporic, or Zygotic

At times, the algal literature describes a structure, mode of reproduction or process as quite different, unique, peculiar, and so forth. How many

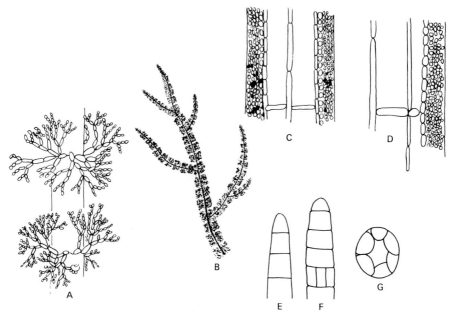

Figure 6-20.
A. *Batrachospermum,* with a large central filament and branches with beadlike cells.
    Microscopic.
B. *Batrachospermum.* Note the branch clusters. Measured in millimeters.
C–G. *Lemanea.* Microscopic.
C. Longitudinal section through the plant body. Note the central strand, and the parenchyma-
    like outer portion.
D. Detail of the interior near a radiating filament. Note the ascending and descending strands,
    just under the exterior.
E. The tip, demonstrating that the plant body originates in a uniseriate fashion.
F. Similar to E, but showing the first multiseriate portion.
G. Cross section of the multiseriate portion shown in F.

times can this be so? Even in taking a quite conservative view in the use
of these terms, one such adjective would have to be used for the life
history of *Lemanea,* and perhaps other freshwater red algae. But first a
description of the organism.

Lemanea (Figs. 6-20 and 6-21) is a freshwater form found attached to
rocks in cold, fast-moving streams. The organism is macroscopic,
unbranched, or sparingly branched, coarse, and about 4 to 6 cm in
length. Definite swellings give the strand the appearance of having nodes
and internodes. The plants are frequently colors other than red, with
many having a green or almost blackish appearance.

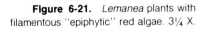

Figure 6-21. *Lemanea* plants with filamentous "epiphytic" red algae. 3¼ X.

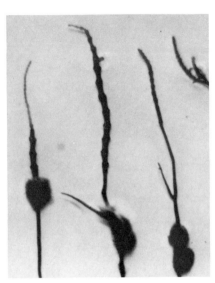

Growth is from divisions of an apical cell (Fig. 6-20), which, in order, cuts off central cells and surrounding pericentral cells (Fig. 6-20). The central cell serves as the axis of the plant. Several divisions and elongations of derivatives of the pericentral cell result in an outer cortex separated from the axial cells by means of spokelike cells, the basal cells (Fig. 6-20). At the distal end of a basal cell, just under the cortex, ascending and descending branches or filaments grow (Fig. 6-20), with the carpogonial branch developing on these threads. Then the trichogyne forms, elongates, and protrudes through the cortex. Spermatangia develop at the nodes from cortical cells.

After fertilization and formation of carpospores, carpospore release can take place by rupture of a portion of the cortex. The carpospores develop into a prostrate and branching form that may produce *Lemanea* plants from budlike outgrowths, or from monospores, developing at the tips of the filaments.

Magne has studied this organism both in the laboratory and the field. Carpospores released from mature plants were germinated on glass slides. The filaments produced resemble *Chantransia* or *Audouinella*. Microscope slides with these small attached, branching forms were studied in the laboratory and others were incubated in nature. Soon, small, erect *Lemanea* plants were seen. Then Magne observed that meiosis took place in the division of the apical cell! In each of the divisions of meiosis a nucleus was discarded, so that one haploid nucleus, and cell, resulted. This haploid apical cell divided by mitoses and cytoplasmic divisions to produce the upper portion of the *Lemanea* plant; the plant is diploid below and haploid above. At maturity it forms spermatia and carpogonia.

Frequently, in some streams, delicate filamentous forms resembling *Audouinella* develop on the *Lemanea* plant body (Fig. 6-21), and also on adjacent vegetation. It would appear that those on the *Lemanea* could result from a poor dispersal of carpospores, or from germination of carpospores before they were liberated. On the other hand, some might be truly an *Audouinella* (Fig. 6-22) growing on a *Lemanea*.

It would not be surprising to find that the reproductive story in *Batrachospermum* (Figs. 6-20 and 6-23), was the same. Magne feels that some cytological information, which is almost 100 years old, would support such an interpretation of the life history.

Some Additional Genera in the Florideophytes

A few general comments can be made for all of the following organisms. In some cases, because there is such a broad range of morphology in a genus, only one species is mentioned. Many species names are not presented. Representatives of crustose and both uniseriate and multiseriate erect forms are presented. Little mention is made of

Figure 6-22. The filamentous red alga sometimes found attached to *Lemanea* plants. 1600 X.

Figure 6-23. *Batrachospermum,*
showing the long central strand and
the beadlike cells of the
photosynthetic portion. 1600 X.

reproductive structures, although certain of the specimens you collect
will have conspicuous, mature reproductive bodies. Cystocarps, whether
on separate appendages or embedded, often cause some inflation of the
thallus. These cystocarps, as well as groups of tetraspores, are often
visible with the naked eye or a hand lens.

Acrochaetium.   This pinkish-red, microscopic (c. 2 mm), branching,
uniseriate epiphyte grows from a simple or disclike base. There are many
species in the genus, most of which reproduce by means of
monospores, and some by the formation of tetraspores. Few have been
reported bearing sex organs. An *Acrochaetium*-like organism has been
shown to be the tetrasporic generation of a species of *Liagora.*

Antithamnion.   This genus contains branching, predominately
attached species that are uniseriate filaments. Small refractory cells, only

a fraction of the size of vegetative cells, are often found between vegetative cells. These are the gland cells. Some species may have rudimentary cortications, for some cells appear to grow loosely around the main axis. Branching is distinctly opposite (hence the name), or four branches may originate from each node.

Batrachospermum.   Batrachospermum (Fig. 6-23) is a freshwater form, commonly found in slow-moving streams or in ponds. As with many of the freshwater red algae, it can be greenish, blue-green, red, or violet. Macroscopically it resembles a small chain of soft beads. The plant body is a uniseriate chain with small branches arising at the nodes (Fig. 6-20). Some branches can parallel the main filament and thus form a cortex around the main axis.

Germination of *Batrachospermum* carpospores results in the formation of a filamentous stage that resembles *Chantransia*. Monospores, formed from terminal cells, or "buds" develop into *Batrachospermum* plants.

Chondrus crispus.   This organism, commonly called Irish moss, has a flattened, repeatedly dichotomously branched thallus quite firm in texture. Plant bodies arise from a rather slender basal portion and are up to 15 cm in height. Although flattened, the pattern of growth and the frequency of branching, especially at the tip, result in a three-dimensional thallus. There is a great deal of variability both in the degree of branching and the color, which can range from yellow green to almost black. There are extensive beds of this organism in the lower littoral and sublittoral in Great Britain, New England, and Canada. The organism can be easily harvested using rakes, even from boats. It is a prime source of carrageenan (Chapter 20).

Corallina.   *Corallina officinalis* is an erect bushy plant, pinnately branched, especially at the tips (Fig. 6-24). *Corallina* is heavily calcified, but not at the joints or nodes (Fig. 6-25). Thus the plant has a conspicuous articulated appearance and is flexible at the nodes. Even internally, cellular organization in the regions between the joints is unlike that at the points of articulation (Figs. 6-25 and 6-26). The organisms are found in the upper sublittoral down to about 20 m below the surface. They usually have a pinkish coloration, partly due to the calcification, but this can vary with the distribution of the organism from shallow to deeper water.

Cystoclonium purpurem.   This red alga is relatively large, up to 60 cm in length. It is frequently and irregularly branched, with a bushy

Figure 6-24. *Corallina* plant
body. Courtesy of N. Proctor.

appearance and a conspicuous main axis. Smaller branches are
numerous and quite firm. It may resemble several other algae on first
encounter, but one variety has many tendril-like proliferations of smaller
branches. It is found in the sublittoral, and large plants are common in
the tidal wash.

Figure 6-25. *Corallina*, showing segments and ''joints,'' as well as
branching. 20 X Courtesy of W. Johanson.

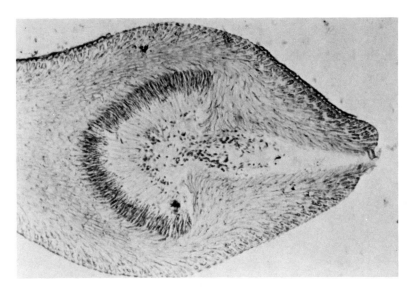

Figure 6-26. *Corallina,* with calcium carbonate removed. Reproductive cells, here spermatia, are formed in such chambers. 325 X. Courtesy of W. Johanson.

Dumontia.   *Dumontia incrassata* is an erect, tubular plant, with long, unbranched, sometimes quite twisted appendages. The branches are fleshy and frequently at maturity are hollow; they may become inflated. Plants are from 20 to 60 cm in length, are often found in groups, and are common in the sublittoral or in tidal pools.

Gelidium.   This organism, from which agar is obtained, is a common Pacific genus. It is frequently a deep red color and plant bodies up to a meter in size may be collected. *G. crinale,* on the other hand, is a smaller form, 5 to 6 cm in height. It is irregularly branched, with more abundant terminal branching. It may be found on rocks in the littoral zone.

Gigartina.   This is a common genus in the Pacific. Some of the members of this genus are among the largest of the red algae, reaching about 2 meters in length (Fig. 6-27). They are frequently quite fleshy and very tough, but some species are smaller and obviously dichotomously branched. *G. stellata,* which occurs on the East Coast in the sublittoral, is commonly found in the same area as *Chondrus.* In fact, some *Gigartina* is frequently collected with the latter when material is gathered for preparation of carrageenan. The novice might at first have some

Figure 6-27. *Gigartina,* one of the largest of the red algae.

difficulty distinguishing the two, but the irregular outgrowths from the stem, containing the reproductive bodies, help in identifying *Girgartina.*

Hildenbrandia.   This crustose form adheres so closely to the rock on which it is found that one has difficulty in determining where the rock ends and *Hildenbrandia* begins (Fig. 6-28). It is best to collect both alga and substrate. The organism is paper thin, only a few cell layers in thickness, and is not calcified. The plant body is composed of a layer of branching filaments which cover the area by growing in a radiating direction. It is found in both littoral and sublittoral zones and can be a most abundantly distributed alga, giving most of the rocks in the area an orange, deep red, or rusty color.

Liagora.   *Liagora* is a common, lime encrusted form found in the Pacific, Caribbean, and the Mediterranean. The plant body is dichotomously and repeatedly branched, forming a rather dense tuft of ropelike branches. The life history of one species has been worked out in culture. The plant we call *Liagora* is the gametophyte phase. The alternate stage, which has been identified as an or *Acrochaetium*-like plant, produces tetraspores.

Lithothamnium.   This genus, like *Corallina,* is one of a large family of calcified plants, the coralline algae. Unlike *Corallina, Lithothamnium* is

Figure 6-28.   *Hildenbrandia*, a firmly attached, crustose red alga.

crustose and lacks erect, flexible branches. It is a pink (or bleached and white, if dead), smooth or knobby, tightly or loosely adherent, epilithic or epiphytic, thick or thin plant. *Lithothamnium* and other crustose coralline genera produce reproductive cells within conceptacles which open by one or, more rarely, by several pores. In certain areas (e.g., off Brittany) crustose corallines develop into free-living nodules that cover miles of ocean floor; this is marl. In tropical areas, corallines act with coral animals in producing massive reefs.

Phycodrys.   *Phycodrys rubens* (Fig. 6-29) is one of the attractive, deep-water red algae with a wide distribution. It is a deep red color, sometimes almost purple. The thallus is just a few cell layers thick and, although membranous, it is really composed of pseudoparenchyma. There is a definited mid-rib and numerous smaller "veins" projecting from it. With the irregular lobed appearance, it deserves the common name, "sea oak" or, "oak-leaf alga." This alga will usually be collected by diving or use of a dredge, but with some luck, plants are found washed up on shore. It often takes patience to collect such plants in shallow water.

Rhodochorton.   This genus has both marine and freshwater species. It is a uniseriate, sparingly branched filamentous form which can be

Figure 6-29. *Phycodrys.* The oak leaf red alga. 1⅝ X.

microscopic to about a centimeter in length. Some species are found attached to rocks, but many are epiphytic. For most of the organisms that have been described there is much more information about the tetrasporic phase than the gametophyte. As a matter of fact, *very little* is known about the haploid phase of any species. It is not at all surprising then that, when an atlernate phase of another red alga is found, a *Rhodochorton*-like form sometimes turns out to be the alternate diploid phase.

Rhodymenia.   This alga is a membranous, dichotomously or palmately branched form up to 30 cm in length. The species *R. palmata,* called dulse (Fig. 6-30), is collected locally in Great Britain and on both shores of North America, and sold as a snack or for chewing (see Chapter 20). This organism is one that thrives in the sublittoral in some areas, but not all stages presumed to be necessary for completion of the life history are known. For example, in some areas on the east coast of North America, no female plants have ever been collected. A name change to *Palmaria* is gradually becoming accepted.

    Many other interesting red algae can be found in the ocean and in freshwater environments. See, for example, Figure 6-31 which illustrates a brackish water alga, and other genera mentioned below.

Figure 6-30. *Rhodymenia,* attached to a kelp stipe. Courtesy of R. Wilce.

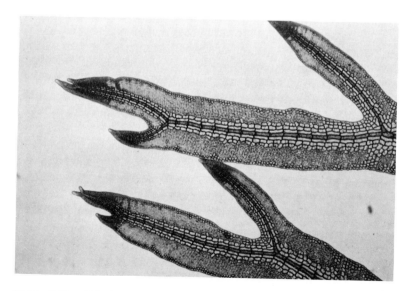

Figure 6-31. *Caloglossa.* With this symmetrical plant body it is possible to determine origin of cell lines, branching, and reproductive cells. 160 X. Courtesy of C. Yarish.

THE COMPLEXITY OF RED ALGAL LIFE HISTORIES

It is very difficult to describe life histories of some red algae. First, information may have been reported only once, or it may be true for just one species in a genus, or the new information might link one organism with a species formerly placed in another genus. The story becomes more involved when we find that not all tetrasporophytes of *Bonnemaisonia* species are *Trailliella*. *Bonnemaisonia armata* has been shown to alternate with *Falkenbergia* tetrasporophytes, another small filamentous plant. Many *Falkenbergia* collected are sterile, and one does not often get viable carpospores from *B. armata*. With just this information, you can see that it is difficult to unravel this life history. A third *Bonnemaisonia* species has as an alternating phase the tetrasporophyte of *Hymenoclonium!*

A further complication is that older literature has presented certain of these life histories in other ways, often indicating that meiosis was zygotic. Thus there could be no alternation of macroscopic plants. Although we have already proven this interpretation to be incorrect in several cases, are there any of these plants with zygotic meiosis? Perhaps there is reduction division when the zygote divides in some organisms that have *direct* development of an erect plant from a small prostrate form. With direct development there is no spore production before the erect plant begins to grow. In other words, an erect stage develops from a prostrate stage without benefit of spores.

We must take into account additional facts. Even in some culture studies, in which the life histories have been completed several times, investigators have not observed all essential reproductive structures. For example, *Gigartina* life histories have been completed without male plants or tetrasporophytes. In some laboratories in which the *Bonnemaisonia-Trailliella* complex has been studied, only female plants were produced, whereas in another laboratory, all gametophytes that had sex organs were male!

Field reports repeatedly show that only plants of one type will be found in certain localities, or when cystocarps are detected, they are sterile. Perhaps some of these organisms are out of their natural range and cannot go through a complete reproductive cycle in that environment, at least not one involving sexual reproduction.

The organisms thus could be relying on alternative methods of reproduction. With production of asexual spores, growth from small fragments, or formation of new individuals from inconspicuous prostrate bases, organisms could grow for several years without ever completing what we think of as a complete life history. A life history involving sexual reproduction could take place in other different environments. With frequent reintroduction of the organism from such areas by spores, organisms growing in several habitats would be maintained at the extremes of their range.

Why do these organisms fail to reproduce the same way in all environments? There is information from culture studies, as well as field observations, that environmental factors are involved. For example, a *Ceramium* would not form gametophytes when the temperature was low. Does this mean that the haploid plants would be found only in very late spring and summer? With some organisms, tetrasporophytes have been located in the field only in winter months. In some areas these were not reported until the 1960s. Although many local floras may have been examined extensively in the past, most observations were in the summer. In the other seasons the professors were back in their inland universities teaching students.

There is now some evidence that day length, or nutrient level, might be involved in the formation of the tetrasporophytes of *Nemalion.* However, they appear to form best under long days, not what one would expect to find in winter months (see above). Perhaps some organisms have evolved a mechanism for completion of the life history, or at least one pathway, only where there is a precise set of environmental conditions. If they were to find themselves in another locality, brought there by human intervention, or a favorable current, they might exist for quite some time by asexual means.

Brief mention should be made of observations of both tetrasporangia and gametangia on one individual, at least in some *Cystoclonium* and *Chondrus* species. How can we fit such plants into the categories we have already discussed? These truly are remarkable organisms!

CALCAREOUS RED ALGAE

There are different amounts of calcium carbonate dissolved in seawater in various parts of the ocean. Tropical waters, which at times are supersaturated, have larger numbers of calcareous red algae; various forms of carbonates are deposited. Some genera are found in the same order with *Nemalion,* where there is aragonite deposition. But the most heavily calcified forms, with calcite deposition, are in another order. The remainder of the calcified red algae are a few species in one genus in a third order. In all there are about 600 species in 39 genera.

Calcareous forms are also found in other groups of algae. The deposition is thought to be connected with photosynthesis and to be a cell surface phenomenon. Recent work shows that other cellular phenomena must also be involved. But why would a *Halimeda* or a *Corallina* have, in general, heavy calcification, with little, if any, at the joints? With the tropical red algae it is possible that the calcified forms have some advantage in those sites where there is heavy grazing. But we also know that calcareous forms may be grazed by other organisms. There is also a filtering of bright sunlight provided by the calcification. Perhaps there is a ballast provided for some green algae found in and on soft and shifting substrata. Such organisms are not easily shifted about.

In atolls the algae are major contributors as both components and also cementing organisms. They join together the main components, the coral animals. In such areas the algae have been found growing down to 200 meters, despite the light-shielding lime over the photosynthetic portion. In borings of atolls some fossil algae have been reported a thousand meters from the surface, indicating some sinking and continuous reef building.

In spite of their hard surface, these algae are also quite brittle. They soon disintegrate into fragments in the herbarium folder unless properly cared for.

REFERENCES

Bird, C., C. Chen and J. McLachlan. 1972. The culture of *Porphyra linearis* (Bangiales, Rhodophyceae). *Can. J. Bot.* 50: 1859–63.

Bourrelly, P. 1970. *Les Algues d'eau Douce.* III. Les algues bleues et rouges. N. Boubee and Co., Paris. 512 p.

Chihara, M. 1974. The significance of reproductive and spore germination characteristics to the systematics of the Corallinaceae: non-articulated coralline algae. *J. Phycol.* 10: 266–74.

Dawson, E. 1966. *Marine Botany. An Introduction.* Holt, Rinehart and Winston, Inc., New York. 371 p.

Dixon, P. 1973. *Biology of the Rhodophyta.* Oliver and Boyd, Edinburgh. 285 p.

Dixon, P. and W. Richardson. 1968. The life histories of *Bangia* and *Porphyra* and the photoperiodic control of spore production. *Proc. Internat. Seaweed Symp.* 6: 133–9.

Fritsch, F. 1945. *The Structure and Reproduction of the Algae.* Vol. 2. The University Press, Cambridge. 939 p.

Hommersand, M. and R. Searles. 1971. Bibliography on Rhodophyta. In J. Rosowski and B. Parker (Eds.) *Selected Papers in Phycology.* Dept. Botany, Univ. Nebraska, Lincoln. pp. 760–7.

Johansen, H. and L. Austin. 1970. Growth rates in the articulated coralline *Calliarthron* (Rhodophyta). *Can. J. Bot.* 48: 125–32.

Kingsbury, J. 1964. *Seaweeds of Cape Cod and the Islands.* The Chatham Press, Inc., Chatham, Mass. 212 p.

Knaggs, F. 1969. A review of Florideophycean life histories and of the culture techniques employed in their investigation. *Nova Hedwigia* 18: 293–330.

Kylin, H. 1956. *Die Gattungen der Rhodophyceen.* Gleerup Fölag. Lund, Sweden. 673 p.

Magne, F. 1969. Meise sans tetrasporocystes chez les Rhodophycees. *Proc. Intl. Seaweed Symp.* 6: 251–4.

Smith, G. 1944. *Marine Algae of the Monterey Peninsula, California.* Stanford Univ. Press, California. 622 p.

Taylor, W. 1957. *Marine Algae of the Northeastern Coast of North America.* University of Michigan Press, Ann Arbor. 509 p.

West, J. 1972. Environmental regulation of reproduction in *Rhodochorton purpureum.* In I. Abbott and M. Kurogi (Eds.) *Contributions to the Systematics of Benthic Marine Algae of the North Pacific.* Japanese Society of Phycology. pp. 213–30.

Phaeophyceae — brown algae

The brown algae are the most complex forms found among the algae. There are no species that exist as unicells, colonies of any type, or simple unbranched filaments. Plant bodies range in size from a millimeter or so to about 70 meters in length. They may be small, branched, attached, filamentous forms, or larger plant bodies with certain portions similar to those found in higher plants. Included are organisms that are filamentous (some of which occur in tufts), folios crustose, elongate cylinders (even hollow), ropelike and bladelike, with both simple and compound blades (Fig. 7-1). They are found most often firmly attached to various substrates, often with elaborate holdfast systems. In addition to these structures, which *resemble,* roots, some forms have stemlike and leaflike appendages. However, they lack the vascular tissues of higher plants. Nevertheless, the "sieve tubes" of some of the complex kelps are even somewhat like the vascular tissue, phloem. They conduct some photosynthate through the plant body.

When we think of seaweed in north temperate waters, we usually think of the brown algae, for they are the most conspicuous of the plants along rocky coasts. They are not as numerous, nor as large, in the tropical waters, but are abundant in colder habitats. Brown algae are most conspicuous in the littoral or intertidal waters, but are quite abundant, especially the kelps, in sublittoral waters. Most are attached, with forms such as the crustose *Ralfsia* attached so well that one might better collect it by gathering the rock *and* the organism. The fingerlike projections of the holdfast of *Laminaria* can provide a firm attachment to a rocky substrate. Some species of *Sargassum* serve as examples of brown algae which are found floating in enormous numbers in the Atlantic between the West Indies and Africa, covering a few million square miles of sea surface.

In addition to the use of brown algae as a source of alginic acid, they have been harvested and used for fertilizer or potash. They can serve as sources of iodine and bromine, concentrating these elements above levels found in sea water (Chapter 20).

BROWN ALGAL THALLUS TYPES

Cytologically there are great similarities among brown algal cells. All are uninucleate (Fig. 7-5) quite a contrast, for example, to the green algae with multinucleate and coenocytic forms. Most cells contain plastids either in discs or bands; however, in at least some forms there can be a large, platelike plastid or one that is stellate. Flagellated zoospores and gametes are also remarkably uniform in their structure, for almost all are

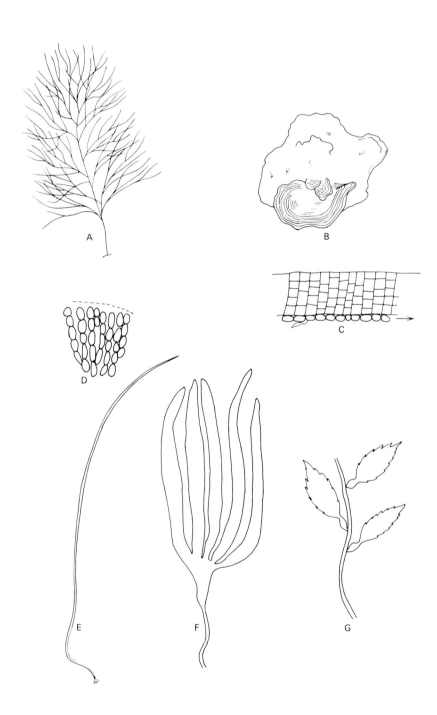

A

B

C

D

E

F

G

Figure 7-1. Morphological types in the brown algae. These include branched filaments (A), crustose types (B), and those which are whiplike (E), blades (F) or have leaflike appendages (G). In cross section (C) the crustose type frequently has a filamentous base from which rhizoids may develop (below), as well as erect, cuboidal cells of the upright filaments (above). Certain forms are pseudo-parenchymatous, with filaments closely appressed and surrounded by a common wall material (D).

laterally biflagellated. In contrast, in the green algae there are many, many zoospore types and there is such a range of structure found in plastids that plastid form is useful in classifying organisms.

The brown algae can be most complex, both internally, divided into a number of regions and cell types, and externally, possessing a wide range of morphological features not at all common to other algal groups. Certain species remain as free filaments, while others become quite entangled, for example, the basal portion of *Elachistea* filaments. In some organisms filaments are closely appressed into a pseudoparenchyma. Most brown algae are more highly evolved and possess true parenchyma.

The main site of cell division can be at the apex of the plant body (apical), at the base of a hair (trichothallic), in a region between the base and the apex of a filament (intercalary), among surface cells or epidermis of the plant body (meristoderm) or cell division may occur in almost any cell in the thallus (diffuse)(Fig. 7-2). In some cases a single cell may be involved, as with apical cell division in some filamentous forms. This would occur in the basal attached system of *Ectocarpus*. In other organisms, such as the trichothallic growth of *Desmarestia*, there may be division of just a few cells of a filament.

In the higher brown algae, including *Dicytosiphon*, *Laminaria*, and *Ascophyllum*, the plant body is composed of a true parenchyma. Cell division may be diffuse, as in *Dictyosiphon*, or more commonly it is restricted to certain regions in which many cells are capable of division. In such a case there is a dividing tissue, the meristem. Meristems may be located at the apex, they may be intercalary, or there may be a meristoderm (Fig. 7-3). *Fucus* species develop from the activity of a single apical cell, but close relatives have an apical meristem. Intercalary meristematic activity is found at the junction of the stipe and blade of *Laminaria;* the meristem activity provides some length to the stipe and continually adds to the length of the blade (Fig. 7-3). Cell division at the meristoderm can provide greater diameter to the stipe of *Laminaria*, or provide some thickness or increase in size to the plant body of a *Fucus*.

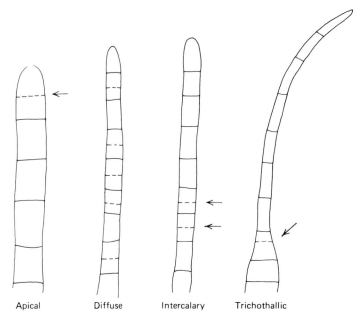

Apical   Diffuse   Intercalary   Trichothallic

Figure 7-2. Sites where cell division can take place in brown algae. Division may be restricted to the apical cell (apical), or it may be in almost any cell of the filament (diffuse). Also figured are types in which divisions are restricted to a zone removed from both the base and apex (intercalary), and division at the base of a hair (trichothallic).

Those divisions of a tissue, such as the meristoderm, that are parallel to the surface of the plant body are said to be periclinal. They initiate a process which would add thickness to a plant. However, as the outer diameter increases, occasional divisions in a plane perpendicular to the surface, also perpendicular to the periclinal divisions, are necessary. These are called anticlinal divisions.

With certain organisms the focal point of cell division may change with age. First it may be throughout the short filament (diffuse), then trichothallic for a short time and finally apical growth may take over. In maturing plants only the apical division would be noted.

Pseudoparenchyma can be developed from close aggregations of both branched and unbranched filaments, as in *Ralfsia,* or from loosely arranged filaments with considerable common wall material, such as *Leathesia* (Fig. 7-19). The latter organism is hollow at maturity, but even forms with a parenchyma can develop a hollow center.

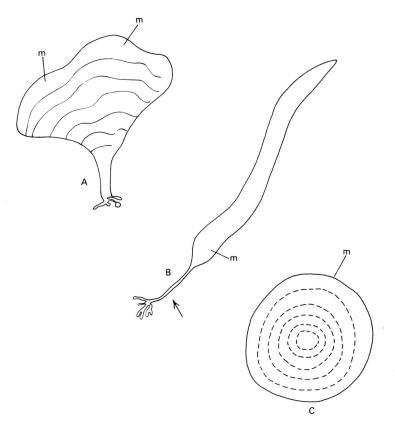

Figure 7-3. Position or localization of meristems in the advanced brown algae. Meristems (m) may be found at the apex, as in *Padina* (A), may be intercalary as in *Laminaria* (B), or may be in the surface cells and be called a meristoderm (C). The cross section (C) was taken from the stipe of *Laminaria* (arrow in B).

In those forms that are filamentous or pseudoparenchymatous, the plant body is often heterotrichous (Fig. 7-4). Advanced forms do not exhibit heterotrichy. The basal portion, from which erect filaments project, may be composed of branches filaments, or it may be pseudoparenchymatous. The erect plant body can be either uniaxial or multiaxial with considerable intertwining of filaments. The base is frequently perennial, allowing regeneration of the erect structure yearly, after mechanical loss, or grazing.

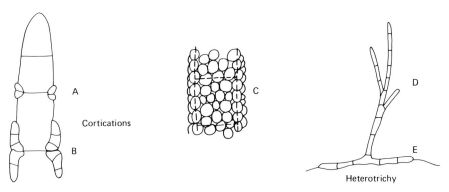

Figure 7-4. Pattern of development in which cells are cut off from the central filament (A), and grow around the latter (B), producing a cortex (C). Heterotrichous growth, with a basal system (E) from which erect filaments (D) develop.

Secondarily, cortications may appear when filamentous growth from the main axis envelopes the latter (Fig. 7-4). With cortications around the uniaxial or multiaxial system a mature plant body of some complexity can develop. The cell structure of some organisms then resembles a true parenchyma, but a study of development would show the filamentous pattern of growth. The external cells, the cortications, provide some supporting strength to the erect filaments, as well as additional photosynthetic tissue. They may have an external origin, as in other groups of algae, or an internal origin, for example, *Chordaria* and other advanced brown algae.

Almost all the phaeophytes possess excellent mechanisms for attachment, including both the prostrate system of heterotrichous forms and rhizoids which might even penetrate the substrate. The larger brown algae have a parenchymatous holdfast system. This is composed of fingerlike projections, which by growing around the substrate (such as a rock) and by some branching provide another very effective attachment system. In some forms the system resembles an inverted woven basket. With this attachment system organisms can withstand the forces in the intertidal zone. For some organisms, the flexible stipe provides some resilience, while for others, elongation of the stipe places photosynthetic segments of the plant body in favorable positions for trapping light energy.

BROWN ALGAL CHARACTERISTICS

1. Pigmentation

The brown algae have chlorophylls *a* and *c,* alpha and beta carotene, and several xanthophylls, such as flavoxanthin and lutein. The abundance of the xanthophylls fucoxanthin and violaxanthin provides the distinctive brownish color. For the most part, the brown algae are true to color, brown to brownish green, and do not have the many color exceptions found in other groups. Pigments are in plastids with the thylakoids in groups of three, with girdling lamellae present.

2. Storage Products

The food reserves are laminarin, mannitol, and fat. Laminarin is a beta 1:3 linked glucan with 1:6 branch linkage. In addition, for many of the brown algae algin or alginic acid is a cell wall component.

3. Motility

As far as is known there are no unicellular brown algae. Thus, it is not surprising that there are no flagellated types. However, reproductive cells, both zoospores and gametes, are flagellated. Most are laterally biflagellated, with an anterior flagellum with hairs (sometimes called the tinsel type) and a trailing flagellum. The latter has no hairs and is called the whiplash type (Fig. 7-5). Unlike other groups of algae in which there is a great deal of flexibility in the number, size, and position of flagella, there is remarkable uniformity in the phaeophytes. However, the sperm of *Dictyota* are uniflagellated, but presence of two basal bodies suggests that one flagellum has been lost.

4. Wall Components

With one exception we usually do not pay much attention to brown algal walls. That exception would be in the manufacture of commercial products from alginic acid. The latter is found in greatest abundance in the types called kelps and in fucoids; it is not found in other photosynthetic organisms, including other algae and higher plants. It is composed of both mannuronic and guluronic acids. The sodium salt of the acid is widely used as a stabilizer in foods and commercial products, so that all of us have both used and eaten this product of seaweeds (Chapter 20). In all the phaeophyte walls that have been examined there is a microfibrillar portion obvious after preparation and examination with the electron microscope. This is the cellulose portion; in some cases it may amount to only a small fraction of the total wall bulk. The orientation

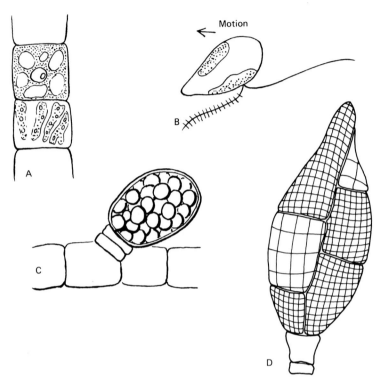

Figure 7-5. *Ectocarpus,* some morphological features. Microscopic.
A. A portion of an *Ectocarpus* filament. The bandlike plastids are shown in the lower cell, with the central nucleus and vacuolate protoplasm evident in the upper cell.
B. A laterally biflagellate zoospore of a brown alga. Note the two plastids.
C. A unilocular structure, frequently a zoosporangium, on a short stalk. Reduction division is thought to typically occur here.
D. The plurilocular structure, which can form zoospores or gametes. Again, there is a short stalk.

of the microfibrils is random, except in some elongating cells. In some cases, the orientation in one layer may be unlike that in another. Most brown algal walls contain sulphated polysaccharides, including alginic acid and fucoidins. A few species of *Padina* have calcium carbonate, but this material is much more common in red algae and some green algae.

REPRODUCTION
Some brown algae can reproduce asexually by fragmentation, while most forms have an asexual phase through the production of motile spores. In

the majority of the brown algae there are separate sporophyte and gametophyte generations, which may be isomorphic or heteromorphic. At one time we thought that the gametophyte generation was always the smaller phase in a heteromorphic life history, but now we know of several exceptions. Basic laboratory and field life history studies are still needed for a few organisms.

With *Fucus* and its allies there is no free-living haploid vegetative phase. The organism reproduces sexually and at maturity releases gametes from inflated reproductive tips. In the brown algae there is a variety of life histories, and even various interpretations of some of them.

The reproductive cycles of brown algae are known to be affected by photoperiod, temperature, available nutrients, and salinity.

SOME REPRESENTATIVE GENERA

Primitive Brown Algae

Ectocarpus (Figs. 7-6 and 7-7) here represents one of the morphologically most primitive types. At one time it was considered to have the simplest and most primitive life history in the brown algae. The organism we recognize is a uniseriate, freely branched, erect filamentous plant found in the sublittoral and littoral. It can be only a few millimeters or up to half a meter in length. In at least some species the plant

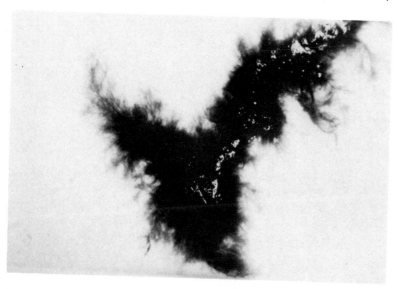

Figure 7-6. *Ectocarpus* plants attached to a hidden coarse alga. 1⅝ X.

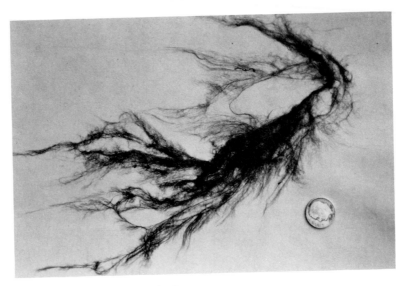

Figure 7-7. *Ectocarpus* plant bodies.

develops by trichothallic growth, but in other cases cell divisions can be accomplished at almost any point in the filament. There are several bandlike plastids per cell (Fig. 7-8). The organism attaches to the substrate, frequently another alga, by a prostrate filamentous portion in which there is apical growth. In some *Ectocarpus* species there can be

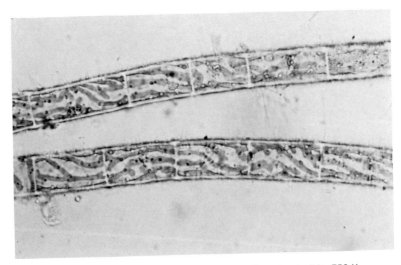

Figure 7-8. *Ectocarpus* vegetative cells, showing elongate plastids. 750 X.

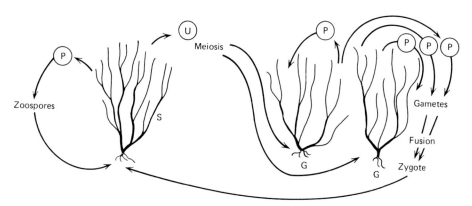

Figure 7-9. Life history of *Ectocarpus*, with several possibilities. The "typical" cycle. The sporophyte (S) produces biflagellated zoospores by an initial reduction division, in unilocular structures (U). The zoospores develop into plants that externally are identical to the original. These are depicted as G. These form plurilocular (P) structures that also have biflagellated swarmers. But the latter act as gametes, and individuals from different plants fuse. The zygote develops into the original sporophyte (S). (See the text for modifications from the above.) The diploid plant (S) can also bear plurilocular structures and thus reproduce asexually. Also haploid swarmers from haploid plants (G) can germinate and form similar plants.

the formation of an outer cortex, which develops by growth of some branches around the main axis.

Reproduction is by formation of swarmers (Fig. 7-5) from sporangia or gametangia. The gametophyte forms ellipsoidal or conical, sessile or stalked, multicellular gametangia (Figs. 7-5 and 7-10). A single biflagellated gamete (Fig. 7-5) arises from each cell of this plurilocular structure (Fig. 7-10). Zygotes, produced by fusion of gametes from two individual plants, grow into sporophytes (Fig. 7-9). Occasionally, gametes develop without fusing and thus they reproduce the gametophyte. One can think of these swarmers as either gametes or zoospores; environmental conditions determine which pathway will be followed.

The sporophyte produces two reproductive structures. One is a *diploid* plurilocular structure (Fig. 7-9) which forms diploid swarmers, here zoospores. (Sometimes plurilocular structures are formed only with elevated temperatures.) Upon germination the zoospores develop into new sporophytes (Fig. 7-9). The second reproductive structure, the unilocular body, is oval or spherical, and unicellular (Fig. 7-5 and 7-11). Following meiosis and several mitoses, 30 to 60 haploid swarmers develop (Fig. 7-9); the latter produce the gametophyte phase (Fig. 7-9). Thus, in the broad sense, the life history is similar to that of *Ulva;* there are identical haploid and diploid, free-living generations.

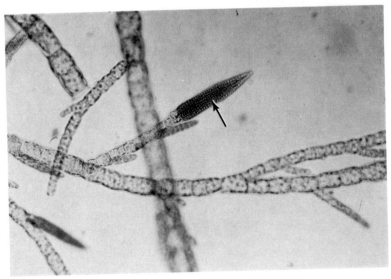

Figure 7-10. Plurilocular reproductive structure. 300 X.

Recently some variations in this "typical" life history have been reported. With *Ectocarpus siliculosus* the gametophytes and sporophytes are not always isomorphic. The sporophytes can be smaller, or sometimes even more robust. In some cases the alternate plants have

Figure 7-11. Unilocular reproductive body. 400 X.

been described as separate species. For example, *E. draparnaldiodes* is the gametophyte phase and *E. fasciculatus* the sporophyte in one life history. Baker and Evans recently demonstrated a *Myrionema* stage, a cushionlike plant, in the life cycle of *Ectocarpus*. With *Ectocarpus siliculosus,* Müller demonstrated a very complex life history, involving gametophytes that may be haploid or diploid as well as haploid, diploid, and tetraploid sporophytes! Obviously phenomena such as a doubling of chromosome number and parthenogenesis are involved in the reproduction of such an organism.

Thus, not all members of this genus reproduce in a similar fashion; in some localities there is no sexual stage. Some species have the potential of alternation of isomorphic generations, but under other conditions might have a heteromorphic alternation. These differences in life histories could be related to temperature. Other forms of *Ectocarpus* apparently are typically heteromorphic, with the complication that two "genera" might really be involved.

With the phaeophytes this is the first example of a life history in which there are two phases, one called *Ectocarpus* and the other in the genus *Myrionema*. (We will discuss other examples later.) We can use both names only temporarily. There are rules of priority in attaching the correct name to the organism studied. But changes in other categories, even at a level just below the class name, are also in order. For example, in the brown algae organisms are grouped into three categories, based on the type of life history. All organisms with isomorphic alternation of generations are placed together. With a recent accumulation of data on life histories, there has been a reconsideration of the classification of the brown algae. There should always be a rethinking of the criteria used at all levels in taxonomy.

Scytosiphon

We shall examine this organism in some detail, for it is a common littoral zone alga that has been known for over 160 years. Nevertheless, there was always some doubt about the life histories of most *Scytosiphons* (Fig. 7-12) examined. For a number of years only plurilocular structures were reported whether one looked at young or mature plants, or at plants in quite different habitats. If one recalls the life history of *Ectocarpus* (Fig. 7-9), it is clear that this fact could not provide sufficient evidence to determine whether the plant was haploid or diploid. Without evidence for the fusion of flagellated cells, except in one unconfirmed report, investigators believed that *Scytosiphon* was a sporophyte, lacking a gametophyte stage.

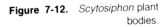

Figure 7-12. *Scytosiphon* plant
bodies.

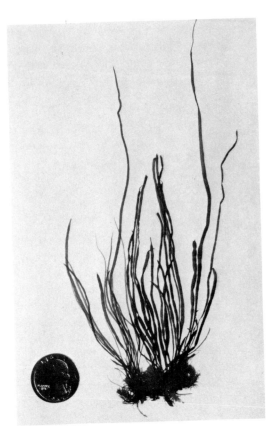

The organism is widely distributed, a winter–spring annual and occurs in rather dense aggregations in the intertidal zone. Plants up to a meter in length are commonly found attached to rocks. Plants are elongate with a narrow (often about a centimeter or less) diameter (Fig. 7-12).

At maturity they are tubular and regularly constricted. In his book on the seaweeds of Cape Cod Professor Kingsbury says they remind him of a series of sausage links. The constriction is normally not quite that pronounced.

When *Syctosiphon* is young, it is a solid cylinder. Soon there is a differentiation into a cortex with smaller photosynthetic cells and a medulla with larger, nonpigmented cells. As the organism ages there is a differentiation into areas that might be loosely called nodes and internodes. At the nodes (Fig. 7-13) there is limited increase in diameter; these points are the constricted portions of the plant body. In the

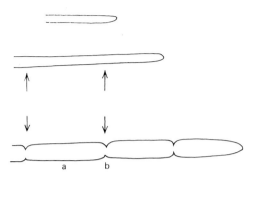

Figure 7-13. The development of form in *Scytosiphon*. As the cylindrical organism elongates, growth at several points (arrows) is limited. Increase in diameter is restricted. As the cylinder increases its diameter in the internodal area (between the arrows), constrictions become obvious. A cross section (A), taken at a, reveals the tearing of the medulla, whereas the cross section (B), taken at b, reveals a definite cortex and medulla.

internodal area there is more rapid growth of the cortical cells, not only providing a broader plant body, but also pulling the cortex away from the central cells of the medulla. Thus the organism becomes hollow at maturity. In some protected areas there are no constrictions in plants of an entire population. Growth rate is apparently involved in this morphological alteration.

Recently Nakamura and Tatewaki independently completed the life history of *Scytosiphon* in culture (Fig. 7-14). The organism is haploid and heterothallic. When mature, plants produce plurilocular gametangia which release typical brown algal gametes. Sexual reproduction has been observed. The zygote attaches to a substrate and elongates. One cell division follows another, producing a short filament of a few cells, bearing a terminal hair (Fig. 7-15). The cells then begin to divide, forming branches, and soon, because of the radiating pattern of growth, a disc (Fig. 7-15) is formed. The latter is the prostrate system from which short, erect filaments are formed. This stage will grow well at 13°C in a 14 hr L/10 hr D cycle. If the temperature is lowered 3 degrees and the cycle reversed to a 10 hr L/14 hr D cycle, unilocular sporangia develop at the base of the erect filaments. Spores produced from the unilocular sporangia develop into *Scytosiphon* plants. The initial stages in the germination of the spore, until a filament of a few cells is formed, are similar to those observed in the germination of the zygote. Then a small,

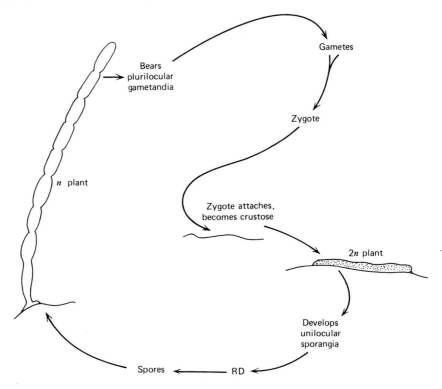

Figure 7-14. The life history of the organism, *Scytosiphon-Ralfsia*. Reduction division (RD) takes place in the crustose *Ralfsia* stage.

irregular disc is produced and many uniseriate filaments radiate from it. These photosynthetic filaments become multiseriate, and eventually a hollow thallus resembling the mature *Scytosiphon* described is produced from each filament (Fig. 7-14).

Thus the alternate phase in this life history is smaller than *Scytosiphon* and is crustose. It resembles one species of *Ralfsia* (Fig. 7-16). This life history is thus remarkably heteromorphic. We will see other heteromorphic cycles later, but in those cases the size relationship is reversed. The sporophyte is considerably larger (page 205).

The elucidation of this life history proved to be quite interesting and other discoveries soon followed. Several investigators in the United States and Great Britain have confirmed the *Scytosiphon-Ralfsia* relationship. However, they have not detected sexual reproduction. It appears then that the erect *Scytosiphon* merely arises vegetatively from a *Ralfsia* base. There are further complications:

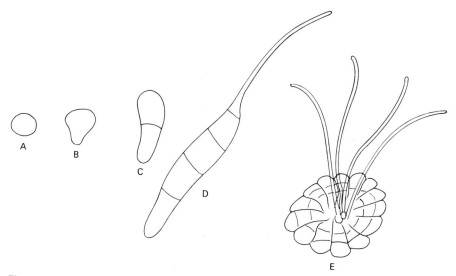

Figure 7-15. The development of the zygote of *Scytosiphon* into the crustose stage. The zygote (A) elongates and divides transversely (B, C). After reaching a stage resembling a short filament (D), a radiating pattern of growth becomes apparent because of frequent branching of the original filament (E). Numerous hairs project from the center of the disc.

1. In Denmark there is a *Scytosiphon* crustose relationship, but the prostrate form resembles *Microsporangium,* a relative of *Ralfsia.*
2. *Ralfsia*-like alternates can be found with other genera. The *Petalonia-Ralfsia* life history is also described by some as lacking a sexual phase.

Additional Related Genera
The following are brief descriptions of some genera that you might encounter in the field. For identification of specimens it is best to consult one of several published keys or treatises on brown algae. The following accounts give some idea of the range of structure seen in the brown algae.

Desmarestia.   *Desmarestia* (Fig. 7-17) species, erect plants in the sublittoral, are widely distributed in temperate waters, as well as in the arctic and antarctic. Plants may be up to 4 meters in length and are filamentous or bladelike. Branching is opposite and frequent, with a pinnate habit developing in some species. Some forms are rather delicate and greatly branched, where others are more coarse or sometimes bladelike. In this case, the blade develops initially by trichothallic growth.

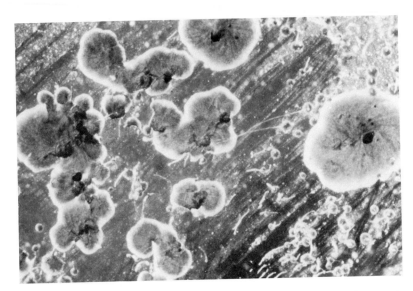

Figure 7-16. *Ralfsia*-like brown algal crusts. Life size. Courtesy of R. Wilce.

Some species of *Desmarestia* are quickly known to the collector because of their release of sulphuric acid from their vacuoles. This occurs shortly after collecting, as the organism cannot survive for long in the collector's bucket or plastic bag. The result is the discoloration or destruction of other algae in the container. The pH in the vacuole has been measured at 2 and lower! After one disaster the collector learns to keep *Desmarestia* in a separate container. The storage of sulphur-containing compounds could be a mechanism for removing an excess from protoplasts.

Dictyota.   This organism is a parenchymatous, dichotomously branched alga. The cells of *Dictyota* originate from divisions of an apical cell. The thallus is composed of narrow, strap-like ribbons. The reproductive cells are grouped in patches or sori on the plant. Spores are formed in tetrads in unilocular sporangia; thus only the products of meiosis are spores. The organism is frequently cited because of its form, the fact that sperm have only one flagellum, *and* because sexual reproduction is controlled by light and tidal rhythms. Experimental work has shown that the formation and release of the gametes is regulated by phases of the moon; thus gametogenesis takes place monthly with the spring tide.

Figure 7-17. *Desmarestia.* A filamentous brown alga well known to collectors because of the acids stored in the vacuole.

Elachistea.   *Elachistea* is an epiphyte found in small but dense clumps on coarse algae. Many epiphytes will usually be found on the same plant. Some forms are apparently found only on certain plants, especially fucoids. *Elachistea* filaments branch only in basal portions. The basal portions are either growing on the host plant, or have penetrated the host. The emerging portion of the filament is unbranched and it may be photosynthetic or sterile. The photosynthetic filaments are larger, and typically quite straight. The smaller filaments, called paraphyses, are sometimes curved. Reproductive structures are found associated with the latter. In some species of *Elachistea* no haploid phase has been detected. Spores produced by all individuals germinate directly into plants that are identical in both appearance and chromosome number.

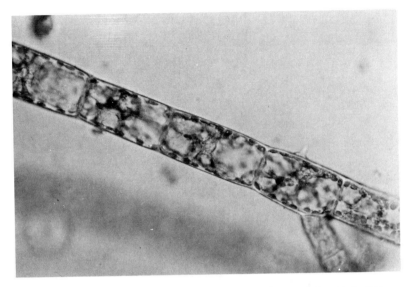

Figure 7-18.   Ectocarpoid alga. Cells with a single nucleus and discoid plastids. 750 X.

Giffordia.   This brown alga is a branching filamentous, bushy plant, not unlike *Ectocarpus* in outward appearance. However, the cells contain numerous disc-type plastids (Fig. 7–18). Many species were formerly classified as species of *Ectocarpus,* which has bandlike plastids. Some phycologists feel that these differences in plastid structure do not provide sufficient evidence for the separation of these two genera. *Giffordia* species grow on a variety of substrates, especially other algae. Reproductive structures are sessile, that is, not on a short branch, and there may be several in a series.

Myrionema.   These plants are small cushionlike growths of about a millimeter in size epiphytic on numerous other algae whether green, brown, or red. There is a heterotrichous pattern of growth, with the prostrate radiating base adhering to the host because of the composition of the wall material. The upright filaments are sparingly branched and quite densely packed. No gametangia have been described for most species. The organism has been described as the alternate phase in the life cycle of at least one *Ectocarpus*. As with other organisms we have seen or will see, our knowledge of reproductive structures was incomplete merely because we looked for them on *Mryionema*-like plants. Now that we know a heteromorphic life cycle is possible, this lack of information makes sense. *Myrionema*-like stages are probably found in the life histories of several algae.

Padina.   *Padina* species are the only calcified brown algae. This is a very common tropical form that is fan-shaped and has conspicuous growth lines. When young, there is a single apical cell, which soon develops into a marginal apical meristem. The latter can provide new cells as the fan continues to develop outward. One *Padina* has been examined in culture where there was an isomorphic alternation of generations. But when the same investigators examined field material, only the sporophyte phase of the species was found. Perhaps this will be another example of distribution of one phase of a life history limited by the environmental factors at specific locations. This phenomenon is also observed in the red algae.

Petalonia.   In the northeast, this organism is a spring-winter annual found in the high littoral. It produces a ribbonlike blade with a gradual tapering base, short stipe, and small disclike holdfast. Initially one can confuse the genus with young *Laminarias* and some members of the genus *Punctaria*. To be certain of the genus, examine the holdfast carefully and make sections of the blade. Only *Laminaria* will have a holdfast with fingerlike projections and a prominent stipe. In *Petalonia* cross sections there are an outer photosynthetic cortex and an inner colorless medulla with enlarged cells. In cross section the blade of *Punctaria* is only a few cells in thickness, and all cells are of the same diameter.

More recently both Wynne and Edwards have pointed out that there is a *Ralfsia*-like (see below) stage in the life history of *Petalonia*. The *Ralfsia* stage can release spores that develop into the erect stage, or *Petalonia* plants can be seen arising directly from the crustose form. Apparently the alternation of stages here is not obligatory. Both day length and temperature have been reported to determine the form achieved. The blade is formed when the temperature is cool and days are short, such as in fall to winter months.

Punctaria.   This brown alga is a flattened bladelike form that is attached by a disc holdfast to a variety of substrates. The blade, which develops by diffuse growth, is just a few cells thick, with no differentiation into medulla and cortex. Sometimes it is essential to cut a cross section in order to confirm the identification of this genus, because it can be confused with *Petalonia*. Blades of this annual can be 25 cm in length with extremes of 50 cm. Apparently some species reproduce asexually and for at least one species there has been confirmation of microscopic, filamentous gametophytes.

Pylaiella.   *Pylaiella,* a winter–spring annual, at first glance can resemble some species of *Ectocarpus* or *Giffordia,* for there are discoid plastids. Although the branching filamentous forms are mostly uniseriate, there are at least occasional divisions to make a few regions multiseriate. These regions might be detected only when the plant is mature. Some forms might be only a few centimeters in length, but plants up to 50 cm have been recorded. Frequently filaments twist together so that the main axis can resemble a strand of twine. Reproductive cells are in a series, and are in an intermediate position in the filament. They are said to be intercalary. The unilocular sporangia, which have bulging lateral walls, may be in a series of up to 30 sporangia. The plurilocular structures are oblong and cylindrical. In *Pylaiella,* zoospores from unilocular sporangia can be formed without an initial meiosis. These swarmers settle to produce new individuals identical to the parent. This is called the direct type of life history.

Ralfsia.   The fleshy crustlike forms, represented by *Ralfsia,* (Fig. 7-16) are found in the sublittoral zone, but are sometimes exposed at low tide. They are good examples of a heterotrichous pattern of growth and the formation of pseudoparenchyma. *Ralfsia* thalli can reach a diameter of 10 cm, with a growth pattern that might remind one of a single bracket fungus growing on a rock in the ocean. With dense growth one thallus might develop and overgrow the pioneers of that season. Many of these organisms were not well known until recently simply because they were not extensively studied. Identification of species is difficult because one must observe cytological detail as well as reproductive structures. The bulk of photosynthesis takes place in the upper filaments of the pseudoparenchymatous mass. Among the upright photosynthetic filaments are sterile hairs and the reproductive structures. The crusts adhere to the substrate by simple rhizoids or by the adhesive properties of the prostrate filament walls.

At one time it was believed that *Ralfsia* plants had an alternation of isomorphic generations. Now some life histories have been completed in culture and fusion of gametes was not observed in several of these studies. Plants can bear both unilocular and plurilocular structures; the interpretation of the life history is not easy. Other forms have recently proved to be of considerable interest to phycologists. One *Ralfsia*-like

Figure 7-19. *Leathesia* plant. Courtesy of N. Proctor.

species is known to be the sporophyte phase in the life history of *Scytosiphon* (see earlier). Another *Ralfsia*-like plant is the alternate phase in a *Petalonia-Ralfsia* life history.

The Kelps

Laminaria.   We shall consider *Laminaria,* (Fig. 7-20) which can serve as an example of the kelps, in some detail. The term kelp actually has two meanings. Most phycologists in the United States think of it being used for the larger, parenchymatous brown algae—those with some external differentiation into parts similar to those of vascular plants. However, for some investigators and in some European countries, kelp means the ash obtained from the burning of seaweed (Chapter 20). It is these larger, harvestable seaweeds which are collected, dried, and burned to obtain algal ash.

 Laminaria can also exemplify those organisms that have an intermediate type of life history. (Fig. 7-30). They have an alternation of heteromorphic generations with a large sporophyte phase and a much smaller, inconspicuous gametophyte stage. Since the gametophyte phases are microscopic, considerable experimentation is sometimes needed to detect them. It is not surprising then to find that some details of the haploid phase are only known from culture studies.

Figure 7-20. *Laminaria* blade with ruffled edges. Courtesy of N. Proctor.

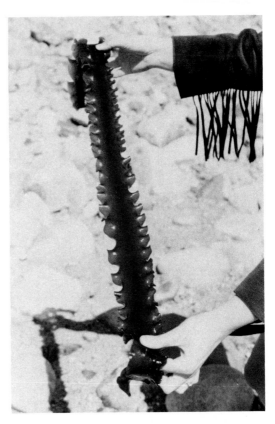

Laminaria (Fig. 7-21) is a large kelp found attached to rocks in the sublittoral zone (Fig. 7-22). The main portion of the plant is a large blade or blades (Fig. 7-23) which arise from a stemlike stipe. The organism is

Figure 7-21. The kelps.

A. *Laminaria*, with holdfast, stipe, and blade. Size: meters in length.

B. *Nereocystis*, with an elongate stipe. Size: many meters long.

C. *Chorda*, the devil's shoe string. Size: many cm. long.

D. A cross section of a *Laminaria* stipe. Note the growth rings and central medulla.

E. One of the trumpet hyphae from a kelp stipe. Diagrammatic. Microscopic.

F. Section of a *Laminaria* blade in the region where unilocular structures are formed. Note the smaller unilocular structures, and the paraphyses (with numerous plastids inside). Microscopic.

G. The male gametophyte, with terminal antheridia (arrow). Microscopic.

H. The female gametophyte, with the contents of terminal cells acting as eggs. Note the basal rhizoids (r). Microscopic.

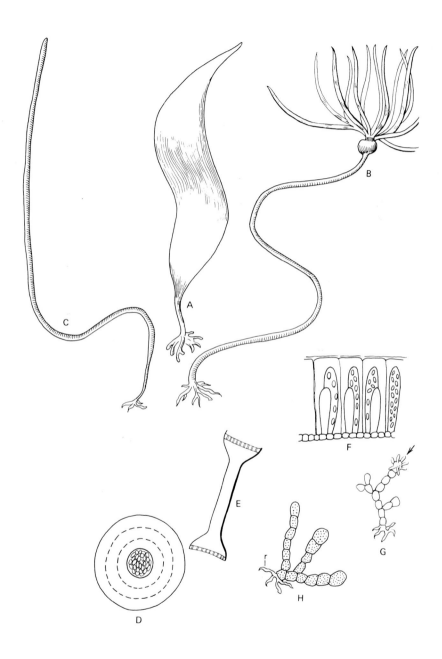

attached by means of fingerlike projections, the holdfast system (Fig. 7-22). Growth takes place at the end of the stipe and base of the blade so that organisms up to 14 meters long, with a stipe of about 5 meters, are found. The organisms thrive in colder waters and are not usually exposed at low tides. The structure of the holdfast and stipe is used in delimiting species. Whether the stipe is solid or hollow is no longer considered a reliable characteristic for distinguishing species of *Laminaria*. The blade erodes at the tip and in some forms is lost completely each year. *Laminaria digitata* has a dissected blade (Fig. 7-23). Many species of *Laminaria* have served as food for animals, and as a source of alginic acid, potash, and iodine.

Beginning an examination of the *Laminaria* life history with the zygote, we see the conspicuous part of the life history, the sporophyte, and the plant we all recognize as *Laminaria*. In early stages it is microscopic and filamentous. But first let us examine the zygote, which elongates and then divides. After division the lower cell may become a rhizoid, but in other cases the formation of this cell type is postponed for several divisions. There are further divisions and growth, with divisions in two planes giving rise to a flat sheet, one cell layer thick. At this time there

Figure 7-23. *Laminaria* species with dissected blade. Courtesy of N. Proctor.

are rhizoids coming from the lower end. These develop into the attaching cells and filaments, and eventually organize into a holdfast. Cell divisions are soon restricted to the base of the flat sheet or blade, so that a stipe and a blade eventually develop (Fig. 7-21). The dividing tissue is called an intercalary meristem.

A cross section of a young stipe shows a superficial layer and a central medulla, while a cross section later shows a superficial epidermis, a cortex that appears to be composed of two regions and a medulla, with all regions having cells quite distinct for the region. Cells toward the exterior of the cortex, as well as the epidermal layer itself, are capable of dividing (meristoderm) so that thickness may be added to the stipe (Fig. 7-21). More medulla is supplied when internal cortex layers differentiate into medulla.

Some of the medullary cells have a longitudinal orientation and have enlarged portions where cross walls form. Because the terminal portion reminds one of the open end of a horn, the cells and filaments are called trumpet hyphae (Fig. 7-21). The end walls of each cell in a trumpet hypha can be perforate. The end wall reminds one of the sieve plate in phloem cells in a vascular plant, and the hypha itself of a sieve tube of phloem. In a related form, chemical and staining analyses have shown that the material accumulated at the ends of the hyphae is the substance callose, just as in the sieve tubes of phloem. Experiments have also shown that there may be some passage of material through the medulla from one portion of a kelp to another. For example, tracer experiments with *Nereocystis* (Fig. 7-24), in which mucilage ducts were severed, show that translocation of material from photosynthesis occurs at a rate of 37 cm hr^{-1} in the medulla, and likely this translocation takes place in the hyphae. Some *Laminaria* species have mucilage ducts, but they are found in the blade. The duct, a tubular structure, is surrounded by specialized secreting cells.

Groups of unilocular sporangia (groups called sori) are found in irregular patches on both sides of the *Laminaria* blade. There are thousands of the unilocular sporangia, along with sterile filaments or cells, the paraphyses. Since the latter are a little taller, they *appear* to offer a protective surface for the reproductive structures. One

Figure 7-24. *Nereocystis*, the bull kelp. Here a portion of the stipe has been coiled. Note the many "leaves" displayed at the end of the bladder. Courtesy of N. Proctor.

sporangium and one paraphysis arise together, as division products of one superficial cell (Fig. 7-21).

Since meiosis occurred in the first divisions of the protoplast of the unilocular sporangium, the 16 to 64 laterally biflagellated swarmers are all haploid. Most of the remainder of the observations, those concerning the gametophyte stage, are from culture studies. Investigations of this stage in the life history have been repeated with many of the brown algae in this group, and the details of the small, microscopic gametophytes are remarkably similar.

The zoospore swims for a short while, settles down on a substrate, and forms a thicker wall. A small (even as few as two dozen cells) and freely branched microscopic gametophyte develops. Each cell has a single nucleus and several discoid plastids. The sexes are separate; thus the organism is heterothallic. One might find the gametophytes close to the sporophyte, near the base, or even attached to some portion of the sporophyte, as lining the holes of *Agarum*. Since few sperm are formed by each gametophyte, a crowding of gametophytes would insure fertilization.

Oogamy is displayed by the gametophytes (Fig. 7-21). Egg cells may be formed at the tips of the short branches, or be intercalary. Gametophytes need blue light for egg and sperm formation. This information came from a laboratory study, but inasmuch as gametophytes are found in the sublittoral, blue light is certainly available. As it is formed, the protoplast of the potential egg rounds up and at least partially escapes from the original wall of the oogonium. This serves as the egg. It may still be attached to the original walls, or be extruded into the ocean. There is frequent branching of a cell at the tip of the male gametophyte (Fig. 7-21), and a single laterally biflagellated sperm arises from each chamber. The sperm are released through a pore, and fertilization takes place either while the egg is still attached to a female gametophyte or external to the gametophyte; the process is oogamous.

The zygote then divides mitotically and gives rise to the large plant we recognize as *Laminaria*, or in a similar fashion to one of the other kelps one might be studying. Limiting factors in reproduction include both survival of the young sporophytes and the chance distribution of gametophytes. (Male cells cannot swim very far.) This life history is said to be heteromorphic, because the haploid and diploid phases are so different in size and outward appearance. It is interesting to recall that we can recognize both gametophytes and sporophytes of an organism such as *Ulva*, and both would be called *Ulva*. But only the sporophyte plant here is recognized as *Laminaria,* and one would have extreme difficulty even trying to determine which kelp gametophyte belonged to which genus.

As one might expect with these organisms there is considerable variation in the pattern of reproduction so far discussed with kelps. In some cases the alternate phase in the life history is unknown. Thus additional laboratory and field investigations are needed in order to gather this basic information.

The Rockweeds

Fucus.   This group of brown algae contains a number of organisms that are smaller than the kelps (Fig. 7-25) and have no immediate visible alternation of generations. Some phycologists believe that they lack the ability to produce spores and are thus unlike *Ectocarpus* and *Laminaria,*

Figure 7-25.   *Fucus* (right) and *Ascophyllum* plants. ⅓ X.

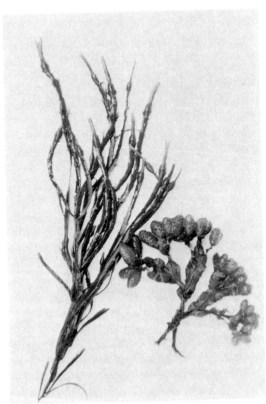

but we will see later that the cytological facts might be interpreted in two ways. *Fucus* and related genera are widely distributed, frequently found in the intertidal zone where the organisms are subjected to periodic, and at times drastic, changes in salinity, moisture, temperature, and so forth. *Sargassum* (Fig. 7-27) may be attached, or free floating.

Growth is by division of an apical cell, and then enlargement and differentiation. The plant body is made of parenchyma cells, but in some cases the internal portion is filamentous.

Fucus (Fig. 7-26) and many of its relatives are common on both coasts of North America, but mainly in colder waters, and definitely where they can attach to rocks in the intertidal zone. *Fucus* plants are about 30 to 50 cm long (Fig. 7-26), composed of straplike, dichotomously branched portions, with an apparent midrib. Some species have rather prominent bladders, which may occur in pairs, along the thallus. There are microscopic cavities just under the epidermis, with hairs projecting through holes or cryptostomata. At the apices structures similar to these cavities become apparent at time of reproduction. The tip becomes inflated (Fig. 7-25), swelling to four to five times normal size, and taking on a lighter yellow appearance. Inside each of the cavities, called conceptacles (Fig. 7-26), reproductive cells may be seen (Figs. 7-28 and 7-29). Both homothallic and heterothallic *Fucus* species are known. The large swollen tip, with numerous conceptacles, is called a receptacle.

Plants that are called female contain reproductive bodies which resemble unilocular sporangia, and are supported by a stalk of one cell. Many of these arise from the base of the conceptacle, and sterile cells, called paraphyses, originate with them from the base; the paraphyses emerge from the opening in the conceptacle (Fig. 7-29). The reproductive structure undergoes meiosis (Fig. 7-26), and with *Fucus* there follows one mitotic division. Then the protoplast is divided, so that eight spherical cells are formed (Fig. 7-26). The latter are called eggs, and they are eventually released, along with the numerous other eggs, from the same conceptacle. The eggs are then located in the water outside the plant.

In a heterothallic form there would be inflated tips on other plants, with conceptacles, and with paraphyses, but in addition, there would be fertile filamentous bodies bearing antheridia (Figs. 7-26 and 7-28). In some cases, the antheridium is borne on the conceptacle wall. When just forming, there is a single diploid nucleus in one antheridium, but meiosis and four successive mitotic divisions follow, resulting in 64 laterally biflagellated sperm. The sperm are also released, along with the contents of numerous antheridia from one conceptacle. They are attracted to the

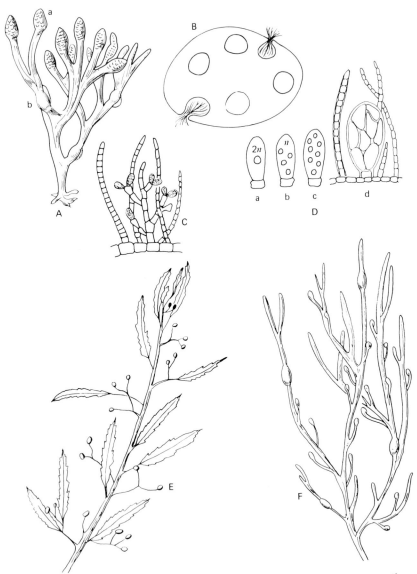

egg and, eventually, one of the several sperm in the area fertilizes the egg. The zygote, as well as the egg, is nonmotile.

Fucus eggs and zygotes are easily collected by manipulating fertile plants in the laboratory. They have been used extensively in studies of polarity, with emphasis on the role of sperm penetration, pH, light, and geotrophism on the first divisions and the formation of the rhizoid.

Figure 7-26. *Fucus* and related genera.

A. *Fucus*, with the inflated tips, the receptacles (a) and bladders, one of which is shown as b. Length measured in cm.

B. Cross section of an inflated receptacle, with several conceptacles, and in two cases the section has gone through the pore in the conceptacle. Note the emergent hairs. Microscopic.

C. In a male plant one would find these filaments in the conceptacle. The antheridia are at the tips of the branched filament. Note unbranched paraphyses. Microscopic.

D. Reduction division and formation of the eight eggs in *Fucus*. The filaments, as in C, are paraphyses. There is meiosis (a–b) followed by a mitotic division and cytoplasmic division. The eggs are released when there is a break in the oogonium wall. Microscopic.

E. *Sargassum.* Note the leaflike appearance of the appendages, and the bladders. Natural size.

F. *Ascophyllum,* with several bladders. Length: c. 0.5m.

Embryos that are apolar have been produced. It appears that there is some orientation of cellular organelles, and then some external manifestation of polarity can be seen.

In one interpretation of the life history *Fucus* does not exhibit free-living forms of two types, a gametophyte and a sporophyte, but exists as a diploid plant which produces gametes, after meiosis.

But upon closer examination this might not be the case, because the immediate four products of meiosis in *Fucus* do not act as either eggs or

Figure 7-27. *Sargassum* plants recently washed up on the shore after a storm. Plant at the right is attached to a small stone ⅓ X.

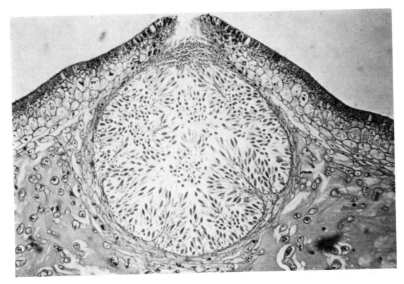

Figure 7-28.   *Fucus.* Cross section of conceptacle that will release sperm. 60 X.

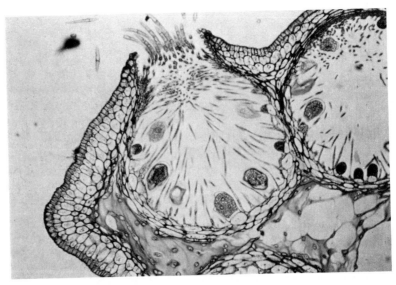

Figure 7-29.   *Fucus.* Cross section of conceptacle that will release eggs. 60 X.

sperm. In comparison with other brown algae we have discussed, this would be the best way to interpret the life history. The chamber in which the eggs are formed would not be called an oogonium, but a sporangium, at least initially. The four products of meiosis are spores, and then with one additional division, the four spores yield eight eggs. In the antheridium, or one might also want to call this structure a sporangium, at least initially, the four products of meoisis do not act as sperm, but there are four mitotic divisions *before* sperm result.

Thus there might be a haploid phase in *Fucus,* but it is extremely reduced and is not free-living. If one were to examine the haploid phase of *Ectocarpus,* it would be at least as large as the diploid phase. In the kelps it would be reduced to a microscopic form. Could it be that the few cells just prior to egg and sperm are in reality the remainder of the gametophyte phase in *Fucus?*

An examination of *related* genera provides some additional interesting information. In some forms reduction division in an oogonium (here that is the correct term) produces four eggs (*Ascophyllum*), while in other genera there may be just two eggs (*Pelvetia*), or one egg (*Sargassum*). One egg would be produced if there was loss of one of the products from each of the divisions in meiosis. In these organisms there are apparently still some mitotic divisions of haploid cells in the development of sperm.

After examination of the changes that can take place in the development of eggs in these genera, one can erect an evolutionary line in which the haploid vegetative phase has been eliminated, with even the number of eggs produced in an oogonium reduced. Have some species of *Sargassum,* which apparently lack sexuality, evolved just a step further?

A summary statement concerning life cycles in the brown algae is in order. Previously we had thought that organisms fell into three categories, based on an alternation of either isomorphic or heteromorphic generations, or merely cytological changes. Now it appears that this framework for the classification is based on examples of too few organisms. Also several organisms are found to deviate from the accepted and "expected" pattern. There is more flexibility. Thus a position on classification must remain fluid, and we should expect some exciting developments in the next few years. Figure 7-30 presents a scheme that shows the relationship of various brown algal life histories we have examined. The gametophyte phase is much reduced in the advanced forms, and it is apparently lacking in some species of *Sargassum.*

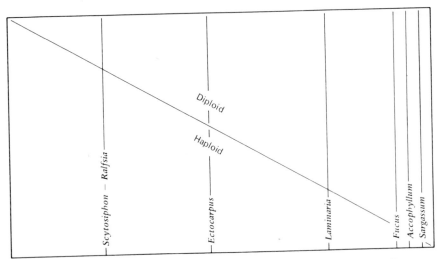

Figure 7-30. A diagram helpful in visualizing various life histories of brown algae. The dominance of a certain phase is indicated by the length of the line. More primitive organisms would be to the left.

Some Additional Genera of Kelps and Fucoids

Agarum. *Agarum cribrosum,* commonly called the devil's apron, is one of the kelps that is preferred as fodder. It gets the common name because of its broad sometimes ruffled blade, but also because of the numerous holes. As a result of the fire which he tended, the apron worn by the devil had numerous holes burned in by flying sparks! Initially, these holes are just weak spots in the blade, which are gradually eroded and enlarged. In some cases one can find gametophytes germinated at the margin of the holes. It is a deep-water form, which resembles, in some ways, a *Laminaria* (with holes) but there is also an obvious midrib.

Alaria. This is another cold-water form, with plants up to 3 m in length, and like *Agarum,* in the northeast it is found only in the colder waters north of Cape Cod. With the southern shore of the Cape warmed by the Gulf Stream, and the water north of that point cooled by water from the Gulf of Maine, there is a sharp break in the distribution of at least a few genera. Many of these organisms are found in the Pacific. *Alarias* have a holdfast, stipe, and blade and when young resemble *Laminaria.* But at maturity the prominent flattened midrib and the numerous deciduous short blades arising laterally along the margin of the stipe make this organism distinctive. Sporangia are formed only on these smaller blades, which can appropriately be called sporophylls. The latter are deciduous, and thus sporangia can be carried distances before spores are released. Like *Agarum,* this perennial is also gathered from the sublittoral for fodder. In Great Britain the stipe and sporophylls are eaten by some people.

Ascophyllum.   This very common rockweed is frequently found attached by a disc-like holdfast in the intertidal zone. The novice often confuses *A. nodosum* (Fig. 7-25) with one or more species of *Fucus,* but the absence of both a midrib and dichotomous branching in *Ascophyllum* can be used to distinguish them. Size and color differences may also be used. *Ascophyllum* branches are straplike and not as flat as in *Fucus* (Fig. 7-26). On occasion one can see numerous small clusters of young plants originating from a portion of a large plant. *Ascophyllum* has remarkable powers of regeneration and thus active growth of new individuals occurs as a broken branch or wound is healing. The air bladders, inflated portions of the main axis of a branch, are elongate and larger than in *Fucus.* They are obviously wider than the branch on which they occur. They not only provide buoyancy, but also enable the numerous straplike branches to fully extend in the moving water of the intertidal zone.

Ascophyllum receptacles are enlarged portions of the branch (here a lateral branch) and are distinctly lighter in color than the main thallus. The small lateral branches bearing sporangia are deciduous and thus this portion of the plant can be carried some distance before eggs and sperm are ready for release.

In the northeast one does not find many kinds of macroscopic epiphytes on *Ascophyllum.* However, one might find a population densely epiphytized by one species of *Polysiphonia.* This is one example of a host-epiphyte specificity.

Chorda.   The most primitive kelp, *Chorda,* bears a close resemblance to a rawhide shoe string and is commonly called devil's shoe string (Fig. 7-21). The holdfast is in the form of a disc and from it come the stipe and blade. However, in this case, the blade is almost a continuation of the form of the stipe. Although commonly about a meter in length, some forms more than five times that length have been collected. The surface of the blade is covered by numerous hairs. In some species these are elongate and very conspicuous.

Durvillea.   This organism is called the bull kelp of the southern hemisphere. The same common name is used for other brown algae, for

Figure 7-31. *Durvillea*, a South American fucoid. Courtesy of R. Wilce.

example, *Nereocystis. Durvillea* (Fig. 7-31) is a large brown alga, up to 10 m in length, forming kelp beds in the South Pacific and even the Antarctic. The plant is anchored by a holdfast (Fig. 7-32), which may reach a diameter of 60 cm, and has a stipe and fan-shaped blade. The blades are internally dissected into many chambers bordered by thin partitions. Gas may be trapped in these chambers, and thus the blade floats quite well. At maturity conceptacles are formed over the surface of the blade. This organism is harvested locally, as in South America (Fig. 20-9), for chemical products, for cattle feed and fertilizer.

Macrocystis. *Macrocystis* (Fig. 7-33) grows in 20 to 30 meters of water and commonly produces plants many meters in length. Plants in deeper water would have to be 35 meters in order to reach the surface. The largest size recorded is difficult to establish, but it was probably around 70 meters long! Fresh weights of from 120 to 250 kilograms are

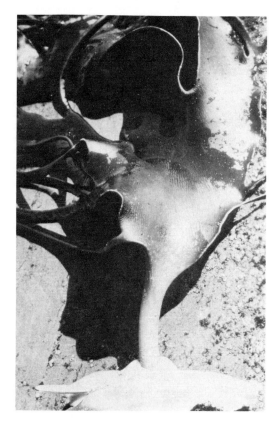

Figure 7-32. *Durvillea* holdfast and stipe. Courtesy of R. Wilce.

reported. This organism, called the giant kelp, grows in colder waters, sometimes almost in dense forests. On a rocky substrate there will be a massive holdfast, but organisms are found on sandy bottoms. Some species are common in the colder waters of the southern hemisphere.

Those plants that begin developing in deeper water grow extremely rapidly by elongation of the stipe. There is usually no branching to the stipe. Numerous blades are produced and each has a basal bladder, a hollow bulbous growth that aids in floatation. The organism is a perennial. It is harvested extensively and regenerates quickly after harvest.

With one species of *Macrocystis* experiments were performed to demonstrate the movement of the bound carbon from photosynthesis within the plant. Even though kelps have elongate cells that are compared with vascular plant phloem, would they actually utilize

Figure 7-33. Portion of a
Macrocystis plant body displayed out
of water. Courtesy of N. Proctor.

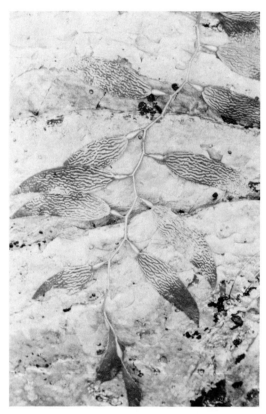

conducting cells? The photosynthate, identified by carbon labeling, was shown to move at rapid rates both up and down the stipe.

Nereocystis.   The bull kelp (Figure 7-24) is a Pacific form that can grow in dense algal forests. The organism, an annual, grows in both shallow and deep (30 m) water and can reach a length of 40 m, with a long 35 m stipe. As an annual it must achieve this growth in a year! Thus, some rather remarkable growth rates have been recorded for stipe elongation. There is an extensive holdfast and a hollow stipe which is quite branched at the tip. Immediately below the point of branching or stipe dissection is an inflated bulblike bladder which can have a diameter of 6 to 14 cm. Distal to the bladder, arising from numerous branches of the stipe, is the display of photosynthetic tissue, with from 30 to 60 simple blades, up to 5 m in length. The blades hang down from the water surface where the large bladder is floating. Sporangia are

Figure 7-34. *Postelsia* habit, at the lowermost edge of the littoral zone. Courtesy of N. Proctor.

produced in patches, which, when near maturity, actually drop out of the blade, leaving a large gap. All sporangia in that sorus then are transported to a new environment on the ocean bottom. Gametophytes are produced on the bottom substrate.

Some *Nereocystis* plants can be among the kelps harvested in the Pacific. It is quite visible from the water surface since the bladder floats and the leaves are displayed near the surface.

Postelsia.   This is an erect annual kelp found in the Pacific northwest, (Fig. 7-34), commonly called the sea palm. It is exposed only with extremely low water, firmly attached by a holdfast with fingerlike projections. The stipe is firm, erect, and stemlike, up to a meter in length (Fig. 7-35). It is hollow and quite rubbery. A number of blades, often not as long as the stipe, project in a flat plane from the tip of the stipe. With

Figure 7-35. *Postelsia*, the sea palm, with holdfast, stipe, and display of leaves. Courtesy of N. Proctor.

its elongate structure and serrate edge when the blades are displayed, the organism looks like a miniature palm tree.

The organism is found in very rough water and withstands the crashing of waves. *Postelsia* grows poorly in some protected localities.

Sargassum.   Commonly called gulf weed, this genus is composed of many species, most of which are tropical. One commonly thinks of the *Sargassum* of the Sargasso Sea, the planktonic form found in the mid-Atlantic, east and south of Bermuda. Few species are found in the plankton. Most are attached to rocks and shells (Fig. 7-27) in the low littoral and sublittoral by a rather extensive holdfast. Several erect branches, a meter or two in length, may be attached by an elaborate disc-type holdfast. Externally, *Sargassum,* with its numerous leaflike appendages, is about as complex a plant body as one can find in the algae. The leaflike structure will serve as an example of the great degree of differentiation. The leaves have a "petiole," midrib, serrate edge and possess numerous hairlike appendages. The planktonic species have numerous small stalked, globular bladders, which *superficially* resemble small fruits of a higher plant. Attached forms have fewer bladders.

It is interesting that such a large form thrives in the plankton. Most benthic algae do not survive very long unless they remain attached to the substrate. For the majority, being torn loose by a storm means sure death within a few days.

The planktonic forms of *Sargassum* are seldom as plentiful as one might imagine from the popular accounts of the flora of the Sargasso Sea. At times only a small patch of the brown alga will be found, perhaps the equivalent of a few plants aggregated together. Then one might find a windrow of *Sargassum*. It is never found in such dense populations that commercial harvesting could be attempted.

For *Sargassum* there is a life history involving production of eggs and sperm from a diploid plant, similar to *Fucus*. However, planktonic forms apparently reproduce only asexually, at least for a period of many years. By extensive growth and fragmentation they have been able to maintain populations in the open Atlantic.

EVOLUTION

In one postulated evolutionary sequence emphasizing reproduction and some vegetative features there are five lines in the brown algae:

Line 1.   The ancestral forms were filamentous, isogamous, and isomorphic. The other lines originated from this stock.

Line 2.   The life history remained isomorphic, but the plant body became parenchymatous. In the larger forms, oogamy was possible.

Line 3.   Here the life history was heteromorphic, but the gametophyte was larger, and pseudoparenchymatous. Sporophytes were smaller, encrusting forms.

Line 4.   The life history was no longer isomorphic, because the sporophyte plants increased in size. Through the intertwining and compaction of filaments a pseudoparenchymatous form developed, or parenchyma developed. The gametophytes became reduced filamentous forms. Sexual reproduction was ultimately oogamous.

Line 5. Life histories were decidedly heteromorphic in this group; in some cases a free-living haploid phase was completely lost, as is true of at least some of the relatives of *Fucus*. Oogamy was the rule. The sporophyte phase was parenchymatous, with the possibility of some filaments visible in the medulla.

REFERENCES

Bourelly, P. 1968. *Les Algues d'eau Douce.* Initiation a la systematique. Tome II. Les algues jaunes et brunes. N. Boubee and Co., Paris. 438 p.

Bird, N. and J. McLachlan. 1976. Control of formation of receptacles in *Fucus distichus* L. subsp. *distichus* (Phaeophyceae, Fucales). *Phycologia* 15: 79–84.

Buggeln, R. 1976. The rate of translocation in *Alaria esculenta* (Laminariales, Phaeophyceae). *J. Phycol.* 12: 439–42.

Dawson, E. 1966. *Marine Botany: An Introduction.* Holt, Rinehart and Winston, Inc., New York. 371 p.

Druehl, L. and M. Wynne. 1971. Bibliography on Phaeophyta. In J. Rosowski and B. Parker (Eds.) *Selected Papers in Phycology.* Dept. Botany, Univ. Nebraska, Lincoln. pp. 791–6.

Edwards, P. 1969. Field and cultural studies on the seasonal periodicity of growth and reproduction of selected Texas benthic marine algae. *Contrib. Mar. Sci.* 14: 49–114.

Fritsch, F. 1945. *The Structure and Reproduction of the Algae.* Vol. II. The University Press, Cambridge. 939 p.

Kingsbury, J. 1964. *Seaweeds of Cape Cod and the Islands.* The Chatham Press, Inc., Chatham, Mass. 212 p.

Kylin, H. 1947. Die Phaeophyceen der schwedischen Westküste. *Lunds Univ. Arssk. N. F., Avd.* 2, 43: 1–99.

Müller, D. 1964. Life-cycle of *Ectocarpus siliculosus* from Naples, Italy. *Nature* 203: 1402.

Müller, D. 1975. Experimental evidence against sexual fusions of spores from unilocular sporangia of *Ectocarpus siliculosus* (Phaeophyta). *Br. Phycol. J.* 10: 315–21.

Nygren, S. 1975. Life history of some Phaeophyceae from Sweden. *Bot. Mar.* 18: 131–41.

Smith, G. 1944. *Marine Algae of the Monterey Peninsula, California.* Stanford Univ. Press, California. 622 p.

Stewart, W. 1974. *Algal Physiology and Biochemistry.* Univ. Calif. Press, Berkeley. 989 p.

Taylor, W. R. 1957. *Marine Algae of the Northeastern Coast of North America.* University of Michigan Press, Ann Arbor. 509 p.

Wynne, M. 1969. Life History and Systematic Studies of Some Pacific North American Phaeophyceae (brown algae). Univ. Calif. Publ. Bot. 50. 88 p.

Wynne, M. and S. Loiseaux. 1976. Recent advances in life history studies of the Phaeophyta. *Phycologia* 15: 435–52.

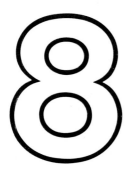

Bacillariophyceae — diatoms

Diatoms are common organisms found in fresh water and marine habitats, as well as in or on the soil. In the ocean they are common in the plankton and are frequently the bloom organisms, especially in colder months. In addition, they are in and on sediments, or attached to larger benthic algae, seagrasses, animals, and rocks. In fresh waters they could be either the dominant planktonic organisms or attached to rocks in small streams. In lakes and large rivers, diatoms are frequently the most important photosynthesizers in winter months, whereas in small streams the attached diatoms could be the main primary producers throughout the year.

DIATOM THALLUS TYPES

Thallus types are simple, so that organisms exist as unicells, colonies, or filaments. Unicells may be free-floating or attached (Fig. 8-1). The attached cells could adhere to the substrate by a short stalk or remain in contact along most of the cell surface, perhaps attached by an extruded mucilage. At times stalks become branched (Fig. 8-1) so that a dendroid or fan-shaped colony results.

After cell division the new cells may adhere and after subsequent divisions a colony or a filament result. Both forms remain rather simple (Fig. 8-1). Certain forms, for example, *Amphipleura, Schizonema,* and one species of *Navicula,* remain within a common branched mucilaginous tube. These aggregations produce macroscopic brownish forms that are easily confused with macroscopic seaweeds in other classes.

For almost a century diatoms have been grouped into two categories, pennate and centric forms. In addition to the very obvious symmetry in each group, there are other general characteristics. The centric forms, often with radial symmetry, usually possess many plastids, reproduce sexually with flagellated sperm, and lack a raphe system. Many centric forms are common in the marine plankton, though they are not exclusively marine.

The pennate forms, generally with bilateral symmetry, possess fewer plastids (frequently two per cell), reproduce sexually by autogamy, and may possess a raphe system. A raphe is a longitudinal slit or opening found on one or both walls (Figs. 8-2 and 8-12). The raphe system is involved with movement of vegetative cells. Some pennate diatoms have what at first might appear to be a raphe. This pseudoraphe (Fig. 8-9) is a

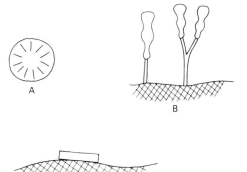

Figure 8-1. Plant body types seen in the diatoms. These organisms exist as unicells (A,B,C), which may be free living or attached. Some are dendroid (B, right), but many are filamentous (D,E) or colonial (F).

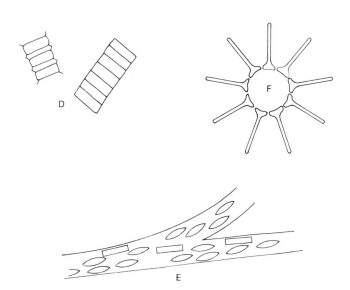

clear, linear type of ornamentation on one or both walls. Actually it is the absence of ornamentation, a place where there are no puncta or striae. It is longitudinal and frequently in the same location where one would find a raphe. However, the pseudoraphe is not a slit in the diatom wall. Pennate forms are common in fresh waters or are attached forms, frequently also found in and on sediment. Many pennate forms are also marine types.

Figure 8-2. Electron photomicrograph of a *Navicula* valve. 7000 X. Courtesy of P. Hargraves.

DIATOM CHARACTERISTICS

1. Pigmentation

Diatoms contain chlorophylls *a* and *c*, alpha and beta carotene, and several xanthophylls, with diatoxanthin as the dominant one. They also contain the xanthophylls fucoxanthin and diadinoxanthin. The pigments are in plastids with the thylakoids in groups of three. There are girdling lamellae which surround the periphery of the plastid. Diatoms contain two to several plastids per cell.

2. Food Reserve

The food reserve is chrysolaminarin or leucosin. The compound is a beta 1:3 linked glucan, with branches linked from the 1 and 6 positions. Many diatoms have conspicuous oil droplets of food reserve within the cell. Many of the oil deposits we now use for heating homes and running automobiles probably were produced by diatoms of the past.

3. Motility

No vegetative cells of diatoms are capable of motion by action of flagella or cilia. However, many of the pennate forms move, without flagella or cilia! When this type of motion occurs, a true raphe must be present on at least one wall. The wall with the raphe is in contact with the substrate. We do not yet have a satisfactory explanation of this type of movement. The subject will be covered later in the chapter.

In the centric group, where sperm may be formed, there is one anterior flagellum on these motile cells. The flagella of *Lithodesmium* and *Biddulphia* lack the two central strands in the normal 9 + 2 flagellum structure. (Fig. 2-6).

4. Walls

The diatom wall is composed of silica impregnated in pectin; cellulose is lacking. An individual cell is composed of two overlapping halves, and in centric forms it resembles the common laboratory petri dish or plate. When we observe a cell from the top, we speak of seeing it in valve view. The view from the side is called the girdle view. In various texts you will see the wall referred to as the frustule, valve, test, or theca. An epitheca would be the larger wall, and hypotheca would refer to the half that fits within the epitheca. Since diatoms can be upside down in nature as well as in the laboratory on a slide, the hypotheca is not always below.

The two wall halves are joined by at least two girdle bands, just as one could join the petri dish halves with a band of masking tape. In many diatoms, the bands add little to the thickness, but in an organism such as the marine *Rhizosolenia,* many additional intercalary girdle bands add considerable dimension to the diatom in girdle view. *Tabellaria* would be a pennate type with several intercalary bands. Wall structure has been, and still is, the basis for classification at all levels.

In the study of diatoms permanent slides are used. Inasmuch as the wall is glass, and classification is based on wall structure, material is ashed on a hot plate (below the temperature which would melt glass), or acid cleaned prior to making a permanent mount.

Some other common diatom terms that refer to the wall ornamentation are:

punctum—a thin area or a depression, sieve area, or hole in the glass wall. With the electron microscope we can see that what we thought was a hole is actually somewhat occluded by a sieve of silica. The plural is puncta.

stria—a linear type of ornamentation, which in reality is a close arrangement of puncta.

costa—internal or external thickening of the valve, frequently elongate.

raphe—a slit in the face of the valve, on one or both valves. It may be median or marginal. The entire structure is quite complex.

pseudoraphe—a clear linear feature, observed because other ornamentation is lacking. It is the space between the ends of adjacent striae or costae. It is not a slit.

nodule—thickening in the wall, either in the central region, or at the poles of pennate diatoms.

septum—internal partition. The plural is septa.

You might find certain other terms helpful for specific diatoms. These may be found in the books, manuals, or literature dealing with diatoms in detail or in keys to local organisms.

REPRODUCTION

Typically the diatom reproduces either by simple division of a cell or by auxospore formation. In the first case, if the organism is unicellular, two new individuals result. However, with colonial or filamentous forms, cell division for a time merely increases the cell number of an individual filament or colony. With fragmentation or dissociation of cells, new individuals can also result.

Resting spores (Figs. 8-3 and 8-4) are formed in a few genera.

Cell Division

At division, mitosis occurs, the wall halves separate, a nucleus moves to each half, and cytokinesis takes place. Then by means of activity of numerous vesicles derived from golgi bodies, new silica walls are deposited. The wall deposition is initiated at the center and develops in a centrifugal pattern, with new walls fitting *within* both old halves. Consider for a moment how a raphe, costa, septum, or spine would be deposited at a precise location, giving the wall such precise symmetry. Perhaps the plasmalemma acts as a template.

McDonald, and also Pfitzer, pointed out in the last century that the formation of new walls within old walls would lead to a gradual size reduction in any population (Figs. 8-5 and 8-6). However, pennate populations may reduce their length extensively, but have little change in width. Inasmuch as diatoms are found within a certain size range for a

Figure 8-3. Centric diatom resting spore. *Leptocylindrus* 7000 X. Courtesy of P. Hargraves and the Journal of Phycology.

particular species, there must also be a minimum size as well as mechanisms that allow cells at specified times to divide without a decrease in the dimensions of the cell. The diatoms thus can and do maintain populations at the lower end of the size range. Some organisms apparently are never observed except at this size.

In order to eliminate size reduction and maintain a uniform population, certain diatoms do not have overlapping halves. With girdle bands precisely shaped and correctly positioned, upper and lower wall components are easily joined. In addition, the wall might not be completely rigid, with hingelike arrangements at the valve margin allowing some flexibility. Many organisms can be maintained in the

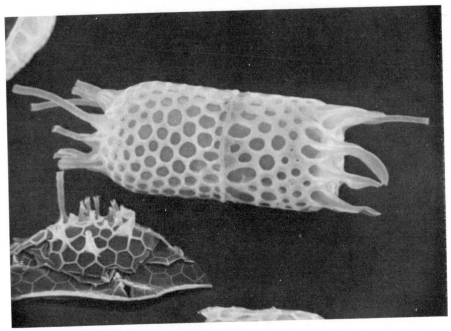

Figure 8-4. *Stephanopyxis* resting spore in girdle view. Note spines and the two components of the spore. SEM. 1500 X. Courtesy of P. Hargraves and the Journal of Phycology.

laboratory at the minimum size for many months, until auxospore formation takes place. The latter is discussed in the next section. In summary, we have now come to understand that a regular decrease in size (Fig. 8-6) is not absolutely necessary.

Auxospore Formation

When cells of any population have reached their minimum size, or are near the lower end of the size range, a diameter increase can occur by

Figure 8-5. Diatom reproduction. After repeated divisions and formation of new walls within old walls, diameter reduction can occur. As observed in girdle view, there can be an increase in cell depth.
The sexual cycle of a centric homothallic population (seen in valve and girdle view) is figured below. Cells are at the lower end of the size range. Under the appropriate environmental stimulus certain cells become eggs and others produce sperm. The cell forming the egg has valves separated and a somewhat expanded protoplast appears between the epivalve and hypovalve. Four sperm are produced from each parent cell. After fertilization the zygote becomes modified into the structure we call an auxospore. The auxospore is formed within the boundary of the fertilized egg, and cell diameter is markedly increased.

means of sexual reproduction. The initial enlarged cell is called the auxospore, and the process auxospore formation (Figs. 8-5 and 8-28).

Although auxospore formation has been reported as an exclusively sexual process, the evidence for this in a number of forms is really lacking. Natural populations of *Ditylum* maintain themselves by vegetative enlargement, usually involving the extrusion of a naked protoplast and

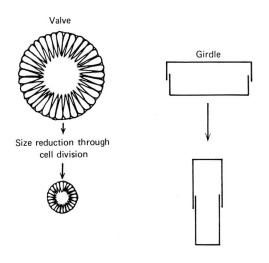

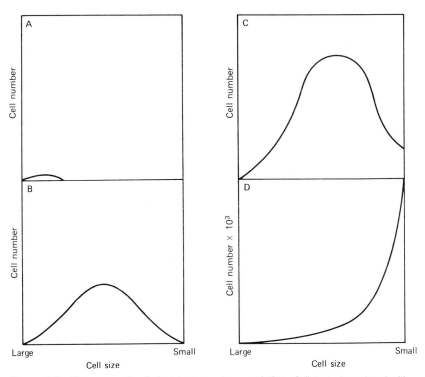

Figure 8-6. The range of cell sizes in a growing population of diatoms inoculated with recently produced auxospores. As the population grows (A to D), the average cell diameter decreases, but a few cells in the large size range persist until the cells die or are grazed.

deposition of a wall with a larger diameter. *Cyclotella* can form auxospores in several ways, one of which is probably asexual.

It is generally *thought* that pennate diatoms reproduce by autogamy (Fig. 8-7) with no flagellated cells, and that oogamy occurs in centric forms. However, both isogamy and oogamy have been reported for *Cyclotella.* For those centric organisms in which sexual reproduction has been observed in culture, for example, *Stephanopyxis,* or in material from the field, there is one typical pattern, with only minor differences among species (Fig. 8-5). Sperm may be anteriorly or posteriorly uniflagellate, and 2 to 64, or more, arise from one parental cell. Four sperm per cell is more typical. A large single oogonium develops from one cell, and fertilization is possible in between separated valves or by way of fertilization pores. However, there have been few reports of the observation of plasmogamy itself, as in *Stephanopyxis.* The zygote

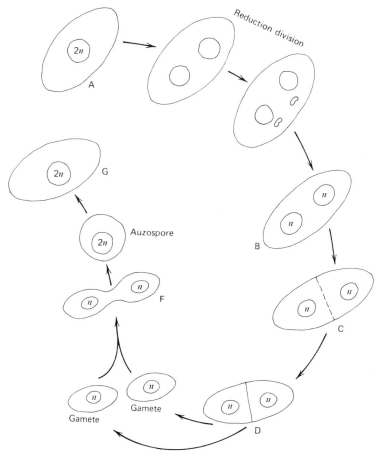

Figure 8-7. Sexual reproduction in pennate diatoms in which two haploid nuclei survive after meiosis. Reduction division takes place (A–B), followed by cytokinesis (C, D). The two gametes which result then fuse (F), forming an auxospore. After a period of dormancy this auxospore enlarges, and in this way a new pennate cell at the upper end of the size range is formed.

produces within its boundaries an ornate, flattened, silica-walled auxospore, with a diameter considerably larger than that of the vegetative cells.

In the pennate diatoms there are two general patterns of sexual reproduction, with many minor modifications. The initial stages resemble stages in any cell preparing to divide. The cells that are going to mate have a close contact at the girdle region. Meiosis occurs, but the number of haploid nuclei that survive in each cell may be either one or two. (Other products of the divisions are lost.)

1. When only one daughter nucleus survives (Fig. 8-8), the resulting cell acts as a gamete. The valves would separate and the protoplast would fuse with another similar gamete.

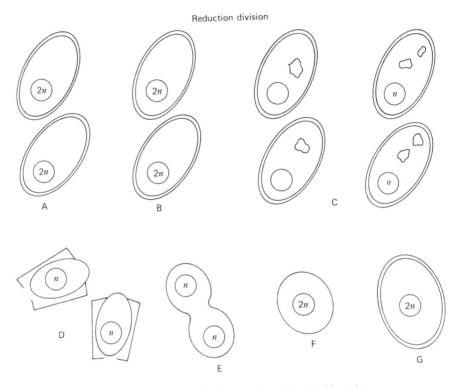

Figure 8-8. One method of reproduction in pennate diatoms. In this series there is only one surviving nucleus after meiosis. Two diploid cells approach each other (A), and in the second pairing (B) each nucleus prepares for a meiotic division, which takes place in the two pairs to the right (C). Polar nuclei are lost and one haploid nucleus remains from each meiotic division. Then in the series on the next line, the cells are first shown in girdle view (D). The valves have separated, exposing the two protoplasts, with a haploid nucleus in each. The protoplasts fuse (E) and the nuclei fuse (F), restoring the diploid state. In the last figure the silica wall has now been deposited around the cell (G). It resembles the original cells which entered into reproduction.

2. When there are two products of the meiotic division (Fig. 8-7), cytokinesis occurs and each nucleus is surrounded by its own protoplast. Then *daughter* protoplasts may fuse. This is an example of autogamy. In either case, a spherical zygote is formed. Then there

is usually a resting period, after which enlargement of the zygote or auxospore takes place. The original wall is ruptured during this process. When a new epivalve and hypovalve are formed they are considerably larger than walls originally around gametes.

There are few reports of meiosis and sexual reproduction in diatoms. The chromosome figures presented are not always convincing. With many possible variations in life histories in other algal groups, it would not be surprising to find that certain diatoms deviated from the life history we have shown to be typical of diatoms, that is, diatoms existing only as diploid organisms (with gametic meiosis).

Some examples of size increases by vegetative means are known. In some cases, as in *Ditylum,* the diatom halves separate, the protoplast is exposed, and an enlarged cell is formed without sexual reproduction. Flagellated cells are seen in other diatoms and auxospores are formed, but fertilization has not been detected cytologically. This phenomenon has been reported for certain species of *Cyclotella.* In similar cases, the organism has the capability of forming sperm, but it can also achieve a size increase without them. We are thus certain that not all auxospores are the result of a sexual reaction.

The range of cell sizes in a growing population, inoculated with auxospores, is depicted in Fig. 8-6.

SOME REPRESENTATIVE GENERA

Diatoms are easily separated into rather distinct categories which makes their study much easier. As we have already seen there are forms with bilateral symmetry, and others with radial symmetry. We will examine the pennate forms (bilateral symmetry) first, initially looking at those diatoms that have a pseudoraphe on both valves.

The Pennate Forms

A. A GROUP OF ORGANISMS IN WHICH BOTH VALVES HAVE A PSEUDORAPHE.

1. Members that can develop zig-zag colonies, with the cells joined by mucilagenous pads at the corners of the cells. Colonies are not observed in most acid cleaned preparations.
Tabellaria (Fig. 8-9). This common freshwater form is frequently seen in girdle view, and in chains. Cells are joined by the mucilage pads. It possesses several septa, which are straight, or sometimes slightly wavy. The latter are best seen in girdle view, parallel to the valve surface. The valve view is distinctive, with polar and median swellings.

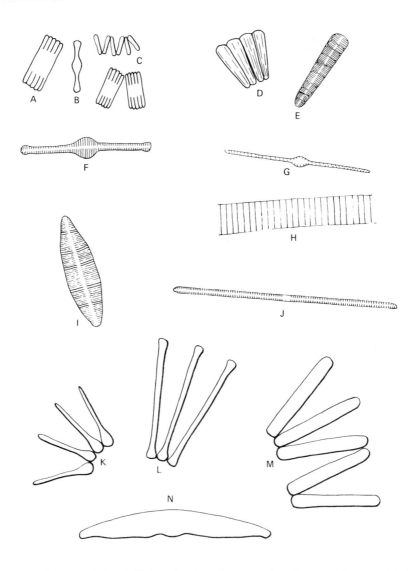

Diatoma (Fig. 8-9) is a freshwater organism frequently seen as a unicell, or at times in chains. It is unlike *Meridion* in that it is not wedge-shaped, and it lacks the septa of *Tabellaria*. Costae are formed on the surface.

Striatella is a marine form with a naviculoid appearance in valve view. The organism has fine puncta and is frequently weakly silicified. It shows many septa and numerous intercalary bands in girdle view.

Figure 8-9. Diagrams of some common pennate diatoms, shown in valve view, girdle view, or both. Microscopic. There is no substitute for examination of the diatoms themselves.

A–C. *Tabellaria.* A. Girdle view, showing the internal partitions, septa. B. Valve view, without the puncta and pseudoraphe. C. Girdle view of two types of colonies, with cells joined by mucilage pads.

D. *Meridion* in girdle view. E. *Meridion* in valve view.

F. *Tabellaria* in valve view with ornamentation and pseudoraphe.

G–H. *Fragilaria.* G. In valve view. H. Colony or ribbon of *Fragilaria* cells, shown in girdle view.

I. Valve view of *Diatoma.* Pseudoraphe is evident; the perpendicular markings, usually at right angles to the pseudoraphe, are ribs or costae.

J. An elongate *Synedra.*

K–L. *Asterionella.* Both are fragments of potential circular or helical colonies. Cells are seen in girdle view.

M. *Thalassiothrix,* a zig-zag marine diatom, which also forms stellate, radiating colonies. Some put the zig-zag form in the genus *Thalassionema.*

N. *Ceratoneis* or *Hannaea.* The latter is now the accepted name for this freshwater form.

Rhabdonema is more heavily silicified. There are numerous elongate septa, which may possess only small apertures. Thus the interior has numerous compartments. Ornamentation can be seen on the girdle.

Grammatophora (Fig. 8-10) is a marine form that resembles *Tabellaria,* but the four septa are undulate as seen in girdle view.

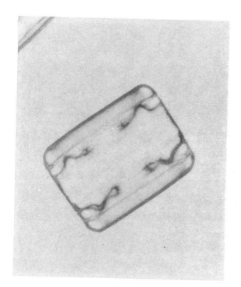

Figure 8-10. *Grammatophora.* Girdle view showing the undulate septa. 600 X.

Thalassiothrix (Fig. 8-9) is an organism with elongate cells and weak ornamentation except for a beadlike margin. This genus contains both zig-zag colonies and radiating colonies. At times others put the zig-zag types in the genus *Thalassionema,* and even use the feature of heteropolarity to support this decision.

2. The remainder do not form zig-zag chains.

Meridion (Fig. 8-9), a colonial type, is found in both marine and fresh waters. The individual cells are wedge-shaped in both valve and girdle views. As the cells divide, daughter cells adhere so that fan-shaped colonies develop. Sometimes the fan forms a complete ring before it fragments. Costae are observed especially in valve view, but can be seen along the margins in girdle view.

Fragilaria (Figs. 8-9 and 8-11), a ribbonlike filament, is found in fresh and brackish waters. When it breaks up, some forms can be confused with the next genus, *Synedra.* In what view would you commonly see a *Fragilaria* (Fig. 8-9)?

Synedra. There is great range of form within the genus, especially with regard to size, with some species extremely long and needlelike (Fig. 8-9). Both marine and freshwater forms are known.

Asterionella (Figs. 8-9 and 8-11). This ornate colony has marine and freshwater representatives, which have cells joined in stellate colonies, or portions of the same. Some of the individual cells look like a bone, with a knob at each end. In what view would you commonly see an *Asterionella*?

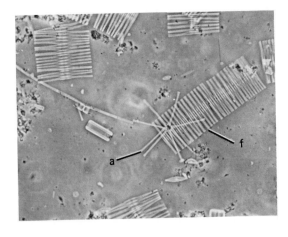

Figure 8-11.   *Asterionella* and *Fragillaria* colonies. Both seen in girdle view. 150 X.

Hannaea was formerly called *Ceratoneis* and is sometimes mistakenly called a *Fragilaria*. This freshwater form (Fig. 8-9) is distinctly, but slightly, bent, with a bump on the inside of the bend. *Eunotia* cells are distinctly, but slightly, bent with an undulate appearance. A short, true raphe appears at each pole.

Raphoneis has large beadlike puncta on a broadly elliptical valve. It is seldom seen in girdle view.

Licmophora is a stalked, epiphytic marine form with cells wedge-shaped in both views. In the girdle view one can see septa and intercalary bands. It does not have the costae of *Meridion*.

B. A GROUP IN WHICH ORGANISMS HAVE A TRUE RAPHE ON ONE VALVE AND A PSEUDORAPHE (OR EVEN THE COMPLETE ABSENCE OF A LINEAR MARKING) ON THE SECOND VALVE.

Achnanthes. This organism (Fig. 8-12) is found in brackish or fresh waters. Many of the species found in the sediment are quite small. In valve view the organism is fusiform or boat-shaped, but in girdle view it is bent or undulating. When attached to the substrate, the valve with the raphe would be in contact.

Cocconeis (Fig. 8-12) species are found in both marine and freshwater habitats, sometimes as epiphytes completely covering the surface of a macroscopic marine alga. Marginal septa are obvious in some species in valve view. Epiphytic forms are most common, but cells free from the host are also encountered.

Rhoicosphenia (Figs. 8-12 and 8-13) is a marine or freshwater form with two internal septa. The latter have large oval holes in the center. In girdle view the organism is comma-shaped with a combined bent and wedge shape.

C. PENNATE DIATOMS WITH A TRUE RAPHE ON BOTH VALVES. IF A KEEL IS PRESENT, IT IS AXIAL (FIG. 8-14). (AS IN A SHIP, THE KEEL IS THE BACKBONE, RUNNING FROM BOW TO STERN). MOST OF THE FORMS LISTED CAN BE FOUND IN FRESH WATERS AND AT LEAST IN BRACKISH WATER, IF NOT IN A COMPLETELY MARINE ENVIRONMENT.

Navicula (Figs. 8-2 and 8-12). This form is frequently listed and figured as a typical pennate diatom. Ornamentation is provided by puncta (Fig. 8-2) which on occasion might be so small that they are visible only with the electron microscope.

Pinnularia (Figs. 8-15 and 8-16). This form has a more rectangular appearance, with the ornamentation provided by costae.

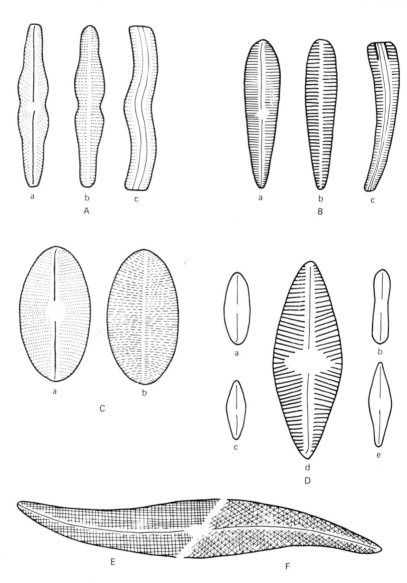

Diploneis has the raphe enclosed by elongate silica ribs, running parallel to the long axis. The ribs are sometimes called horns. *Gyrosigma* and *Pleurosigma* (Fig. 8-12) are two S-shaped types from marine and fresh waters. In *Gyrosigma* the striae have a transverse and longitudinal orientation whereas in *Pleurosigma* there are oblique striae in addition to those which are transverse.

Figure 8-12. Pennate diatoms. Microscopic.

A. *Achnanthes*, seen in valve view, a and b, and girdle view, c. In a, the raphe is evident on the lower valve, and the pseudoraphe is not apparent in b.

B. *Rhoicosphenia* seen in valve view, a and b. In a, one can see the raphe on the hypovalve, and the pseudoraphe on the epivalve in b. A view of the girdle area shows the bent shape (c).

C. *Cocconeis* hypovalve in a, and epivalve in b. The pseudoraphe is not always as evident as shown on the epivalve.

D. Various shapes of *Navicula* species. As shown in d, the raphe is evident and can be seen on both valves: a, b, c, and e show some of the various shapes possible.

E. Along with F there is represented a sigmoid diatom. On the left portion, as in E, the markings, puncta, are arranged parallel and perpendicular to the raphe. E thus would be *Gyrosigma*.

F. The right portion of the sigmoid diatom is a diagram of a *Pleurosigma*, with cross hatching in two directions and a third orientation at right angles to the raphe.

Gomphonema (Fig. 8-15) may be a stalked (attached) form. It is asymmetrical about the transapical plane—front to back—but otherwise symmetrical (Fig. 8-17). A common type is the so-called Coke-bottle diatom. It is unique with its H-shaped plastid.

Cymbella (Fig. 8-18) is symmetrical about the transapical or transverse plane, but asymmetrical about the apical or longitudinal plane. On opposite sides of the raphe the asymmetry is apparent (Fig. 8-19).

Mastogloia is a pennate form with a septum on each valve. The latter resembles a shelf running parallel to the upper surface, completely around the diatom. Transverse septa originate at the surface of the longitudinal septa and attach to the valve. The interior is thus compartmentalized.

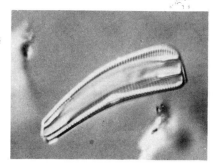

Figure 8-13. *Rhoicosphaenia*. Girdle view. 600 X.

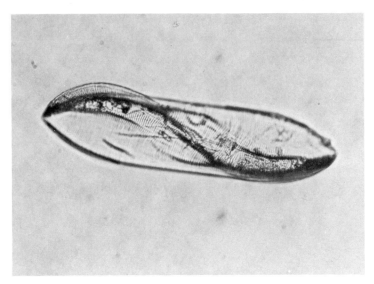

Figure 8-14. *Entomoneis.* Note the sigmoid keel. 600 X.

Epithemia has an arched appearance, transverse septa, and the raphe in an excentric V-shaped position. The angle between valves may be about 150°.

Amphora at first appears to resemble *Cymbella,* but in girdle view one sees that the valve surfaces are not parallel, rather far from it! Thus in girdle view one can see both valves, both with true raphes. Marine

Figure 8-15. Pennate diatoms. Microscopic.

A. *Pinnularia,* with costae.

B. *Stauroneis,* with a thickened central area.

C. *Gomphonema,* which is asymmetrical in both valve (a) and girdle views (b).

D. *Cymbella,* with an excentric placement of the raphe.

E. *Surirella,* with a marginal keel. Any median ornamentation is not a raphe.

F. *Nitzschia* or *Hantzschia.* The keel is along the margin, and the raphe is in the keel. The two genera could be similar to a or b, when seen in valve view. But in cross section, it is apparent that in *Hantzschia* (c), the keels are above each other and in *Nitzschia* (g) the keel on one valve is diagonally opposite that on the other valve. If an organism such as c would divide and the division was equal and parallel to the surface, as in d, two *Hantzschias* would be formed. But if the division was not equal, as in e or f, then one *Nitzschia* and one *Hantzschia* would result. If a *Nitzschia,* as in g, were to divide as in h, then each daughter cell would be either a *Nitzschia* or a *Hantzschia.* However, if g divided as in i, then two *Nitzschias* would be formed. On the other hand, g dividing as in j would give two *Hantzschias.* The plane of division, and position of the keel, are interrelated.

and freshwater types are known. Recently, it has been proposed that cellular fragments produced during formation of resting cells of *Amphora* could be the olive green "cells" frequently found in the marine aphotic zone.

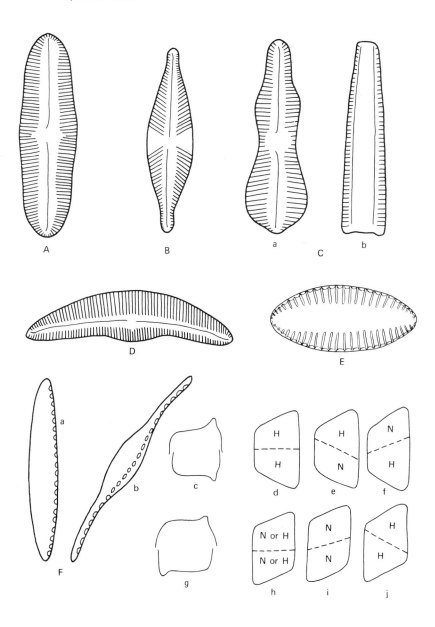

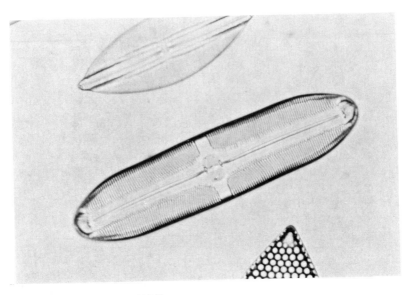

Figure 8-16. *Pinnularia.* 600 X.

Entomoneis (Fig. 8-14), formerly called *Amphiprora,* is also found in marine and freshwater habitats. It possesses a naviculoid shape and a sigmoid keel, in the axial position. It has an hour-glass shape in girdle view. Frequently it is weakly silicified. We will later see that other forms with keels have the latter at the margins.

Many other genera of pennate diatoms can be found (Figs. 8-20 and 8-21).

Figure 8-17. *Gomphonema* in valve view. Note slight asymmetry. 600 X.

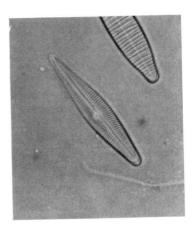

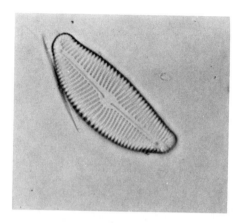

Figure 8-18. *Cymbella*, viewed from outside the valve. 600 X.

D. PENNATE FORMS WITH A TRUE RAPHE IN THE MARGINAL KEEL, IDENTICAL ON BOTH VALVES. THE FORMS WITH AN AXIAL KEEL ARE PLACED IN THE PREVIOUS GROUP. AGAIN THE FORMS CHOSEN ARE PRESENT IN MARINE AND FRESH WATERS.

Nitzschia (Figs. 8-15 and 8-22) has a marginal keel that is apparent because it has circular or elongate openings (Fig. 8-23), as well as

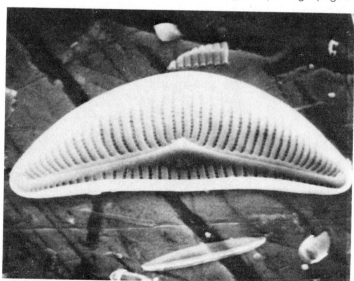

Figure 8-19. *Cymbella.* Scanning EM view from inside the valve, with the raphe positioned toward one edge of the valve. 1600 X. Courtesy of B. Rosen.

Figure 8-20. *Frustulia.* 600 X.

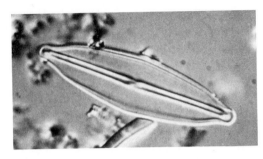

the raphe. The keel on one valve is diagonally opposite that on the other valve. Ornamentation is provided by striae.

Hantzschia (Fig. 8-15) is quite similar to *Nitzschia,* even with the circular openings in the keel, but the two valves have the keel on the same side. In valve view they appear to enhance each other.

Surirella (Fig. 8-24) is not always naviculoid, and has a marginal keel along the entire edge of both valves. Transverse folds of the valve appear as costae or ribs (Figs. 8-24 and 8-25). Striae appear between the folds, and they are lacking along the median line; hence there is a pseudoraphe.

The Centric Forms

Although raphes and pseudoraphes are not found in these organisms, they possess structures found useful in distinguishing genera. Unicellular and filamentous genera are commonly collected.

Figure 8-21. *Podocystis,* an epiphyte in the marine environment. 400 X.

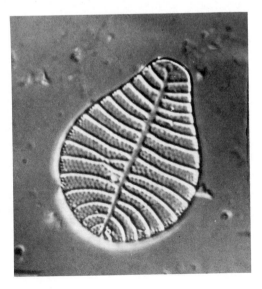

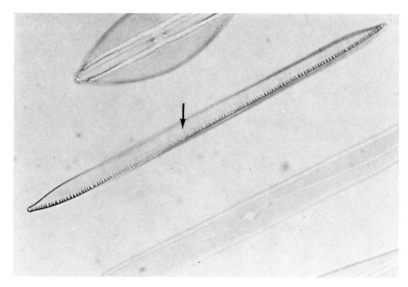

Figure 8-22. *Nitzschia.* 600 X.

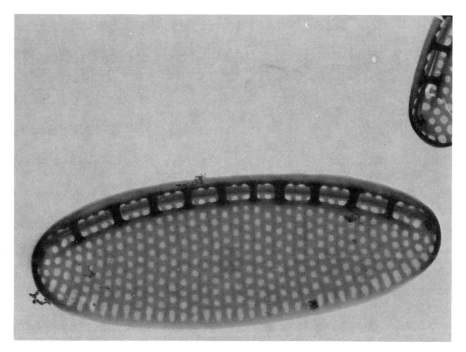

Figure 8-23. Electron photomicrograph of a *Nitzschia* valve, showing the marginal keel. 8700 X. Courtesy of P. Hargraves.

Figure 8-24. *Surirella* cells viewed under Nomarski interference microscopy. Note marginal keel and pseudoraphe. 500 X.

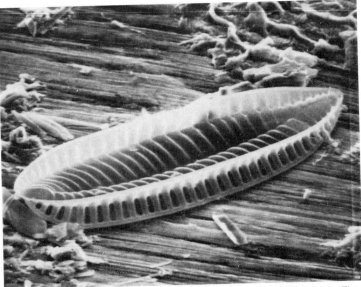

Figure 8-25. *Surirella*. Scanning EM view from inside one of the valves. The organism rests on the marginal keel. 2000 X. Courtesy of B. Rosen.

Cyclotella (Fig. 8-26) This genus, as well as the next, are seen equally often in both valve and girdle views. When a centric diatom is seen mostly in valve view, we know that the organism is shaped like a petri dish (Fig. 8-27). Organisms with the shape of a pipette container would be found mostly in girdle view, for they would not remain with the end up on a microscope slide (Fig. 8-28).

Both marine and freshwater forms are known and polymorphism has been reported in the genus, that is, some organisms can produce forms resembling at least two different "species." Ornamentation is provided by radiating costae (Fig. 8-27).

Coscinodiscus (Fig. 8-29). Some forms are among the largest of diatom single cells. Areolae provide the ornamentation (Fig. 8-26), but marginal spicules may be present. The areolae, closely positioned holes with perforate membranes placed within the holes, are in uniseriate rows, or grouped to form facets. These species are found in marine and freshwater habitats.

Rhizosolenia (Fig. 8-26). This and the next form are seen as frequently as unicells as they are in chains. Marine and freshwater forms are known. The organism is seen in girdle view (Fig. 8-30). Not much of the girdle view of the diatom is taken up by the margin of the valve, but *numerous* intercalary bands are present. These are easily broken apart during slide preparation. Each valve gives rise to one long spine, which aids in linking cells into filaments.

Biddulphia (Fig. 8-31) has two blunt horns on each valve. They are found at the margins opposite each other. The girdle bands are quite ornamented and thus distinct (Fig. 8-31). Chains may be straight or zig-zag. Some forms are triradiate.

Melosira (Fig. 8-26) is a very common chain-forming centric, which as a result is mostly seen in girdle view. Some have little apparent ornamentation, but others have a complete groove near the juncture of the valve and girdle band—the sulcus.

One marine form often called *Melosira* is really a *Paralia*.

Chaetoceros (Fig. 8-32) is a spine-bearing genus that has mostly marine forms. Each cell possesses four corner spines which, because of their great length, overlap with adjacent spines. In some types the spines are interwoven (Fig. 8-26).

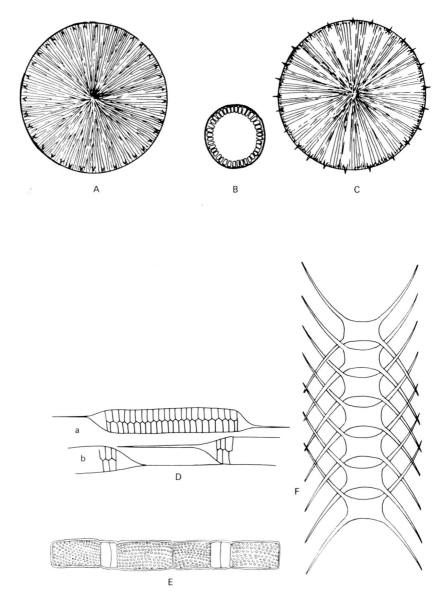

A   B   C

a

b

D

E

F

Skeletonema (Fig. 8-33) is a marine form with silica rods joining the valve of one cell to the next. It is a dominant diatom along the northeastern coast of North America. The cells are small when compared to the above genera. It has a delicate appearance.

Figure 8-26. Centric diatoms. Microscopic.
A. *Coscinodiscus*, with short spines on the edge of the valve and areolae in uniseriate rows. Valve view.
B. *Cyclotella*, with marginal costae, and in this species a clear central area. Valve view.
C. Valve view of *Stephanodiscus*, with alternating multiseriate puncta and clear regions.
D. *Rhizosolenia* seen only in girdle view. In a, the valve to the left has an elongate spine and the valve to the right an additional spine. The two valves are held together by girdle bands and numerous intercalary bands. The junction of each band can be seen in this particular view. In b, the valves of two adjacent cells are seen opposing each other, and in this manner filamentous forms can result.
E. Girdle view of a *Melosira*, showing two complete cells of a filament.
F. *Chaetoceros* seen above in girdle view with intertwined spines. Some cells have well-developed statospores within.

Thalassiosira. This chained form has disc-like cells which are joined by a single central mucous strand. Areolae provide the ornamentation.

Actinoptychus is one centric form with cells single or united by mucilage strands. In valve view there are alternately raised and depressed triangular segments, which provide a distinctive pattern. The central area

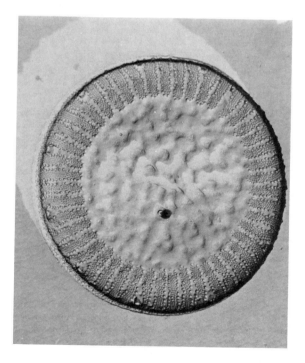

Figure 8-27. *Cyclotella.* Electron photomicrograph of a valve (SEM). 6000 X. Courtesy of P. Hargraves.

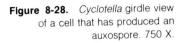

Figure 8-28. *Cyclotella* girdle view of a cell that has produced an auxospore. 750 X.

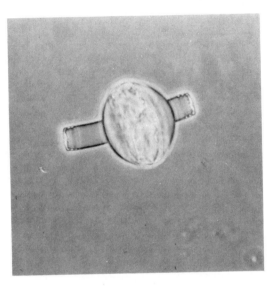

is devoid of puncta, so that this organism appears to have a donut shape. This latter character helps distinguish it from similar genera.

These are a select few centric forms. You might encounter more when looking at plankton samples (Figs. 8-34 and 8-35). They can be identified by consulting monographs or keys to phytoplankton.

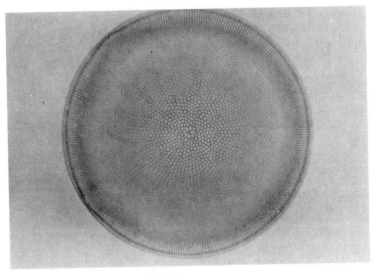

Figure 8-29. *Coscinodiscus.* 200 X.

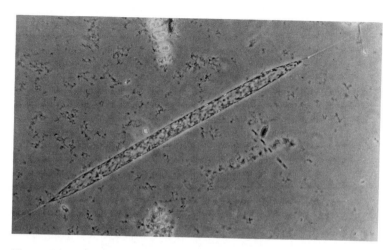

Figure 8-30. *Rhizosolenia.* Girdle view of living cell. 325 X.

THE DIATOM CELL

For convenience we have first considered wall ornamentation, in order to get to know a few diatoms and to enable the student to function in the initial laboratory periods. The organisms can be identified by examining the features discussed thus far. All wall details can be seen on ashed specimens. But diatoms are living entities and some additional cell features are of interest.

As stated earlier the diatom cell is composed of a typical protoplast surrounded by a glass wall that is composed of two overlapping halves.

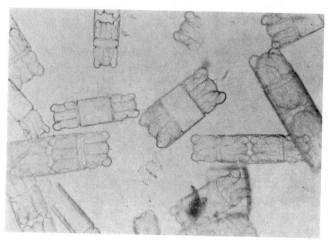

Figure 8-31.
Biddulphia. Cells observed in girdle view, some recently divided. 450 X.

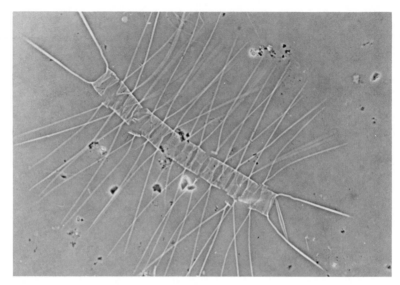

Figure 8-32. *Chaetoceros.* Filament with numerous spines. 200 X.

The latter are joined by girdle bands, and at times are further spaced because of extra bands, the intercalary bands. In addition to the possibility of having a raphe or pseudoraphe, puncta, striae, costae, septa, nodules, or other similar types of ornamentation, some forms have

Figure 8-33.
Skeletonema. Cells of filament seen in girdle view. Note connecting strands of silica. 600 X.

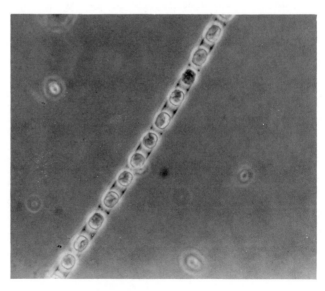

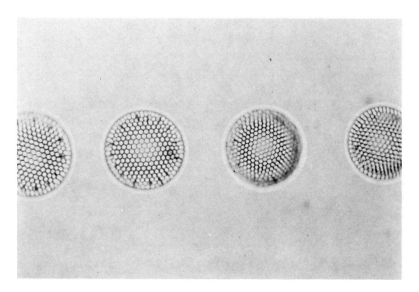

Figure 8-34. *Stephanopyxis* in valve view. 600 X.

spines or thinner elongate appendages, the mucilaginous or chitan strands. In the classification of diatoms, little attention has been paid to internal characters, such as with other algae where great emphasis is placed on plastid structure and number, number of nuclei, and so forth.

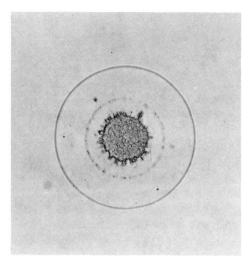

Figure 8-35. *Hyalodiscus.* Valve view showing doughnutlike appearance. 600 X.

It appears that there is great stability in diatom plastid number and structure. Pennate forms frequently have two per cell and centrics have numerous discoid plastids. Even ultrastructural details appear constant. The membrane bound plastid possesses thylakoids in groups of three, a feature common to many of the close relatives of diatoms. In addition, some of the thylakoids completely encircle the plastid. These are the so-called girdle lamellae. With the electron microscope we have been able to confirm certain cell wall details of many diatoms, including those of the raphe system, details of structures such as the single punctum, origin and development of the wall, as well as valve margins, girdle bands and chitan appendages. These wall studies have essentially confirmed previous excellent light microscope studies, in addition to adding some minute detail.

Diatoms are uninucleate. A large diploid nucleus is located in the center of the cell and can be observed with light microscopy, even in unstained material, as in *Pinnularia*. Only small vacuoles are observed. Large oil droplets are at times quite prominent. *Melosira varians* cells, on the other hand, have a large central vacuole, with a conical nucleus always found in the epitheca or upper valve.

DIATOM MOTILITY

Vegetative cells of many pennate diatoms move by gliding. In order to achieve such motion the diatom must have at least one raphe system, and the valve with the raphe must be in contact with the substrate. For a planktonic pennate form, gliding is impossible. If the organism is not too large, does not have much wall thickness, and if the surface area to volume ratio is favorable, it will remain in the plankton. Either a gradual sinking or mass movement of water provides some of the advantages achieved by its motile relatives. Centric forms, which all lack a raphe system, do not move by gliding. These forms are common in the plankton and are widely distributed. With numerous appendages providing an increased surface area, the centric forms rely on excellent flotation to remain in the photic zone as well as to achieve their distribution.

Flagellated cells are not found in the pennate forms. Motion of reproductive cells, sperm, or microspores in the centric group will not be discussed at this time, for the mechanisms are well known. Those cells have typical flagella and flagellar activity. However, some forms lack the central two tubules in the flagellum.

There are several theories with respect to motility of vegetative cells of pennate diatoms. Some excellent information on raphe structure is found in older literature, and more recent information from ultrastructural studies has confirmed most of the older observations. The complete structure of the raphe is exceedingly complex.

In surface view a wall with a raphe system in the median position has two linear markings (a to b, c to d in Fig. 8-36) from the central area toward each pole. The raphe is a slit completely through the wall, but a very elaborate structure, with complex twistings at each end, that is, near the central nodule and the polar nodules. In addition, the raphe normally is not a simple opening perpendicular to the outer surface. Instead, in section, one is reminded of tongue and groove joints (Fig. 8-36).

Many of the attached pennate forms are able to move to a new location on a host by means of this motility, while others migrate in the sediment. The forms with the raphe in a keel are more abundant in the sediment. Apparently with that type of raphe system they are able to make contact with several particles at one time. Keelless types such as *Cocconeis* attach to larger organisms, such as *Polysiphonia*, and, with a curvature of the cell, much of the cell surface is in contact with the host.

As you have seen with living material in the laboratory, diatoms can exhibit several varieties of motion. The most common is the smooth movement to and fro, but this can be equal in both directions, with little or no net gain, or unequal in one direction so that the diatom changes its location. Then there is the series of unidirectional advances. Turning is accomplished by accidental encounters with a structure, because of the contour of the surface, by a twisting of the cell, or a pirouette. Forms

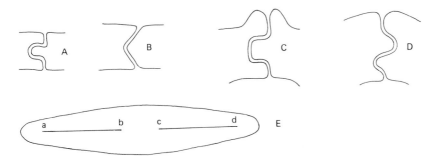

Figure 8-36.   The raphe system in diatoms as seen in cross section (A–D) and in surface view (E). In E the two halves of the raphe system (a–b, c–d) are figured.

such as *Cymbella* always move in an arc. In experimentation with the marine *Bacillaria paradoxa,* springlike activity was noted when the colonial form was restrained for a period and then released. It would appear that some energy was stored and quickly released.

Moving diatoms leave trails. In studies on slides or on agar, a light slime trail is observed. It can be observed more readily if the microscope slide is coated with India ink. Is the slime trail involved in the motion?

Nitzschia can penetrate and then move in 2% agar. (Typically agar plates are about 1.5%.) When discussing such motion by a diatom, the shape must be considered. But movement of even a needlelike diatom into firm agar indicates the presence of force behind diatom motility. A number of diatoms show a rhythmic movement, associated with the tides, in and out of the sediment. This movement demonstrates a certain amount of force. With each tidal cycle the population moves several millimeters up and down through sediments.

The rate of diatom motion has been studied. For example, *Nitzschia putrida* can move faster on glass, 2.7 μm sec^{-1}, than on agar, 0.8 μm sec^{-1}. *Navicula radiosa,* an organism with a 50 μm length, can move 14 μm sec^{-1}, with a maximum rate of 20 μm sec^{-1}. One might wish to compare the movement of a diatom with that of a human, each in its own environment. If each organism were to move its own length each second, the diatom, about 15 μm long, would travel only a few centimeters in an hour. But the human would be moving at about a typical walking speed, about 6 km each hour. For a diatom, that rate is sufficient to enable the species to expand its niche.

Bacillaria paradoxa, now considered a *Nitzschia* by some, is commonly called the carpenter's rule alga. The individual cells, which are joined by a hooklike structure along the length of the keel, slide or glide, extending or reducing the length of the rule. These organisms have an average rate of motion of 8 μm sec^{-1}. There are no protoplasmic connections between cells, but they change direction of their movement synchronously. The timing of the reversal is 80 seconds at 20°C.

Lewin and Guillard believe that there might be an endogenous rhythm with *Bacillaria,* but others feel that some continuity might be achieved since each cell recently divided from an original parent cell could have the necessary message to move synchronously. The lubricant between the cells is both the sea water and any mucilage produced.

Historically there have been many theories to explain all of these types of motion. At first it was felt that there must be cilia or pseudopodia, but these have never been observed. A theory involving streaming cytoplasm, especially flowing through the raphe system, was quite popular for some years, but ultrastructural studies, which show no protoplasm in the raphe, provided little support for such a theory. There has been speculation that gas expulsion or jets of water could make such movement possible, but there have been no observations of any currents around the diatom to lend support to these theories. The secretion of mucilaginous material and the possibility of membrane undulations at present are the bases for current theories.

Drum-Hopkins Mechanisms

Drum and Hopkins confirm the fact that a raphe must be present and add that there must be a solid, or, in the case of keeled types, a particulate substrate. When on a solid substrate, the diatom is only in contact with the substrate through mucilage attachment at the posterior end. When there is a reversal of motion, there is also a change in polarity *and* attachment. In such a case the second part of the raphe system is utilized.

They base their theory on observations of living material, detection of the slime trails, and ultrastructural studies. The latter include not only the detection of the slime or mucilage in the raphe system, but also the presence of two cell organelles, the crystalloid bodies and a fibrillar bundle located adjacent to the raphe system (Fig. 8-37).

Fibrils of mucilage are secreted from the crystalloid body into the raphe system. These could be forced out by pressure from the fibrillar bundle, located under the raphe. The ultrastructure of this bundle closely resembles that of smooth muscle. In contact with water, the mucilage would expand.

Capillarity might also be involved, or independently, it might account for some motion. Because of the shape and dimensions of the raphe, capillary action could account for transfer of the mucilage to the end of the raphe opposite the point of origin. This could be independent of the theory regarding forcing of mucilage into the system, or coupled with it. In either case, the mucilage adheres to the substrate at the end of the raphe, at one point with a solid substrate, and at several sites when moving in the sediment.

If mucilage is always involved in diatom motion, reversal—especially the sudden movements characteristic of some forms—could not be explained easily, if one has to release the mucilage adhesion at one pole and attach the secretion at the opposite pole. At the same time, the reversal of polarity is accomplished within the cell, for material must

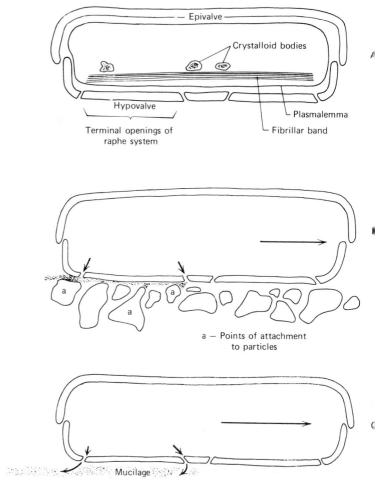

Figure 8-37. Motility in pennate diatoms. Depicted are longitudinal sections of diatoms with raphe systems. A. The organelles implicated in diatom motion, including the fibrillar band and crystalloid bodies. B. The release of mucilage and its attachment to particles (at a) are diagrammed. C. The cell is diagrammed as moving on a solid substrate. After Drum and Hopkins.

come from another crystalloid body, or the position of the body must change quickly. Perhaps then one must also incorporate cytoplasmic streaming, a phenomenon which alone was once thought to affect movement. Now that we know of several phenomena that occur and better understand diatom wall structure and cytology, the time would appear right for an adequate explanation of diatom motility.

DIATOMACEOUS EARTH

Because glass walls are very resistant to destruction, diatoms in sediments give us an accurate picture of at least that component of the algal flora in past time. In some areas, both marine and freshwater, diatom walls have accumulated in enormous numbers. Freshwater deposits may be 10^5 years old, whereas marine material is 5 to 20 $\times$ 10^6 years old. At Lompoc, California, the marine deposit is a few hundred meters thick. The diatomaceous earth is mined and used commercially for filters, insulation, polishing, as a filler, and so forth. It is only 1.25 times the weight of water. Many of the diatom walls are broken, but even the novice can identify a genus or two from a fossil sample (see also Chapter 20).

GOLDEN ALGAE, YELLOW-GREENS, DIATOMS, AND ALLIES

The student will find in older references that these organisms were once placed in one division, the Chrysophyta. In such cases one relied on similarities in pigmentation, especially the presence of fucoxanthin and the absence of chlorophyll *b,* as well as the presence of leucosin or oil as common food reserves. In addition, if silica was not present in the walls of vegetative cells, it was probably found in a reproductive body. Overlapping walls are common. With more recent evidence the evolutionary relationships among these organisms have become less clear, and thus independent origin is frequently suggested.

REFERENCES

Bourelly, P. 1968. *Les Algues d'eau Douce.* Initiation a la systematique. Tome II. Les algues jaunes et braunes. N. Boubee and Co., Paris. 438 p.

Darley, W. 1974. Silicification and calcification. In W. Stewart (Ed.) *Algal Physiology and Biochemistry.* Univ. Calif. Press, Berkeley. pp. 655–75.

Fritsch, F. 1935. *The Structure and Reproduction of the Algae.* Vol. I. The University Press, Cambridge. 791 p.

Hopkins, J. and R. Drum. 1966. Diatom motility: an explanation and a problem. *Br. Phycol. Bull.* 3: 63–7.

Hostetter, H. and E. Stoermer. 1971. Bibliography on Bacillariophyceae. In J. Rosowski and B. Parker (Eds.) *Selected Papers in Phycology.* Dept. Bot., Univ. Nebraska, Lincoln. pp. 784–90.

Lewin, J. and R. Guillard. 1963. Diatoms. *Ann. Rev. Microbiol.* 17: 373–414.

Patrick, R. and C. Reimer. 1966. *The Diatoms of the United States.* Vol. I. Nat. Acad. Sci., Philadelphia. 688 p.

Patrick, R. and C. Reimer. 1975. *The Diatoms of the United States.* Vol. 2, No. 1. Acad. Nat. Sci., Philadelphia. 213 p.

Steele, R. 1965. Induction of sexuality in two centric diatoms. *Bioscience* 15: 298.

Stewart, W. 1974. *Algal Physiology and Biochemistry.* Univ. California Press, Berkeley. 989 p.

Stosch, H. A. von. 1950. Oogamy in a centric diatom. *Nature* 165: 531–2.

Weber, C. et al. 1965. *A Guide to the Common Diatoms at Water Pollution Surveillance System Stations.* R. A. Taft Center, Cincinnati, Ohio. 101 p.

Werner, D. and E. Stangier. 1976. Silica and temperature dependent colony size of *Bellerochea maleus* f. *biangulata* (Centrales, Diatomeae). *Phycologia* 15: 73–7.

Zoto, G., D. Dillon, and H. Schlicting. 1973. A rapid method for clearing diatoms for taxonomic and ecological studies. *Phycologia* 12: 69–70.

Xanthophyceae — yellow-green algae

Many Xanthophyceae were once placed with the green algae. However, the pale green or yellow-green coloration is often a clue that they have a unique group of pigments. These organisms are found more frequently in temperate regions in freshwater and marine habitats, as well as on and in soil. Although marine representatives, found in coastal areas, are often collected, xanthophytes are mainly freshwater and soil forms. Only seldom are they dominant organisms in any environment.

XANTHOPHYTE THALLUS TYPES

Xanthophytes typically exist as unicells, colonies, and both branched and unbranched filaments. Unicellular flagellates are not common and there are no motile colonies. Certain xanthophytes are palmelloid; some are dendroid colonies with elongate strands of mucilage between cells.

In most cases, individual cells are uninucleate, but coenocytic forms are known, for example, *Botrydium* and *Vaucheria*. Although motile genera are not common, many organisms reproduce by motile reproductive cells. These organisms were originally called heterokonts because the flagella are of unequal length.

In general, the plant body types parallel what one finds in the green algae, but there are far fewer species of xanthophytes.

CHARACTERISTICS OF THE YELLOW-GREEN ALGAE

1. Pigments

For some time we believed that the only chlorophyll in the yellow-green algae was chlorophyll *a,* with the earlier report of chlorophyll *e* due to technique. The latter pigment is now recognized as a degradation product. Recently Guillard and Lorenzen pointed out that xanthophytes possess chlorophyll *c.* In looking for other chlorophylls, or in examining other pigments, the chlorophyll *c* was frequently overlooked.

In addition, there are beta carotene and several xanthophylls. The dominant xanthophyll is diadinoxanthin; yellow-greens also contain vaucheriaxanthin. The thylakoids in the plastics are in groups of three. In most genera, as in *Tribonema,* there are girdling lamellae, but they are absent in some forms, such as *Bumilleria.* There may be a pyrenoid within some plastids, an observation confirmed by recent EM studies.

Inasmuch as the xanthophytes, chrysophytes, and diatoms were once classified together, one may wish to compare the pigments found in these organisms (Table 9-1).

Table 9-1. THE MAJOR PIGMENTS OF THE BACILLARIOPHYTES, CHRYSOPHYTES AND XANTHOPHYTES

	Bacillariophytes	Chrysophytes	Xanthophytes
Chlorophyll a	X	X	X
Chlorophyll c	X	Absent in some	Absent in some[a]
Alpha carotene	X	X	
Beta carotene	X	X	X
Major xanthophyll	diatoxanthin fucoxanthin	fucoxanthin	diadinoxanthin

[a] The absence could be due to the fact that investigators have not looked for chlorophyll c.

2. Storage Products

More data are needed on food reserves in members of this group. We do know that many xanthophytes store reserve food as an oil, and *probably* chrysolaminarin is the reserve in a few forms (Fig. 9-1).

3. Motility

In many genera sexual reproduction is not observed, or is a rare phenomenon. Thus it is usually necessary to observe motility in the activity of zoospores. Some genera can form either planospores (zoospores) or aplanospores. Flagella are of unequal length, with the longer type the tinsel type, with hairs in two rows. At the tip of each lateral hair there are smaller filaments. Flagella on *Vaucheria* zoospores lack hairs. The anterior flagellum of *Vaucheria* sperm has lateral hairs, while the shorter posterior flagellum is smooth.

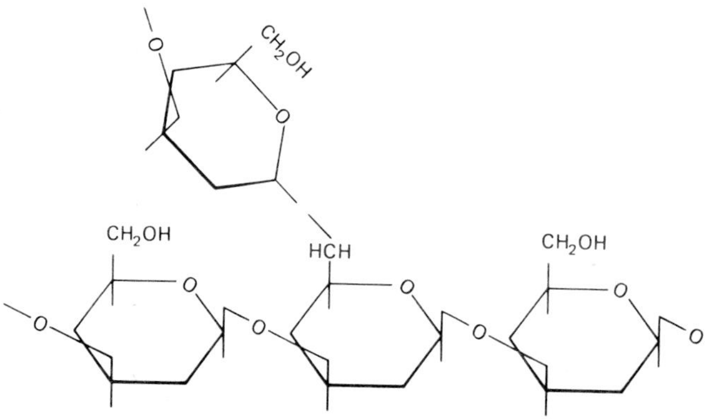

Figure 9-1. Chemical structure of the food reserve, chrysolaminarin.

2. Walls

The xanthophyte walls are typically of cellulose and pectin. In some cases one may observe overlapping walls in living material (*Tribonema*). These resemble the overlapping walls of diatoms, but there are no girdle bands. This type of wall structure could be very common in this group. We must await further studies, especially with the electron microscope, to show us how walls in all genera fit together. Xanthophyte cysts also have overlapping walls.

REPRODUCTION

Organisms reproduce asexually by cell division, by fragmentation of filaments and colonies, or by means of nonmotile and motile spores. Some zoospores are multiflagellated. Sexual reproduction is rarely observed with certain organisms and not known for others. As far as is known, life histories are quite simple; for example, *Vaucheria* has a haploid gametophyte phase, with zygotic meiosis.

SOME REPRESENTATIVE GENERA

Ophiocytium

This genus (Fig. 9-2) with cylindrical cells, may be free-floating or attached (Fig 9-3). Sometimes species are abundant in the sediments. The cells may become very long and twisted. Some have spines, which may be present at one or both ends of the cell. The cell wall is composed of two overlapping halves that are of unequal length. They are most apparent after reproduction by planospores or aplanospores, when empty walls are found.

Figure 9-2. Some members of the Xanthophyceae.

A. *Vaucheria*. The organism is a coenocyte, and has numerous disc-like plastids. A septum is formed when the male (left) or female (right) reproductive structures develop. Macroscopic, but with microscopic reproductive cells.

B. The macroscopic, multinucleate *Botrydium*. Three stages are shown: the small spherical cell, then the formation of the rhizoid, and finally the adult plant with a branched rhizoidal system.

C. *Ophiocytium*, shown in two forms, one of which is more curved. Microscopic.

D. *Botrydiopsis*, with a young and an adult cell. Microscopic.

E. and F. *Tribonema*, shown in two forms. In E, six cells are shown, with the end pieces of two additional cells showing. In F, the cells are obviously bulging.

G. and H. Two types of walls from *Tribonema*. These are the so-called H-pieces, but as one can see the "H" has a third dimension. Remember the H is formed from one-half of a cell wall from two adjacent cells. Microscopic.

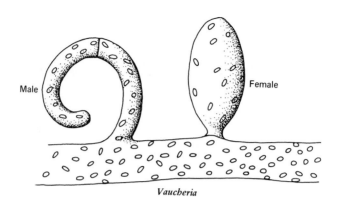

Vaucheria

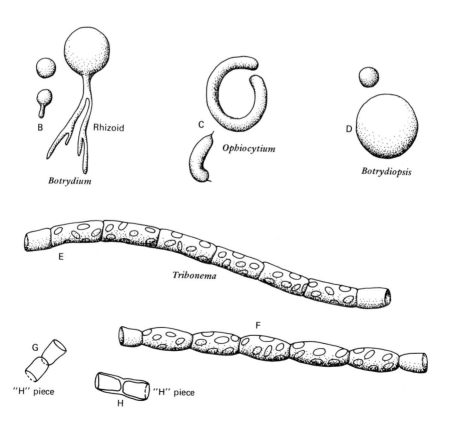

Botrydium   *Ophiocytium*   *Botrydiopsis*

Tribonema

Figure 9-3. *Opiocytium.* Vegetative
cell still attached to parent wall
component. 600 X.

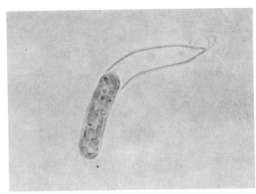

Botrydiopsis

Botrydiopsis (Fig. 9-2) has spherical cells with several to numerous
peripheral plastids, a phenomenon not uncommon in the xanthophytes.
This organism at first resembles a yellow-green *Chlorococcum.* It is
similar to the latter in that vegetative cells can vary extensively in size.
There may be a ten-fold increase in diameter as cells age. Reproduction
is by means of planospores or aplanospores, sometimes with quite
thickened walls.

Botrydium

Botrydium (Fig. 9-2) plants are much larger than either of the previous
genera. The cells are spherical and, except for early development stages,
are multinucleate. In the cooler months in spring and autumn one can
find aggregations of this coenocyte on damp soil, especially at the edges
of a body of water. The *Botrydium* is attached by a basal rhizoidal
system, which also lacks cross walls. The bulbous sphere may be about
a millimeter in diameter, so that aggregations of these plants are visible
to the naked eye. The multinucleate protoplast may divide at times of
reproduction to form planospores or large, thick-walled aplanospores.

Vaucheria

Vaucheria (Figs. 9-2 and 9-4) is a very large, macroscopic filamentous
form. For some time it was classified with the green algae. Even after
pigment analyses, there was still some confusion, perhaps because
vegetative portions and reproductive structures or cells do not have an
identical pigment composition. But when it was realized that chlorophyll
b, as well as true starch, was absent, the filaments could not be green
algae. It was in *Vaucheria* that, under some types of pigment extraction,

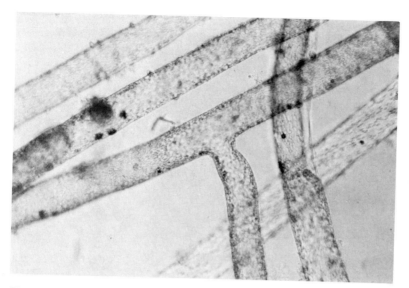

Figure 9-4. *Vaucheria.* Portion of filamentous coenocyte, with branching. 200 X.

an additional chlorophyll band, chlorophyll *e*, was reported. Now this compound is recognized as a degradation product and not a true chlorophyll.

The filaments, forming a felty mass, are coenocytes. In addition to the many nuclei there are thousands of ellipsoidal chromatophores or plastids. Septations are formed at the time of reproduction, or on occasion after an accident to the intact filament.

Both marine and freshwater forms of *Vaucheria* are known. The organism grows very well on mud surfaces, or even in sediments in shallow water. The photosynthetic filamentous portion is exposed, and the colorless branches, of narrow diameter, penetrate the substrate. Some species deposit calcium carbonate on the exterior.

Zoospores are formed singly in terminal sporangia. A cross wall is formed a little distance back from the tip and the protoplast becomes flagellated. The large multiflagellated body is called a synzoospore. Apparently it arose by loss of the ability to undergo cytokinesis. Flagella are paired, subequal in length, and are both of the whiplash type; they are inserted in pairs all over the zoospore surface.

Sex organs are large and develop as side branches (Figs. 9-2 and 9-5). Sperm are formed in elongate, curved spermatangia. The flagella are laterally inserted. The egg is formed singly within an erect oogonium

Figure 9-5. *Vaucheria.* Male and female (2) reproductive structures. Stained specimen. 200 X.

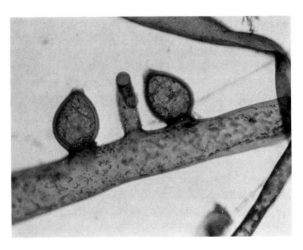

(Fig. 9-2). It is uninucleate and sperm get within the wall through a pore. Meiosis is zygotic.

Tribonema

This form (Fig. 9-2) is a more typical, unbranched, filamentous type. The cells are quite elongate, more than two times the cell diameter, have a single nucleus and several parietal discoid plastids. The walls are composed of overlapping halves. When a filament disintegrates, the so-called H-pieces are seen in the collection. This portion of the wall, formed from the walls of the bottom half of one cell and the upper half of the adjacent cell, is clearly visible because the halves adhere at the position of the initial wall septum. Planospores or aplanospores may be formed singly within the cells.

Members of this genus can be found in running water, in small ponds, ditches, or swamps (Fig. 9-6). Typically, the organisms are found in clean water.

CHLOROMONADOPHYTES

This small and poorly known group of freshwater and marine flagellates, often called chloromonads, cannot be placed with convenience with any algal class, but might be allied with the Xanthophyceae. In the past chloromonads have been classified in a separate division, or class, and have been placed with the cryptophytes. More data are needed concerning their cytology and physiology, on several organisms, before they can be placed accurately. The large, approximately 100 μm,

Figure 9-6. *Tribonema.* Dense growth in a drainage ditch. Note higher plant leaves (arrows).

biflagellated cells (with flagella of unequal length, one tinsel and one whiplash) lack a wall, pyrenoid, and eyespot, are uninucleate, and often have groups of trichocysts at the plasmalemma. The many chloroplasts have a girdling lamella and thylakoids in groups of three. They contain chlorophylls *a* and *c,* beta carotene, and xanthophylls, giving most healthy organisms a bright green color.

Chloromonads reproduce by cell division.

Gonyostomum, Vacuolaria and *Hornellia* are among the better known genera.

REFERENCES

Bourrelly, P. 1968. *Les Algues d'eau Douce.* Initiation a la systématique. Tome II. Les algues jaunes et brunes. N. Boubee and Co., Paris. 438 p.

Fritsch, R. 1935. *The Structure and Reproduction of the Algae.* Vol. I. The University Press, Cambridge. 791 p.

Guillard, R. and C. Lorenzen. 1972. Yellow-green algae with chlorophyllide c. *J. Phycol.* 8: 10–14.

Heywood, P. 1977. Chloroplast structure in the chloromonadophycean alga *Vacuolaria virescens. J. Phycol.* 13: 68–72.

Leedale, G., B. Leadbeater, and A. Massalski. 1970. The intracellular origin of flagellar hairs in the Chrysophyceae and Xanthophyceae. *J. Cell. Sci.* 6: 701–19.

Massalski, A. and G. Leedale. 1969. Cytology and ultrastructure of the Xanthophyceae. I. Comparative morphology of the zoospores of *Bumilleria sicula* Borzi and *Tribonema vulgare* Pascher. *Br. Phycol J.* 4: 159–80.

Norris, R. and W. Blankley. 1971. Bibliography on Chrysophyta (Excluding Bacillariophyceae). In J. Rosowski, and B. Parker (Eds.) *Selected Papers in Phycology.* Dept. Bot., Univ. Nebraska, Lincoln pp. 777–83.

Whittle, S. and P. Casselton. 1975. The chloroplast pigments of the algal classes Eustigmatophyceae and Xanthophyceae. II. Xanthophyceae. *Br. Phycol. J.* 10: 192–204.

Chrysophyceae — the golden algae

Chrysophytes are aquatic forms found in colder regions, such as northern latitudes or a mountain stream, or are found more frequently in winter months. These organisms can grow well under ice in freshwater ponds. They are usually not abundant in the plankton; however, motile forms *may be* the dominant organisms in the nannoplankton of the North Sea, or some freshwater ponds. They may be key organisms in some salt marshes.

Synura, a freshwater colony, often imparts a fishy smell and taste to water in reservoirs (or in other bodies of water we wouldn't normally taste), even when cell numbers are only in the tens of thousands per liter. However, although almost any sample of fresh water will have one member of this class, most golden algae do not receive much attention because they are not dominant or troublesome organisms.

Many organisms are attached forms or are found on or in the sediments. The student will have to spend some time in observations in order to see many chrysophytes, for most are encountered only after making repeated examinations of collections, even in, among, and on other algae and biota.

With some members of this group, as in several others we will consider, it is difficult to determine the exact types of nutrition. Some forms rely exclusively on photosynthesis, while others approach animal-like nutrition. Certain relatives of chrysophytes are holozoic. There can be quite a range of nutrition within the same genus.

A great deal is currently being learned about many of the Chrysophyceae. There are probably two main reasons for this increased knowledge. When forms are not common, any new detailed study adds considerable information, much of which is quite basic. In addition, organisms with external appendages are especially good subjects for ultrastructural studies, and certain of the fine structure is found to be essential in classification, or at least in clarifying observations made with light microscopy. Thus, new insights can greatly affect classification, even at the division or class level.

CHRYSOPHYTE THALLUS TYPES

Organisms in this group exhibit a number of different forms. Some are unicellular, amoeboid, colonial, or palmelloid (Figs. 10-1 and 10-2). The unicells and colonies, the majority of the chrysophytes, are either motile or nonmotile. There are very few parenchymatous forms, or few branched or unbranched filaments. Most organisms have a flagellated stage, which in this group is called a monad. Those with one flagellum are called chromulinomonads, and those with both a long and a short flagellum are the ochromonads. There are a number of colorless forms that are not treated in this text.

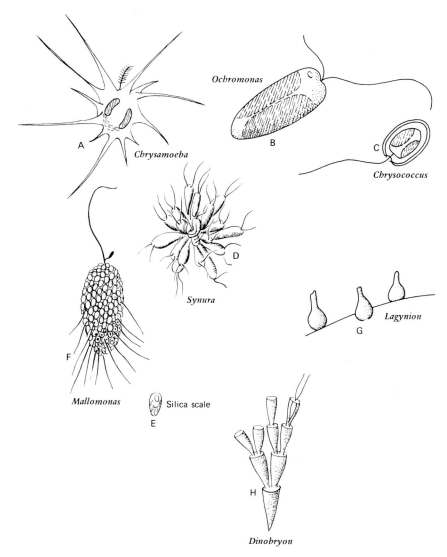

Figure 10-1.   Some Chrysophyte genera and allies.
A. *Chrysamoeba*, an amoeboid type, with an emergent flagellum.
B. The biflagellate *Ochromonas*, with two plastids.
C. *Chrysococcus*, with the protoplast in a small mineralized lorica.
D. The colonial *Synura*, with an individual silica scale shown in E. The silica scales (E) are found over the surface of the plant body.
F. The spiny *Mallomonas*, sometimes seen in a naked state, minus spines; species may lack spines. Two flagella at the apex.
G. The epiphytic *Lagynion*, with individuals in flasklike loricas.
H. The dendriod colonial form, *Dinobryon*. A living cell is depicted at the upper right of the colony.

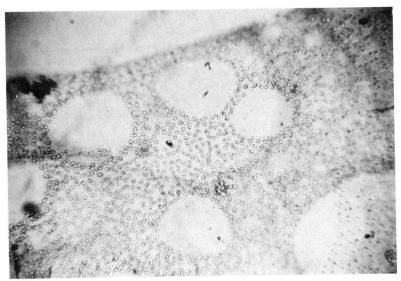

Figure 10-2. *Tetrasporopsis,* an example of a colonial chrysophyte. 325 X. Courtesy of L. Whitford.

Cells are typically uninucleate and have one or two plastids. It is surprising that coenocytic representatives, advanced types in other algal classes, are not found in this class.

There is a great deal of ornamentation to be examined as part of a cell, colony or filament. Ultrastructural studies have proved very informative because of the minute detail seen on many walls.

GOLDEN ALGAL CHARACTERISTICS

1. Pigments

All members have chlorophyll *a* and some possess chlorophyll *c*. In addition, in the plastid one will find alpha and beta carotene and several xanthophylls. In the latter group is the dominant pigment, fucoxanthin. The pigments are in organized plastids, and two parietal plastids are commonly seen. Ultrastructural studies have shown that these forms have the thylakoids organized in groups of three, with a girdling lamella system. Pyrenoids are not always present.

2. Storage Products

Chrysolaminarin and lipoidal material are the food reserves. The former is a branched carbohydrate, with beta 1:3 linkage of glucose residues, and 1:6 linkage for branching.

3. Motility

Vegetative cells or colonies are frequently flagellated, or have a flagellated reproductive stage. One or two flagella are seen. When there are two flagella, they are typically unequal in length, with the longer possessing lateral hairs. This is called the pluronematic or tinsel type. *Hydrurus* is a form with one flagellum. We once thought some other forms had the same arrangement, but closer study, especially with the electron microscope, showed that the second flagellum is held closely to the cytoplasm as with *Chrysococcus,* or embedded in a cell surface depression. These flagella are *quite* unequal in length. Some, for example, *Synura* and *Mallomonas,* have small scales on the flagella. These scales are not composed of silica, as are body scales.

4. Wall Structure

A number of organisms are naked, or in other words, lack a wall. Others possess just cellulose, as a wall component, but there may be silica present either in the vegetative cell or in reproductive stages. Scales that appear on the surface can be organic, or composed of silica or calcium carbonate. Individual cells of *Dinobryon* have a lorica around them. The lorica (Figs. 10-1 and 10-3) is a firm case that is not completely in contact with the protoplast. In the silicoflagellates there are elaborate silica skeletons, external to the plasmalemma.

Figure 10-3. *Dinobryon* colony caught under a *Spirogyra* filament. 325 X.

REPRODUCTION IN CHRYSOPHYTES

Typically, reproduction is by cell division, fragmentation of larger forms, or production of monads. In some genera, cysts or statospores form within the wall of the vegetative cell. The statospore is a flask-shaped structure, with a bulbous body and constricted neck. There is a plug, or stopper, in the opening of the neck. The statospore cells are impregnated with silica. They usually cannot be used as a diagnostic feature when identifying organisms, for many genera have similar statospores.

Sexual reproduction is a rare phenomenon. Isogamy is the rule. In certain cases, the statospore is in reality a walled zygote.

SOME REPRESENTATIVE GENERA

Chrysamoeba

This unicell (Fig. 10-1) spends the majority of its life in the amoeboid stage. A number of long rhizopodia radiate from the central mass of protoplasm. Obviously a wall is lacking, but one plastid, a single nucleus, and a very short flagellum are visible. Reproduction is by division of the organism.

Chrysococcus

Chrysococcus (Fig. 10-1) cells are spherical flagellates, each surrounded by a spherical lorica that is not much larger than the vegetative cell. The lorica has a fibrillar component that may or may not be mineralized (with at least iron componds and carbonates). There are a few holes in the lorica, typically three. The anterior long flagellum projects through one of these openings. Although we see but one emergent flagellum, with this genus ultrastructural studies reveal two, one of which is extremely shortened and does not emerge through the lorica.

If a collection with *Chrysococcus* is ashed, as with diatom preparations, the loricas remain because of mineralization.

Epichrysis

This unicellular coccoid form may be in a stage in which one sees just the nonmotile spherical organism. Reproduction can be by formation of zoospores. However, the products of division may remain close to each other and form large palmelloid masses.

Lagynion

Lagynion (Fig. 10-1) cells are sessile unicells in flask-shaped lorica. The base of the lorica has been greatly flattened. A single protoplasmic strand projects from the long neck of the lorica.

Kephyrion

This organism and *Pseudokephyrion* have the flagellated unicell in a lorica with a broad apical opening (Fig. 10-4). The fibrillar component of the lorica is probably cellulose, but there is also some mineralization. In certain forms in each genus the lorica may have an external spiral thickening. In ashed preparations of some freshwater diatoms these distinctive loricas can be numerous. The genera are distinguished by their types of flagellation.

Ochromonas

Ochromonas (Fig. 10-1) is a commonly occurring unicellular flagellate with flagella markedly unequal in length. The latter are inserted a little off center at the apex. There are two distinctive plastids in the photosynthetic forms. Nutrition in *Ochromonas* ranges from autotrophy to a holozoic response. One can find members of this genus that exhibit photoautotrophy, (with additional requirements for a vitamin), or photoheterotrophy, heterotrophy, or phagotrophy. One should consult texts or references on protozoology for close relatives.

Mallomonas

This organism (Fig. 10-1) has numerous silica scales on the large elliptical or oval cells. The scales frequently bear long spines. Some species have spines, projecting toward the posterior, all over the cell surface; others have just polar spines, while a third group has just

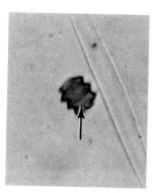

Figure 10-4. *Pseudokephyrion* lorica. 1000 X.

posterior spines. The spines can be easily broken or removed from the scales, and thus one should be careful in making diagnoses without carefully checking many members of the population. Cells have two flagella, one short, one long.

With these rather large forms, scale structure can be studied with the light microscope, but ultrastructural studies have revealed considerable additional detail.

Synura

Synura (Figs. 10-1 and 10-5). These colonies will be among the first members of the class Chrysophyceae a student will encounter. Unlike the colonial *Volvox* and its relatives, the cluster of cells is not surrounded by a common mucilage. In addition, there is vegetative cell division, which results in increase in colony size. Asexual reproduction of *Synura* is by means of fragmentation (Fig. 10-6).

The colonies are quite golden brown because of the two prominent plastids in each cell. Cells also bear numerous scales (Fig. 10-6), which must be examined under magnification of oil immersion light microscopy in order to see detail. Species distinctions can then be made. Flagella are of unequal length, and they bear small scales! In at least one form the flagella can be markedly unequal in length.

Figure 10-5. *Synura* elongate colony. 600 X.

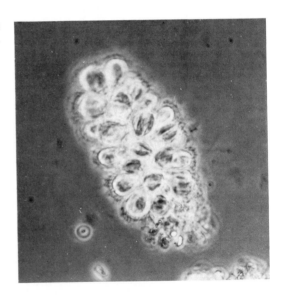

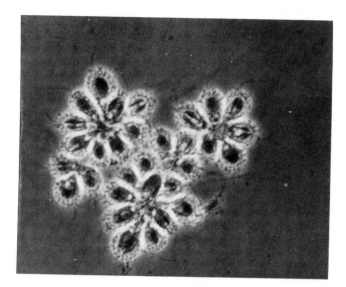

Figure 10-6. *Synura.* Several colonies with scales visible on individual cells. Note also flagella on some of the cells. 500 X.

Chrysosphaerella
These colonies at first appear to be members of the more common genus *Synura.* However, one soon observes that each cell bears two long silica spines. On closer examination with the light microscope, a common mucilage around cells of the colony, *and* only one long flagellum, are seen on each cell.

Dinobryon
Dinobryon (Fig. 10-1) colonies are very distinctive because they are dendroid and composed of numerous cells in urn-shaped loricas. The base of the lorica tapers to a point; at that point it attaches to the rim of the adjacent lorica (Fig. 10-3). Each protoplast has two flagella of unequal length. The entire colony swims freely in the water, with movement in the direction of the open end of the lorica. One or both products of cell division may move from the lorica to the rim, and there a new lorica is produced around the protoplast. Reproduction is by fragmentation.

SILICOFLAGELLATES
These are marine, uniflagellated unicells with golden chromatophores, and an external star-like silica skeleton (Fig. 10-7). Portions of the

Figure 10-7. The structure of two of a variety of silicoflagellate skeletons.

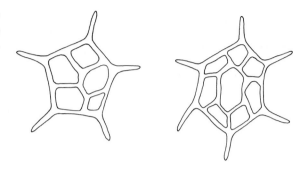

skeleton are often hidden by pseudopodia. As with diatoms these skeletons are seen in ashed or acid-cleaned material and they provided the basis for early taxonomic treatments. They are frequently found in diatomaceous earth, for silicoflagellates were more plentiful in earlier geologic time.

Classification was based on the number of spines on the outer ring of the skeleton and the number of open spaces (windows) in the frame *not* in contact with the basal ring (Fig. 10-7). The spaces were called apical windows. In one system of classification *Dictyocha* had no apical windows, *Distephanus* had one, and *Cannopilus* had two or more.

Recent studies of clonal cultures (started from single individuals) show that the pattern of the skeleton is not stable, for from zero to two apical windows were found on various individuals and spine patterns were not consistent. No new system of classification has been proposed. Other researchers find little variation in natural populations. Silicoflagellates are apparently quite temperature sensitive.

REFERENCES

Allen, M.B. 1969. Structure, physiology and biochemistry of the Chrysophyceae. *Ann. Rev. Microbiol.* 23: 29–46.

Blum, J. 1972. *North American Flora.* Vaucheriaceae. N.Y. Bot. Garden. 64 p.

Bourrelly, P. 1957. Recherches sur les Chrysophycées: morphologie, phylogénie, systématique. *Rev. Algol., Mém. Hors. Sér.* 1: 1–412.

Bourrelly, P. 1968. *Las Algues d'eau Douce.* Initiation a la systématique. Tome II. Les algues jaunes et brunes. N. Boubee and Co., Paris. 438 p.

Hibberd, D. 1971. Observations on the cytology and ultrastructure of *Chrysamoeba radians* Klebs (Chrysophyceae). *Br. Phycol. J.* 6: 207–23.

Ignatiades, L. 1970. The relationship of the seaonality of silicoflagellates to certain environmental factors. *Bot. Marina* 13: 44–6.

Leadbetter, B. and I. Manton. 1971. Fine structure and light microscopy of a new species of *Chrysochromulina* (*C. acantha*). *Arch. Mikrobiol.* 78: 58–69.

Manton, I. and G. Leedale. 1969. Observations on the microanatomy of *Coccolithus pelagicus* and *Cricosphaera carterae,* with special reference to the origin and nature of coccoliths and scales. *J. mar. biol. Assoc., U.K.* 49: 1–16.

Norris, R. and W. Blankley. 1971. Bibliography on Chrysophyta (excluding Bacillariophyceae). In J. Rosowski, and B. Parker (Eds.) *Selected Papers in Phycology.* Dept. Botany, Univ. Nebraska, Lincoln, pp. 777–83.

Van Valkenburg, S. and R. Norris. 1970. The growth and morphology of the silicoflagellate *Dictyocha fibula* Ehrenberg in culture. *J. Phycol.* 6: 48–54.

Haptophyceae

PLANT BODY TYPES

HAPTOPHYTE CHARACTERISTICS
1. Pigmentation
2. Food Reserve
3. Motility
4. Walls

REPRODUCTION

SOME REPRESENTATIVE GENERA   Representative Coccolithophorids

CLASSIFICATION

REFERENCES

These organisms which were recently removed from the Chrysophyceae have no common name, except haptophytes. One might wish to call them the haptonema-bearing organisms. The haptonema is not found in other classes. This external, filiform appendage bears a superficial resemblance to a flagellum; it may be straight, and protrude forward, or it might be tightly coiled. It can be shorter than flagella, or many times longer. In cross section, observed under the high magnification of the electron microscope, one sees three concentric membranes with six to eight, typically seven, single fibers in the center. This is in contrast to the flagellum substructure, with the nine double and two central microtubules.

Haptophytes fall into two groups. The first includes forms that resemble simple chrysophytes, but the motile stage has a haptonema. The second group includes the coccolithophorids which have two types of scales. The first type has a calcite thickening and such scales are called coccoliths. The other scales are unmineralized.

Many haptophytes in the first group are in the small size range of algae, that is, under 20 μm. They were formerly not collected in many studies, for they easily pass through a plankton net. Most are quite delicate and were easily destroyed by the typical fixatives used in early studies. Thus, today many fundamental questions about these organisms remain unanswered.

Haptophytes can be found most often in the marine plankton, where they may be the key primary producers, but attached forms and freshwater genera are known. Up to 45% of the total plankton in parts of the south Atlantic, or frequently as much as 75 to 95% of the volume of the Mediterranean plankton can be coccolithophorids. The latter are not common in freshwater. We have much to learn about coccolithophorids, especially their morphological plasticity.

PLANT BODY TYPES

The group includes unicellular, motile and nonmotile types, as well as some colonies and filaments. Because of the loose arrangement of cells, some researchers might call certain organisms colonies, others filaments. The majority of the haptophytes are unicellular, and some unicellular forms, described as distinct genera, have been shown to be motile stages of more advanced forms. These zoospores would have a haptonema.

HAPTOPHYTE CHARACTERISTICS

1. Pigmentation

Pigments are found in plastids (usually two per cell) which have thylakoids in groups of three. There are no girdling lamellae. The pigments are quite similar to those of the chrysophytes, with chlorophylls *a* and *c,* beta carotene, and fucoxanthin as the dominant xanthophyll. In at least one haptophyte, *Hymenomonas,* some of the xanthophylls can be located in a coiled lamellar system, which is not part of the plastid. Most motile cells lack an eyespot (and thus the associated pigments).

2. Food Reserve
The storage product of photosynthesis is probably chrysolaminarin, but we certainly need good chemical evidence from several genera.

3. Motility
Reproductive cells, as well as motile vegetative cells, have flagella of equal length, of slightly unequal length, or flagella that are markedly dissimilar in size. Both flagella lack hairs. The flagella, as well as the haptonemata, are inserted close to each other, either at the cell apex or somewhat laterally.

The haptonema is not responsible for motion, except in a very limited way. An organism may attach at first at the tip of the haptonema, and then use this organelle to adjust the position of the unicell.

Recently one organism has been placed in this class without observations of the haptonema, for there were no motile cells. In this case the similarity in scale structure (discussed below) enabled the scientist to make such a decision.

An amoeboid stage is known for *Chrysochromulina.*

4. Walls
Most haptophytes have external, unmineralized scales, which are composed of cellulose, with a mucilaginous matrix, as well as protein. The pattern of fibrils in the scale has been shown to be radiate on the inner surface and in concentric rings on the outer surface (Fig. 11-5). In most cases there are several layers of scales, and thus in *Chrysochromulina* (Fig. 11-1) one sees a layer of cuplike scales outside the layer of platelike scales. This arrangement of two types of scales is quite common. Rarely are scales found in only one layer, but naked types have been seen, especially in culture.

Coccoliths, calcium carbonate scales in the form of calcite, formed on many cells of coccolithophorids (Fig. 11-2), were first described as carbonate discs from the Cretaceous and were thought to have a nonbiological origin. They were found in the Atlantic when the Atlantic Cable of 1858 was being put down and also on the *Challenger* cruise.

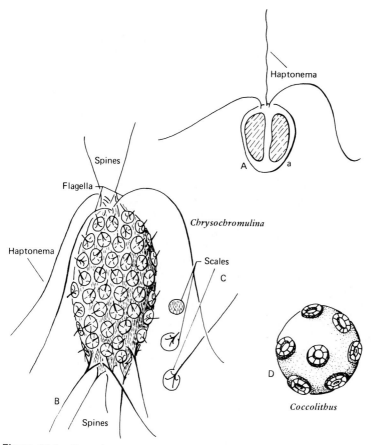

Figure 11-1. Some haptophyte genera. Microscopic.
A.–C. A. *Chrysochromulina*, shown in diagrammatic form as it might appear in the living condition, with two chromatophores (a).
B. and C. *Chrysochromulina*, diagrammatic, as seen with the electron microscope. The cell surface is covered with scales, and there are three types, with the terminal, spine-bearing form seen below (C). There are two flagella and a haptonema adjacent to the second flagellum on the left. In addition, this type shows two protruding anterior spiny scales.
D. The coccolithophorid, *Coccolithus*. Diagrammatic, with just a few of the many coccoliths found on the surface.

The coccoliths found were not known to be algal wall components until 50 years later.

Coccoliths are formed within the vesicles of the golgi, in a manner similar to scale formation in the packet-forming haptophycean alga,

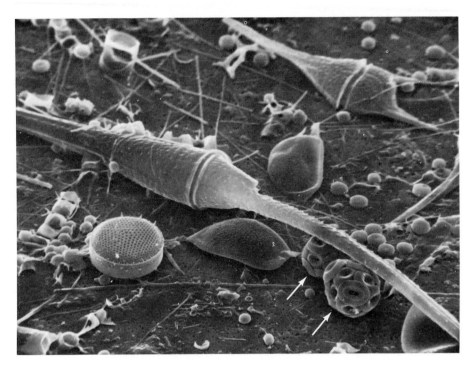

Figure 11-2. SEM of coccolithephorid, showing numerous scales with overlapping edges. (arrows)

Pleurochrysis. Brown has shown that in the scales of the latter there is a cellulose component, the concentric rings of fibers, and a spokelike component which is probably protein. The covering is a glucuronic acid polymer. There is one large golgi apparatus in each cell, and it is apparently connected to the nucleus and plastid. A scale, produced in the golgi, passes through the plasmalemma and is released at one point on the cell surface. During subsequent scale formation, the entire plastid-golgi-nucleus complex rotates, so that scales may be deposited on the entire surface from this one golgi apparatus. With the light microscope this movement of organelles within the cytoplasm has been observed and photographed in several closely related organisms. Coccolithophorids deposit calcite on the basic scale, which for them merely serves as a framework. Bald types, that is, forms lacking coccoliths, have neither the cellulose-protein, nor the calcite components of scales.

REPRODUCTION

Haptophytes reproduce by cell division, zoospore formation, or rarely, as least as we presently understand these organisms, by sexual reproduction. In at least one organism the life history involves forms previously placed in different genera, in which sexual reproduction has been confirmed.

SOME REPRESENTATIVE GENERA

Prymnesium

This well-known organism (Fig. 11-3), which grows in marine or brackish water, has been very troublesome in fish ponds in Israel because *Prymnesium* may release a toxic substance. Growth of this haptophyte is controlled by using ammonium; mature cells then rupture and die.

Cells are ovoid, with a somewhat blunted anterior and they possess two plastids. It is at the anterior end that both flagella and haptonema are

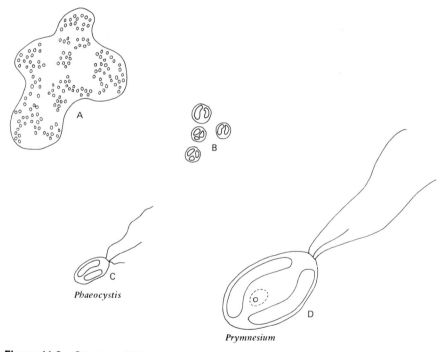

Phaeocystis

Prymnesium

Figure 11-3. Structure of *Phaeocystis*, as a colony (A), unicells (B), and a flagellate (C). D. The brackish water organism, *Prymnesium*.

inserted. The flagella, of equal or slightly unequal length, may be more than twice the length of the cell. The haptonema is permanently extended.

Phaeocystis

Members of this planktonic genus (Fig. 11-3) are colonial, and can be seen with the naked eye. Individual cells are nonmotile and spherical, but zoospores are haptophycean.

Phaeocystis has been studied in culture. A form from the south Atlantic produces acrylic acid. In nature the Phaeocystis is grazed by invertebrates; the latter are eaten by marine birds. The acrylic acid then reduces the number of bacteria in the intestinal tracts of these birds. Phaeocystis is also found in the plankton of the north Atlantic. Fish migrations can be affected by blooms of members of the genus.

Chrysochromulina

This flaggellated haptophyte (Fig. 11-1) has been studied extensively by Parke and Manton in England. Species may be found in a variety of shapes, including spherical, pyriform, and cylindrical unicells. Actually there are three or four types of scales on the surface. One layer of platelike scales is surrounded by a series of cuplike scales. At the cell apices there may be scales with elongate spines. In recent studies Manton found that some truncated spines are in reality deep cups, and thus there is an exposed open end. Scales, even those as long as the cell itself, are formed internally in golgi vesicles. The coiled haptonema can be extended, and probably serve as an organelle of attachment. Haptonemata, when extended, can be up to 20 times the cell length.

In the plankton of some English lakes there may be blooms of Chrysochromulina with 10^5 cells per milliliter. Marine and brackish forms also exist.

Representative Coccolithophorids

Hymenomonas.   In this marine organism haploid and diploid stages alternate (Fig. 11-4). The nonmotile stage has been called Apistonema in the past. This stage, found attached to rocks in the splash zone, is filamentous, with irregular branching (Fig. 11-5). Apistonema can also form coccoid unicells or simple colonies. These plant bodies can reproduce asexually by swarmers. The latter lack coccoliths. The Apistonema-like stage can also form gametes. The zygote develops into the Hymenomonas stage. These diploid cells are flagellated and have a

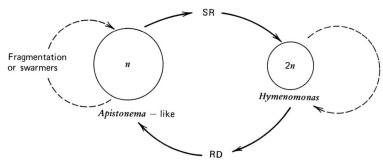

Figure 11-4. Life history and reproduction in *Hymenomonas–Apistonema*. There is alternation of generations between the filamentous and flagellated stage (Fig. 11-5). SR = Sexual reproduction; RD = reduction division.

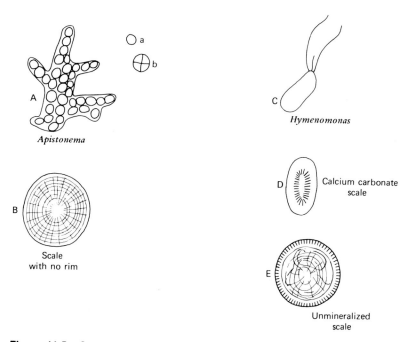

Figure 11-5. Structure of *Apistonema* and *Hymenomonas*. Microscopic. Scales with the electron microscope. *Apistonema* (A) may be a filamentous mass, or exist as either a unicell (a) or a small colony (b). Scales (B) have the typical radiating structure and concentric rings. The margin is irregular. *Hymenomonas* (C) has two flagella and a haptonema. The flagellated cells have calcite (D) and unmineralized (E) scales. The latter have a regular margin, with a rim. After Leadbeater.

short haptonema. On the surface is a layer of unmineralized scales covered by a layer of calcified scales. This flagellated stage can reproduce asexually, and thus the diploid phase can be maintained. Under certain conditions meiosis occurs and the swarmers produced develop into the *Apistonema* stage.

Coccolithus. Some species may have both motile and nonmotile stages and both can be dominant plankters in marine waters. The motile stage has two whiplash flagella and a haptonema. In the past, the two stages were incorrectly determined to be different genera. *Coccolithus* has been studied in culture, and either form can be found with or without coccoliths (Fig. 11-2).

CLASSIFICATION

Early classification of coccolithophorids was based on morphological features, with great emphasis on scale morphology. But information from laboratory studies show that:

1. Some flagellated cells, at one time put in a particular genus, were really reproductive stages of coccoid or filamentous genera.
2. Some organisms have a "naked" or bald stage and a scale-bearing stage. (Apparently some cultures have lost the ability to form coccoliths. Good nutritional studies of coccolith formation are needed.)
3. Some genera will form a certain type of coccolith in the motile stage, and a second type in a nonmotile stage.

Identification is further complicated by the fact that coccoliths range from 1 to 35 μm in diameter. At the lower end of the scale, the size of most coccoliths, one cannot see sufficient detail with light microscopy. The electron microscope is used to reveal additional structural information.

REFERENCES

Black, M. 1965. Coccoliths. *Endeavour* 24: 131–7.
Brown, R., W. Franke, H. Kleineg, H. Falk, and P. Sitte. 1969. Cellulosic wall component produced by the Golgi apparatus of *Pleurochrysis scherffelii. Science* 166: 894–6.
Bourelly, P. 1968. Les Algues d'eau Douce. Initiation a la systématique. Tome II. Les algues jaunes et brunes. N. Boubee and Co., Paris. 438 p.

Klaveness, D. 1972. *Coccolithus huxleyi* (Lohm.) Kamptn. II. The flagellate cell, aberrant cell types, vegetative propagation and life cycles. *Br. Phycol. J.* 7: 309–18.

Leadbeater, B. 1970. Preliminary observations on differences of scale morphology at various stages in the life cycle of *'Apistonema-Syracosphaera'* sensu von Stosch. *Br. Phycol. J.* 5: 57–69.

Loeblich, A. and H. Tappan. 1966. Annotated index and bibliography of the calcareous nannoplankton. *Phycologia* 5: 81–216.

Manton, I. and G. Leedale. 1969. Observations on the microanatomy of *Coccolithus pelagicus* and *Cricosphaera carterae,* with special reference to the origin and nature of coccoliths and scales. *J. Mar. Biol. Assoc., U. K.* 49: 1–16.

Paasche, E. 1968. Biology and physiology of coccolithophorids. *Ann. Rev. Microbiol.* 22: 71–86.

Eustigmatophyceae

PLANT BODY TYPES

EUSTIGMATOPHYTE CHARACTERISTICS

1. Pigmentation
2. Food Reserve
3. Motility
4. Walls
5. Eyespot
6. Flagellar Swelling
7. Pyrenoid

REPRODUCTION

REPRESENTATIVE GENERA

RELATIVES OF EUSTIGMATOPHYCEAE

REFERENCES

This new algal class was first proposed in 1971 by Hibberd and Leedale, who initially based their decision on ultrastructural features of zoospores. Until 1971, Eustigmatophytes were placed with the yellow-green algae.

The organisms have a distinctive eyespot, a flagellar swelling, finely lamellate reserve photosynthate, and other features which we discuss below. One might initially rebel at placing so much emphasis at the cellular level, especially on features the beginning student or field workers collecting these organisms might never see. Inasmuch as certain of these organisms are not widely distributed, the average student is not apt to see representatives such as *Pleurochloris, Polyedriella, Ellipsoidion,* and *Vischeria.* Other scientists might object to basing distinctions at the level of class on ultrastructural features. But one cannot ignore the fact that these organisms do indeed possess some rather unique structural features. Thus we know that they must be separated from other organisms, and we have to decide at what level of classification one should make the important distinction. Pigment analyses now lend support to the erection of this new class.

PLANT BODY TYPES

Eustigmatophytes may be unicellular or filamentous. Unicells are spherical, angular, or elongate; some are attached. Zoospores, which exhibit most of the features distinctive for this class, are naked.

EUSTIGMATOPHYTE CHARACTERISTICS

First we will examine the four characteristics considered for each of the other classes, and then look at additional characteristics.

1. Pigmentation

These organisms contain chlorophyll *a* and no additional chlorophylls. They possess beta carotene and several xanthophylls. Violaxanthin is the dominant xanthophyll. The eustigmatophytes and xanthophytes are the only algae known to possess vaucheriaxanthin. Pigments occur in plastids that lack girdling lamellae. In some forms the thylakoids are in groups of three.

2. Food Reserve

We lack good data on the chemistry of reserve foods. However, some phycologists believe that the storage product is chrysolaminarin. This decision is based on the fact that chrysolaminarin is found in organisms supposedly related to eustigmatophytes. The composition of the finely lamellate material observed with the electron microscope is not known. This material has not been seen in other algal classes.

3. Motility

Vegetative cells are not motile, but most organisms reproduce by zoospores that lack walls (Fig. 12-1). One species of *Monodus* lacks zoospores, and thus is difficult to identify as a member of this group. Most eustigmatophytes described have one long tinsel flagellum but ultrastructural studies reveal two basal bodies. Thus a second flagellum was most certainly present in ancestral forms. In those genera with a second flagellum it might be quite short, or of moderate length. Zoospores are elongate, fusiform, or pear-shaped.

4. Walls

More information is needed on wall structure of vegetative cells. We know that it may be modified to form a stipe, but it is typically quite thin and lacks ornamentation. Some bipartite walls are known.

Figure 12-1. Diagrammatic presentation of the ultrastructure of a eustigmatophyte zoospore. After Hibberd and Leedale.

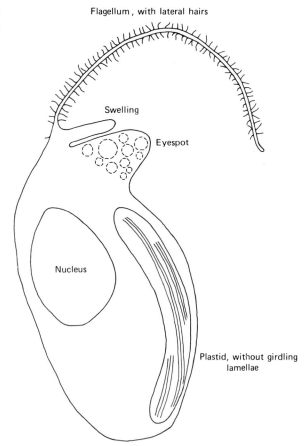

Flagellum, with lateral hairs

Swelling

Eyespot

Nucleus

Plastid, without girdling lamellae

5. Eyespot
The eyespot is orange-red, at the anterior end of the zoospore, and is independent of the single plastid. It is composed of droplets that lack membranes. Also, there is no membrane around the group of droplets.

6. Flagellar Swelling
There is a flagellar swelling near the base of the long flagellum, adjacent to the eyespot.

7. Pyrenoid
The pyrenoid may have a polygonal shape, visible with the light microscope. It may also be on a short stalk.

REPRODUCTION
Reproduction is either by cell division or release of zoospores from the parent cell. As we get to know more about these organisms no doubt we will learn of other possibilities.

REPRESENTATIVE GENERA

Pleurochloris, Polyedriella, and Vischeria
Because the vegetative cells of each of these genera are quite similar, they will be described as a group. The vegetative cells are in the 5 to 15 μm size range, and in any one species may be either spherical or polyhedral, or both. The number of angular or spherical cells in a population depends on the conditions of culture. There is a cell wall, which is distinctly layered, when observed under the electron microscope. There is a single nucleus, golgi apparatus, and plastid. The plastid is parietal and has many deep lobes. In the plastid is the polygonal pyrenoid, which can be detected with the light microscope. This feature is quite helpful in making preliminary identifications.

The zoospores are also quite similar in the three genera. All have a single flagellum, with the second flagellum represented only by the basal body. The zoospores lack a pyrenoid, and some lack a golgi apparatus. The remaining features of the zoospores are characteristic of the group.

Ellipsoidion
This genus has fusiform or ellipsoidal cells that may have terminal papillae. The cells may be up to 20 μm long, with one to several plastids. The zoospores have two emergent flagella, but one is short.

Chlorobotrys

C. regularis is a coccoid alga that does not have a flagellated stage. It lacks the girdling lamellae in a much lobed plastid. The pyrenoid is typical of members of this class. Thus it is placed here, even though a flagellated stage is lacking.

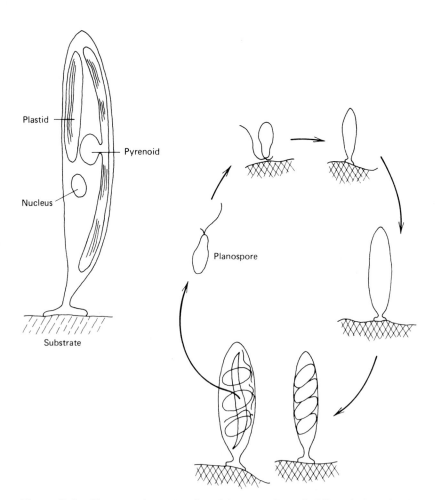

Figure 12-2. Diagrammatic presentation of the vegetative cell of *Pseudocharaciopsis,* as might be seen when thin sections are observed with the electron microscope. A life history of the organism is also presented. After Lee and Bold.

Pseudocharaciopsis

This organism was described in 1973 as a new genus. After examining EM photos of the vegetative cell, with its stalked pyrenoid, Lee and Bold found that zoospores had eustigmatophyte characteristics.

The organism is an attached unicell with tapering apices (Fig. 12-2). Individual cells have several plastids and are multinucleate when they age. *Pseudocharaciopsis* reproduces by the production of several naked zoospores, which after a period of motility settle on a suitable substrate and produce a short stalk and thin wall.

RELATIVES OF THE EUSTIGMATOPHYCEAE

All the members of this class that had been described prior to 1970 were placed in the Xanthophyceae. Xanthophyte zoospores have the stigma inside one chromatophore. There are two plastids per cell. Each cell also bears two flagella with the swelling on the whiplash type. The whiplash flagellum projects backwards when swimming. In addition, most members have girdling lamellae in the plastids.

Recent investigations of pigmentation support the erection of this new class. Now we know that the yellow-green algae have chlorophyll *c,* but this pigment is lacking in the Eustigmatophyceae. They have different dominant xanthophylls, with violaxanthin the dominant pigment in this new class. However, both classes have the minor accessory pigment, vaucheriaxanthin, which is found only in these two classes!

Hibberd feels that most ultrastructural features place the yellow-green algae closer to the phaeophytes and chrysophytes than to the eustigmatophytes. We will need more information on more genera before the position of this group among the algae becomes clear.

REFERENCES

Hibberd, D. and G. Leedale 1970. Eustigmatophyceae—a new algal class with unique organization of the motile cell. *Nature* 225: 758–60.

Hibberd, D. and G. Leedale 1972. Observations on the cytology and ultrastructure of the new algal class, Eustigmatophyceae. *Ann. Bot.* 36: 49–71.

Lee, K. and H. Bold 1973. *Pseudocharaciopsis texensis* gen. et sp. nov., a new member of the Eustigmatophyceae. *Br. Phycol J.* 8: 31–7.

Whittle, S. and P. Casselton 1975. The chloroplast pigments of the algal classes Eustigmatophyceae and Xanthophyceae. I. Eustigmatophyceae. *Br. Phycol. J.* 10: 179–91.

13

Dinophyceae — dinoflagellates

Dinoflagellates (the first syllable is pronounced "deen," not "dine") are so named because of their twirling motion rather than their morphology. They are perhaps best known to all, and especially to the layman, as the organisms involved in the red tides.

This is another group of organisms that has both plant-like and animal-like characteristics. Both marine and freshwater forms are found, and because of the chemistry of the wall material, which is not easily decomposed, several genera are known from the fossil record. Most organisms are free-living, but others are symbionts. Organisms called zooxanthellae are frequently dinoflagellates. They are found in fish, corals, anemones, and other organisms.

The classification of the dinoflagellates is not as firm as one might expect for such an ancient, widely distributed group. At times the cryptophytes (Chapter 14) are included in this class, but they are classified separately here, because they possess phycobilins and the dinoflagellates do not.

PLANT BODY TYPES

These organisms occur as flagellates or sessile unicells, colonies, and filamentous forms. The cells are typically flattened and have a transverse constriction, the girdle, *usually* around the cell equator (Fig. 13-1). Distinctive features of dinoflagellates are the insertion of flagella in the girdle and the flagellar arrangement, with one encircling the cell and one trailing. One group of armored forms has horns, another has spines on the armor plates, and a third category includes dinoflagellates with large, winglike appendages.

The organisms are usually uninucleate and have visible chromosomes throughout the cell cycle. The nuclei appear as in early prophase. For this and other reasons the nucleus is termed the mesokaryotic type. Some organisms have a mesokaryotic and a eukaryotic nucleus.

How did the binucleate form originate? There is now clear evidence that a photosynthetic endosymbiont is present within certain dinoflagellates. Such organisms have been examined with the electron microscope. If one assumes that the parent dinoflagellate was a colorless heterotrophic organism, all cell organelles can be identified, including the mesokaryotic nucleus. An invaginating membrane separates the chloroplasts, eukaryotic nucleus, and other organelles of the endosymbiont from the host cytoplasm. Thus the photosynthetic

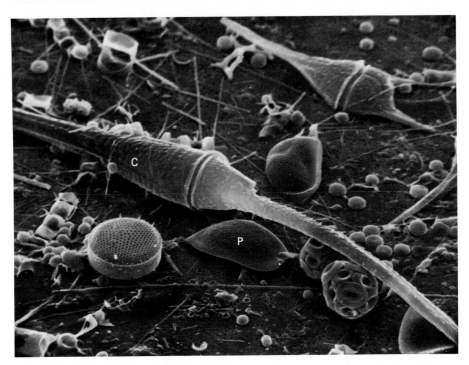

Figure 13-1. SEM of two dinoflagellates, *Ceratium* and *Prorocentrum*, along with coccolitho-phorids and diatoms. 635 X. Courtesy of P. Hargraves.

dinoflagellates can be divided into two groups. One contains membrane-bound pigmented endosymbionts and the second would include dinoflagellates with a single nucleus and their own plastids.

Furthermore, cells with endosymbionts contain the xanthophyll, fucoxanthin. This pigment is usually found in phaeophytes, chrysophytes, and diatoms, and not in dinoflagellates. (The latter have peridinin and dinoxanthin, when without symbionts.) It would appear that fucoxanthin is the pigment of the endosymbiont. A study of *all* accessory pigments of binucleate dinoflagellates supports this view.

Binucleate dinoflagellates usually divide synchronously. It is also now clear that the nucleus of the symbiont is not typically eukaryotic. However, differences in its division cycle are no doubt due to the role of the host nucleus, which governs division of the entire cell.

CHARACTERISTICS OF THE DINOFLAGELLATES

1. Pigments

The organisms contain chlorophyll *a* and *c*, beta carotene, and several xanthophylls, especially peridinin or fucoxanthin, which give most organisms a brownish coloration. The pigments are located in plastids, with thylakoids in three's, and the plastids are not bound by a girdling lamellar system. In contrast to many chlorophytes and cyanophytes, utilization of dissolved organic matter is quite uncommon in the free-living photosynthetic forms. This fact is unexpected because of the range of activity nonphotosynthetic types exhibit in obtaining their carbon. On the other hand, vitamin requirements are not uncommon.

There are numerous plastids in each cell.

2. Food Reserve

The food reserve is either true starch or oils. The starch can be stored in the plastid, but many organisms store starch only outside the plastid.

3. Motility

The dinoflagellates are frequently motile forms, but zoospores can be produced by palmelloid and filamentous forms. Both flagella originate together in the girdle or in the sulcus, a groove perpendicular and posterior to the girdle. The whiplash flagellum trails behind the moving organism. The second flagellum trails behind the moving organism. The second flagellum is quite complex, and should be referred to as the flagellum complex. Close to the cell surface in the girdle there is a striated band. Slightly removed from the striated band is the flagellum itself, which is flattened, ribbonlike, and undulates encircling the organism. Contrary to earlier reports it is not helical, and it does not wind around the striated band. There are hairs projecting from the flagellar surface.

It is claimed that flagellated dinoflagellates can move 100 times their own length each second. For man to move proportionately that fast, he would have to accelerate to about 600 kph!

4. Cell Covering

The exterior surface of a dinoflagellate is usually covered by a plasmalemma, but in some forms there is a typical wall. The most conspicuous portion of the cell covering of many dinoflagellates is within the plasmalemma. This type of covering is composed of thecal plates, and dinoflagellates with them are termed armored. Species without thecal plates are said to be naked. One or more of the following comprise the dinoflagellate cell covering.

1. Cytoplasmic membrane, or plasmalemma.
2. Pellicle. This is *outside* the cytoplasmic membrane. it contains alpha cellulose and is an unornamented fibrous layer.
3. Theca. This layer is the armor, or a series of plates containing alpha cellulose. In various species the number of plates varies from few to many. The theca can be present even if there is a pellicle.

    The thecal plates are divided into anterior or epithecal plates, which can be composed of apical, intercalary (found just below the latter) and precingular plates (found immediately above the girdle), and posterior or hypothecal plates. The latter are postcingular (just below the girdle), intercalary and antapical (at the posterior pole).

    The plates overlap and are held together at the joints. They may extend as projections or horns as in *Ceratium* or as elaborate wings as in *Ornithocercus.* Plates are ornamented and may be rough and fibrous, perforated or merely provided with thickenings. Quite thin and very thick types are known. Many species have pores in the plates.
4. Thecal membranes. These surround the thecal plates and attach to each other at the margins of the plates.
5. Wall. Some forms, for example, *Pyrocystis,* may have a wall composed of several layers. The inner layer is cellulose and the outer amorphous. A layer of sporopollenin, very resistant to decay, is probably present in some dinoflagellates, and certain types can add calcium carbonate to the wall. Fossil dinoflagellates are known from the Ordovician, with over 100 genera reported!

In the marine environment the naked forms are often oceanic. Armored types frequently have a more coastal distribution. Dodge suggests that

Figure 13-2.   Some dinoflagellates. Microscopic.
A. An armored type of dinoflagellate. There are two flagella, one in the girdle groove, the second trailing. Note the plates and girdle.
B. *Gymnodinium,* with numerous plastids. Both freshwater and marine species are known.
C. *Ceratium.* Note the four horns.
D. *Noctiluca.* The cells within are algal symbionts; however, not all species of *Noctiluca* have symbionts.
E. *Pyrocystis* life history. The nonmotile vegetative cell produces two dinoflagellate swarmers. The naked protoplast from each swarmer eventually produces another vegetative cell.

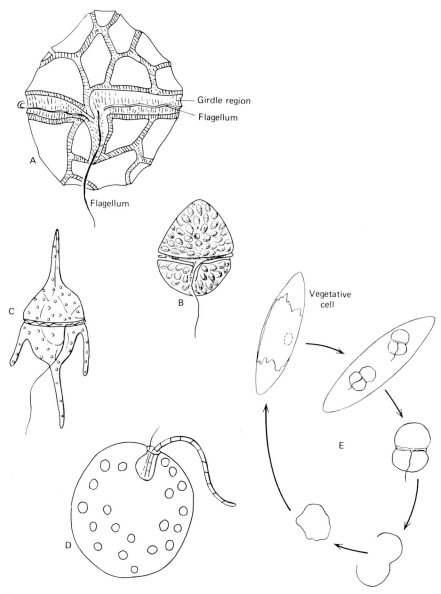

Girdle region

Flagellum

A

Flagellum

B

C

Vegetative
cell

E

D

there are no fundamental differences between naked and armored forms, because intergradations occur. The thickness of the theca, and the length and thickness of the spines or horns, may vary with the season or with the locality. *Ceratium* (Fig. 13-2) species have longer spines in warmer waters. In less dense waters the increased frictional resistance provided by cell processes helps keep *Ceratium* in the photic zone.

DINOFLAGELLATE REPRODUCTION

Asexual reproduction takes place through division of the cell. The cell may divide smoothly along the girdle, or diagonally, with separation along plate margins. When a dinoflagellate divides while swimming, each half gets one flagellum. *Peridinium* releases a naked protoplast from the theca, and division occurs after release. The naked portion has at times been confused with naked genera. Nonmotile types reproduce by forming dinoflagellate zoospores.

Sexual reproduction is not commonly observed. Older reports indicate that the organisms are haploid and have zygotic meiosis. *Noctiluca* (Fig. 13-2) was recently reported to form numerous uniflagellate isogametes from a single cell. The zygote developed directly into a *Noctiluca*. Gametic meiosis would then be expected. Perhaps dinoflagellates have several types of life histories, as do other algae. *Glenodinium,* *Peridinium,* and *Ceratium* (Fig. 13-2) are reported to have zygotic meiosis. The life history of *Crypthecodinium* appears to be complex. Genetic recombination has been confirmed in *Crypthecodinium.*

REPRESENTATIVE GENERA

Pyrocystis

This luminescent form (Fig. 13-2) with a light brighter than that produced by most microorganisms, has nonmotile vegetative cells. This relatively large planktonic genus has a thin, cellulose wall, peripheral cytoplasm, and a large central vacuole. The vegetative cells eventually divide and release two typical dinoflagellate zoospores which closely resemble *Gonyaulax.* The zoospores break open at the girdle and release their protoplasts, which develop into vegetative cells (Fig. 13-2). These findings have been confirmed in culture.

Gymnodinium

Gymnodinium (Fig. 13-2) is a naked marine and freshwater genus. The girdle is transverse and around the cell equator. The slightly flattened cells are ovoid or circular when observed from the dorsal or ventral surfaces. When organisms from colder waters are observed under the microscope, they quickly become spherical. In order to see the normal shape and to keep cells from dying, the microscope slide should be kept cold.

Gonyaulax
This marine and freshwater genus (Fig. 13-1) is armored and is not flattened dorsoventrally. The plates are not very distinct. The girdle is said to be helicoid, inasmuch as both ends are displaced from the transverse plane.

Peridinium
Peridinium is an armored marine and freshwater form (Fig. 13-4) with thick plates. The organism is slightly flattened, and with the thick thecae, it has an angular appearance. There may be short spines on some thecal plates, and the sutures between plates are obvious.

Ceratium
This genus (Fig. 13-2) is also found in both marine and freshwater environments. It is quite distinctive and has up to four elongate horns (Fig. 13-3). One projects from the tip of the upper half of the cell, and as many as three may project from the lower half. Horn length in a population is subject to environmental control, possibly enabling the organism to remain suspended in the water column.

Noctiluca
Noctiluca (Fig. 13-2) is a colorless relative of the photosynthetic dinoflagellates; there are many others. This large, spherical, sometimes macroscopic form has many structural features that are not typical of

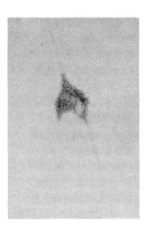

Figure 13-3. *Ceratium.* 250 X.

Figure 13-4. *Peridinium.* SEM of whole cell. Note girdle depression and outline of cell covering segments. 2800 X. Courtesy of P. Hargraves.

dinoflagellates. For example, the transverse flagellum is greatly reduced and modified. There is also a projection of protoplasm near the insertion of flagella. These luminescent organisms are common in the plankton, and they are sometimes collected when there is a red tide. But they are not implicated in red-tide phenomena.

TRICHOCYSTS

Many dinoflagellates, *Gonyaulax,* for example, have trichocysts. These microscopic hairlike projections are released by mechanical means and provide protection or a means of attachment. They probably originate from golgi. They are seen in vesicles and appear homogeneous when first developed, but then develop a neck and a body with a crystalline core. They are released through the cell membrane and pores in the thecal plates. After discharge there is hydration. Electron micrographs of the discharged body show the presence of cross striations, with a repeating unit each five or six bands.

LUMINESCENCE

Dinoflagellates are the only algae that luminesce. *Noctiluca,* *Gymnodinium,* and *Pyrocystis,* among others, are responsible for this "light without heat." *Gonyaulax* luminesces only in the dark period, a process that is under the control of a biological clock. There have been extensive studies of this type of rhythm in *Gonyaulax.*

TOXIC FORMS

There are many toxic dinoflagellates, some of which are involved in shellfish poisoning as well as in fish kills. Filter feeders eat *Gonyaulax* and accumulate the toxin of the dinoflagellate in their tissues. The compound does not affect the shellfish. However, it affects humans in a manner similar to the nerve poison, curare, causing distal paralysis and respiratory failure. Usually around 1% of those poisoned die, and the cause of death is usually respiratory failure.

Red tides are caused by *Gonyaulax* along the California coast, where tens of millions of cells per liter have been found. In the Florida portion of the Gulf of Mexico both *Gonyaulax* and *Gymnodinium* have been found in significant numbers and give the water first a yellow color, then a brownish, and finally a red coloration. These colors are due to pigment changes in the organism; there have been 10^6 and 10^8 cells per liter and as many as 5×10^7 fish killed in one poisoning. The toxin of *Gymnodinium* is released as the cells break up during passage through fish gills. The blooms of red tide now occur almost every summer in Florida and California, but the severe cases do not occur that often. A bloom of *Gonyaulax* occurs yearly in the colder waters of the Gulf of Maine (see Chapter 20).

REFERENCES

Beam, C. and M. Himes. 1974. Evidence for sexual fusion and recombination in the dinoflagellate *Crypthecodinium* (*Gyrodinium*) *cohnii. Nature* 250: 435–6.

Cox, E. and R. Zingmark. 1971. Bibliography on the Pyrrophyta. In J. Rosowski and B. Parker (Eds.) *Selected Papers in Phycology.* Dept. Botany, Univ. Nebraska, Lincoln. pp. 803–10.

Dodge, J. 1971. Fine structure of the Pyrrophyta. *Bot. Rev.* 37: 481–508.

Dodge, J. 1973. *The Fine Structure of Algal Cells.* Academic Press, New York. 261 p.

Jeffrey, S. and M. Vesk. 1976. Further evidence for a membrane-bound endosymbiont within the dinoflagellate *Peridinium foliaceum. J. Phycol,* 12: 450–5.

Loeblich, A. and A. Loeblich. 1974. Index to the genera, subgenera, and sections of the Pyrrophyta, VII. *Phycologia* 13: 57–61.

Swift, E., W. Biggley and H. Seliger. 1973. Species of oceanic dinoflagellates in the genera *Dissodinium* and *Pyrocystis:* Interclonal and interspecific comparisons of the color and photon yield of bioluminescence. *J. Phycol.* 9: 420–6.

Taylor, F. 1975. Non-helical transverse flagella in dinoflagellates. *Phycologia* 14: 45–7.

Tomas, R. and E. Cox. 1973. Observations on the symbiosis of *Peridinium balticum* and its intracellular alga. I. Ultrastructure. *J. Phycol.* 9: 304–23.

Cryptophyceae — cryptophytes

This small group of organisms, containing about 24 genera, has no common name. They are marine and freshwater forms which, when free-living, are often motile. Some members of the Cryptophyceae are symbiotic with animals and are thus called zooxanthellae. At times the group has been classified with the dinoflagellates, and after examining the characteristics, one can see some similarities. However, there are many distinct differences, and I will treat the group as a separate class and division with independent origin.

PLANT BODY TYPES

These are typically naked motile cells with two flagella of slightly unequal length. The cell is oval and flattened. There is an anterior reservoir or sometimes a furrow, which in some species is lined with trichocysts. Some coccoid or palmelloid forms are known, which may have zoospores similar to those just described.

There are one or two plastids, with or without pyrenoids, per cell. Electron microscope studies show that starch is stored outside the plastid, but within the nuclear-plastid complex. The complex itself is surrounded by a membrane.

CHARACTERISTICS OF THE CRYPTOPHYCEAE

1. Pigments

Organisms in this class possess chlorophylls *a* and *c,* and *lack* beta carotene. Alpha carotene is present, along with several xanthophylls. The dominant xanthophyll is alloxanthin.

The pigments are in plastids with thylakoids in groups of two. No girdling lamellar system is present. The cryptophytes are the only *flagellated* algae that have the additional group of pigments, the phycobilins. Incorporation of a blue-green algal cell into a colorless flagellate, and the evolution of the former into the plastid organelle would explain the origin of these pigments in such an algal group. The phycobilins are not found in phycobilisomes, as in the blue-green and red algae. They may well be located in the space between the membranes of the thylakoids, that is, intrathylakoidal (Fig. 14-1).

2. Food Reserve

The food reserve is true starch in some species, as well as fat or oil. We need more chemical information on food reserves in a variety of these organisms.

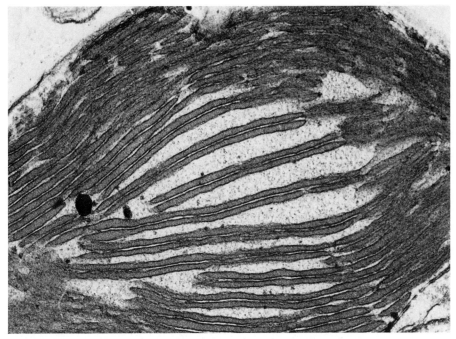

Figure 14-1. *Chroomonas* chloroplast. The phycobiliproteins are perhaps in the intrathylakoidal area. 68,000 X. Courtesy of E. Gantt. et al. and the Journal of Cell Biology.

3. Motility

Many of the organisms present in this group are actively motile during their vegetative phase. The cells are flattened into a dorsal-ventral plane and possess two anterior flagella, which arise from the base of a reservoir or groove. The flagella are of unequal length and both are the tinsel type, with stiff hairs.

4. Walls

Some forms possess a cellulose wall, while both flagellated cells and zoospores have the protein plates of a pellicle beneath the plasmalemma. The pellicle plates have three layers. In some forms lacking a wall there may be a thin fibrillar or granular material loosely arranged outside the plasmalemma.

REPRODUCTION

Reproduction is either by means of cell division or the formation of zoospores or cysts. Sexual reproduction has been confirmed recently in *Cryptomonas*.

REPRESENTATIVE GENERA

Cryptomonas

Cryptomonas (Fig. 14-2), a common flagellate of the group, has a flattened oval shape with two flagella coming from the anterior depression or reservoir. The color of the organism (green, blue-green, or red) varies, depending on its age and physiological state. A contractile vacuole is located adjacent to the reservoir, and trichocysts (or ejectosomes) are observed lining the reservoir. For many years the organism was not accurately classified by phycologists and was included in a miscellaneous group.

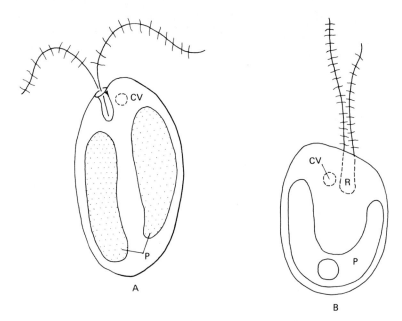

Figure 14-2. Two cryptomonads. Note the anterior reservoir (R), contractile vacuoles (CV), plastids (P), and insertion of flagella. A. *Cryptomonas*. B. *Chroomonas.* The periplast and trichocysts are best seen with the electron microscope.

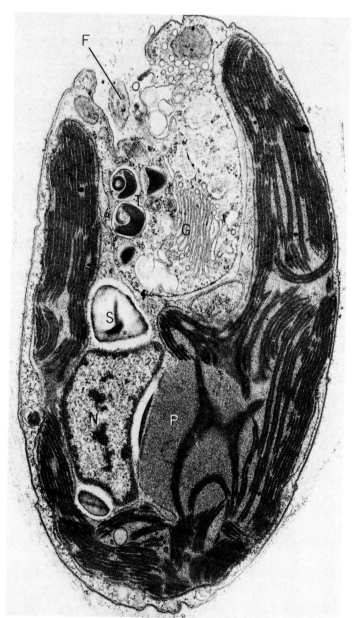

Figure 14-3. Electron photomicrograph of *Chroomonas* cell, longitudinal section. The cup chloroplast encloses the Golgi (G) apparatus, nucleus (N), and trichocysts (T). 40,000 X. Courtesy of E. Gantt. et al. and J. Cell Biology P = pyrenoid, S = starch, F = flagellum.

Chilomonas

This organism is a colorless *Cryptomonas* and is identical to it in most details.

Tetragonidium

This walled nonmotile unicell has an irregular tetrahedral shape. There is a single plastid. Reproduction is by zoospores which typically resemble a cryptophyte.

Chroomonas

These organisms (Figs. 14-1 and 14-3) are flagellated unicells with subapical insertion of two flagella. Their three-layered pellicle is visible with the electron microscope and has been shown to be composed of plate areas. These plate areas, with marginal ridges, cover the entire cell, *under* the plasmalemma.

CLASSIFICATION OF THE CRYPTOPHYTES

In this volume these organisms will be placed in a separate division (Chapter 16). This decision is based on several considerations, with emphasis on the fact that this is the only group of flagellated organisms in which there are phycobilins. If plastids originated by way of endosymbiosis (Chapter 3), cryptophytes would probably have as closest algal allies the red and blue-green algae. The flagellated host cell would have evolved separately, perhaps among the protozoa.

REFERENCES

Chapman, V. and D. Chapman. 1973. *The Algae*. MacMillan and Co., London. 497 p.

Faust, M. 1974. Structure of the periplast of *Cryptomonas ovata* var. *palustris*. *J. Phycol.* 10: 121–4.

Gantt, E., M. Edwards and L. Provasoli. 1971. Chloroplast structure of the Cryptophyceae: Evidence for phycobiliproteins within intrathylakoidal spaces. *J. Cell. Biol.* 48: 280–90.

Lucas, I. 1970. Observations on the fine structure of the Cryptophyceae. I. The genus *Cryptomonas. J. Phycol.* 6: 30–8.

Algal habitats and distribution

When considering algal habitats and algal distribution throughout the world, we first think of either the conspicuous benthic algae, such as the coastal seaweeds, or of the phytoplankton. Phytoplankton are found in almost any body of water, sometimes so abundant that they color the water. There may be millions of cells per liter. Without doubt, the phytoplankton are the most significant category of algae because of their place in the food web.

Although algae are found in a wide variety of habitats, such as arctic lakes, hot springs, or dry soil, certain classes of algae have members that might be found almost exclusively in only one of the habitats described in this chapter.

There are a number of practical reasons for observing, collecting, and studying algae in their natural habitats. Researchers may be interested in obtaining a particular material for experimentation, or in gathering a complete picture of all organisms in a particular habitat. Students or teachers may need to obtain material for the classroom, or they may wish to observe algal distribution first hand. An employee of a government agency may wish to assess the extent of an environmental problem or to conduct a baseline study.

In the following discussion a general overview of a variety of habitats is presented. More information on specific topics is found in Chapters 17 and 18.

AQUATIC DISTRIBUTION

Planktonic Forms

Organisms called phytoplankton (Chapter 18) are important as primary producers. Phytoplankers are found in almost any body of water, floating or passively moving in the water column. Some are able to swim, but could not travel more than a few meters in a day. They are present in coastal and oceanic waters, tide pools, small and large lakes, ponds, and even small temporary bodies of water. Large rivers could have a true plankton, but a number of the organisms found floating might have just recently been scoured from some substrate (see below). In small rivers, and in some small, nonflowing bodies of water, the primary production that occurs is due not to phytoplankton but to attached forms.

The plankton may be sampled by collecting a representative volume of water and observing it directly, or immediately after concentration by centrifugation, filtration, or sedimentation. Ruth Patrick has popularized the use of artificial substrates, even for the collection of organisms floating in rivers and streams. However, in these habitats many organisms probably were recently scoured from rocks, and thus primarily benthic genera are collected.

The plankton could include members of any algal class, but red algae are rarely found there, and *Sargassum* is the one planktonic member of the brown algae. In all classes reproductive cells released from the parent plant would be considered part of the plankton. Planktonic forms from most of the algal classes are microscopic forms typically less than one millimeter in size.

Attached Forms

Attached Microalgae.   In the last decade interest has developed in these organisms in both marine and fresh waters, especially concerning their primary production. These algae can attach to other algae, vascular plants, animals, rocks, or can be located on and in both sand and sediments. In the modern sense of the term *periphyton* refers to algae found attached to many types of substrates, including organisms. In the past *periphyton* had been used to include only organisms growing on submerged vegetation. The term *epiphyte* applies to plants attached to other plants. Then *epipelic* algae are those found growing on and in sediments, *epipsammic* algae are located on or in sand, and *epilithic* algae are located on rocks.

In some habitats attached microalgae are an essential component, or the main component, of the primary producers (Chapter 18). This is especially true in some freshwater lakes (Table 15-1). The figures presented are for primary production in a lake (by attached microalgae) *after* substraction of phytoplankton and vascular plant production.

Organisms found attached to vascular plants frequently grow in dense aggregations. Then they are easily removed from the substrate in

Table 15-1. THE CONTRIBUTION OF ATTACHED MICROALGAE TO THE PRIMARY PRODUCTION OF LAKES IN SEVERAL COUNTRIES (After Wetzel, Foerster)

Lake Site	Percent of Primary Production
Ontario	2
Michigan	25
Poland	40
California	40
British Columbia	60

turbulent waters and are found in the plankton. In calm water they can settle on new substrates. A conspicuous attached community can be found on many marine macroalgae, as well as on marine vascular plants in both shallow water and estuaries.

Members of several algal classes can be found as attached microalgae, but the green algae, diatoms, blue-green, and golden algae are most common.

Attached Macroalgae.   Macroalgae found along the coasts of large lakes, oceans, or bays are stratified vertically, but the zonation is more conspicuous in the marine environment. The most obvious areas are the sublittoral (below the level of exposure to air from tidal activity), the littoral (alternately covered and exposed by the tides), and the spray zones (moistened only by the splashing of waves). Spray zone algae may be covered with water during exceptionally high tides. The benthic algae will occur along a continental shelf, or margin of a lake, to the depth of light penetration. (See Chapter 17 for a discussion of the factors affecting benthic algal distribution.)

Other Forms

A number of both marine and freshwater algae, macroscopic as well as microscopic species, do not conveniently fit into any of the above categories. Certain algae spend portions of their lives in different habitats. Some species are found entangled or entrapped among the appendages of larger forms. In calm water truly planktonic organisms can temporarily land on a substrate. With these forms it is best to follow common practice and to be consistent in the use of terminology developed.

ASSOCIATIONS WITH OTHER ORGANISMS

There is a broad range of associations between algae and other organisms. Some assocations are intracellular, others intercellular or even merely on the surface of the host. The combination is frequently called symbiotic, although supporting data are often lacking. In some cases, it is assumed from observations that each organism is deriving some benefit. Some associations are epiphytic or epizoic, others parasitic.

Animals

The algae growing on or in animals are often called zooxanthellae or zoochlorellae. The latter group is typically found in fresh waters, and includes green algae in addition to *Chlorella,* for example, *Platymonas.* The zooxanthellae are almost exclusively marine organisms and include dinoflagellates and cryptophytes. Symbiotic blue-green algae are found in either marine or fresh waters and are not included in either category.

The majority of associations with animals are found in the tropical marine environment, where the water is nutrient poor. Some researchers speculate that the relationship of alga and invertebrate arose when ingested algae were first only slowly digested. There was a selective advantage, for the alga would continue photosynthesizing until digestion actually killed it. Gradually a system evolved in which the algal partner was protected from destruction. As hosts adapted to the partnership, some organisms have extensively modified the digestive tract, while others have increased surface area of some parts. An example of the latter would be the large mantle of the giant clam, with algae in the surface layers, which offers a broad surface for photosynthesis. Zoochlorellae and zooxanthellae can be found in protozoa, sponges, flatworms, coelenterates, molluscs, echinoderms, ascidians, annelids and jellyfish.

Recently, the hatching of winter flounder was found to be more successful when a particular *Melosira,* and to a lesser degree a second *Melosira* and an *Amphipleura,* were present in the plankton. In Narragansett Bay, R.I., *Melosira* plankters flourish in winter and spring months; the flounder spawning period is from January to April. Eggs naturally clump within minutes of fertilization, and those in the center of the clump often quickly die. However, when these diatoms are present, clumping is not only reduced, but the supply of oxygen is plentiful. More flounder eventually survive.

Plants

Algal partnerships are formed with other algae, with fungi, in some liverworts, associated with ferns, and even with the roots of cycads. Some algae, often parasitic forms, are found in the leaves of vascular plants, for example, *Magnolia.*

Some Marine Associations.   Norris has called the associations among members of the marine plankton consortisms. This general term would include organisms ranging from those that have an epiphytic existence to others living within the cells of hosts. Many of these consortisms are well established, and, surprisingly, Norris found that blue-green algae were often involved. The blue-green *Richelia* lives within the protoplast of the diatom *Rhizosolenia,* and some blue-greens are found within the cell covering of dinoflagellates. In autoradiographic studies with the *Richelia-Rhizosolenia* complex, most of the photosynthesis was detected in the blue-green alga. In addition, in fresh collections the diatom always appeared to be senescent.

Lichens.   Ahmadjian calls the algal partner in a lichen the phycobiont and points out that at least 30 genera of blue-green and green algae can be found associated with fungi in the plants called lichens. Most lichen phycobionts in the green algae are either *Trebouxia* or *Pseudotrebouxia*. Blue-greens found would include at least members of the genera *Anabaena, Nostoc, Calothrix,* and *Scytonema.* However, each lichen species does not necessarily have a specific algal type. For example, *Lichina confinis* has one *Calothrix* species when found in Norway or Spain, but another species is found in the Swedish lichen, *L. confinis.*

Lichens grow in a variety of habitats, and they attract attention because of the adverse conditions under which they are often found. The few marine forms are found in the spray zone. Other lichens grow on soils, buildings, rocks, or the bark of trees, and also pioneer in the extreme cold of the Antarctic. The algal and fungal partners are often easily separated and grown in culture, although individual species often grow slowly. However, attempts to resynthesize the lichen thallus have met with only minimal success. Such a synthesis has been accomplished a few times, but only the initial stages of development are reported.

Identification of the Algal Partner

Taylor has indicated that in spite of the research and reviews on symbiosis, little consideration is normally given to the phycobiont. This has no doubt hindered our progress toward understanding the subject. Identification is quite difficult, and at times impossible, unless the alga is freed from the host and cultured. When there is no flagellated stage, or components of the cell covering are not formed, it is not possible to arrive at the correct generic name. In an extreme case, as in the nudibranch, *Elysia,* the animal digests much of the algal cell structure and retains just the plastids. Inasmuch as *Elysia* feeds on the coenocytes *Bryopsis* and *Codium,* the chloroplasts can be easily freed from the remaining cell components. The plastids then function photosynthetically in the mollusc. Other nudibranches can have similar associations with these organisms, other green algae, or even certain red algae.

At present most taxonomists attempt to place the phycobiont in an established free-living genus, and do not recommend a new taxon for each phycobiont-host association. The dinoflagellate *Symbiodinium,* which probably has a worldwide distribution, can be found in a variety of hosts, mostly coelenterates.

For the alga the association provides a ready supply of carbon dioxide, some organic compounds to supplement photosynthesis, perhaps a needed growth factor, as well as protection from grazing. In addition, if the phycobiont were released along with the reproductive cells of the host, there would be a mechanism for dispersal.

There are disadvantages to a symbiotic relationship. These would include both a reduced light intensity and a different spectrum of inorganic nutrients and nutrient concentrations. The animal would use or tie up some nutrients. With obligate associations there is the problem of frequently reestablishing the partnership when there are offspring.

ALGAE IN HOT SPRINGS

In a number of areas throughout the world, algae are found in thermal springs or in waters that flow from these sites. Such springs are located in some western states of the United States, in Iceland, and in New Zealand. In the region surrounding hot springs there are considerable thermal and water density gradients. Water near the boiling point flows adjacent to rushing waters of mountain streams. Gradual mixing of waters of different temperatures and densities provides a variety of habitats. There is often a variation in water chemistry, for example, water with high sulfur content, because of the geologic conditions under which thermal springs have developed.

Brock points out that green algae are called thermophiles when growing above 35°C, whereas the term is applied to blue-green algae when the temperature is above 50°C. Perhaps it would be best to define a thermophile as an organism growing in and adapted to water warmer than the average for a particular region. Usually only green and blue-green algae have adapted to thermal conditions, but occasionally there is a member of another class. A hot springs diatom, *Achnanthes,* was capable of two doublings per day at 40°C. We can think of the maximum temperature for blue-greens as 75°C, and that for eukaryotic organisms as 55°C. There is no adequate explanation for this ability of prokaryotes to multiply and survive at these temperatures. Evidently their proteins and the mechanism for making protein are more resistant to high temperatures than are those of other forms. In some environments, such as shallow pools and the soil surface, on hot sunny days algae can be exposed to high temperatures daily, at least for a few hours.

Thermophilic blue-green algae such as *Mastigocladus, Oscillatoria* and *Synechococcus* reproduce by means of cell division and fragmentation. Very few thermophiles reproduce by means of spores. Many of these forms probably evolved from the margins of thermal areas where there is a relatively broad temperature range. If thermophiles die because of excessive heat, maintenance populations at the point of a temperature gradient, or in cool waters upstream, provide an inoculum for growth during another season.

Brief mention should be made of the effects of thermal discharges from industry on the biota of a lake or river. Some factories and power plants have discharged heated water into rivers for years, but until recently we have overlooked this practice. There were not many factories with thermal effluents, and the limited column of water, along with the considerable dilution factor, produced little temperature change. However, increasing numbers of atomic power generating plants will inevitably use much greater volumes of water for cooling effluents. In most cases we are only considering a temperature rise of a few degrees into the 20 to 30°C range, hardly producing a truly thermal environment.

Heated effluents from power plants do affect the biota. Some organisms cannot tolerate the temperature increase, and die, whereas others, such as the cyanophytes, often become dominant in the warmer months. In one lake study in which there was a heated discharge from industry the growing season was extended, not only lengthening the spring maximum of diatoms, but also projecting the fall *Stigeoclonium* growth into the winter months. We should approach these modifications of our environment with great caution, for the alteration of one parameter could possibly produce long-lasting and undesirable results in the ecosystem.

SNOW ALGAE

Organisms that grow at around 0°C, and are often found in and on ice, on snow, and in water near that temperature, are called cryophiles. When investigated in the laboratory, these organisms frequently have temperature optima a few degrees higher than the environment from which they were isolated. Perhaps, then, some are located only temporarily in the cold habitat.

The snow algae, most often found in alpine and arctic regions, receive a good deal of attention because of the spectacular nature of their growth. They may color a considerable area of snow red, yellow, brown, or green. The same organism can be responsible for different colors because of the ability to produce a broad spectrum of pigments. Green color would probably indicate good growth conditions with an adequate supply of nutrients. The remaining colors result when secondary pigments are predominant, indicating that resting cells or spores are dominant.

Most often the snow alga is identified as a chlorophyte, for example, *Chlamydomonas nivalis,* but mistakes have been made when investigators relied only on observations of vegetative cells. At times culturing is a necessity. Other green algal genera (*Raphidonema, Cylindrocystis*), chrysophytes (*Chromulina*), and cryptophytes (*Cryptomonas*) are known in this environment. (One green alga reported from red snow was incorrectly identified as a euglenoid.)

It was once felt that the snow flora was similar throughout the world. Now we know that we must learn how to correctly identify all forms. Currently we recognize that there are members of at least a dozen genera in several classes.

The red snow coloration is most intensive and extensive when air temperatures are above freezing. Water droplets in the snow can then habor populations of these organisms. Hoham feels that true snow algae have optimum growth at a temperature lower than 10°C and will not grow above 10°C. He would conclude that certain genera isolated from red snow cannot be true snow algae, but are air borne algae that have recently landed in the snow.

Some cryophiles occur in ice, probably being trapped there as the ice formed. Other forms in lakes grow on the lower ice surface, and may temporarily produce a dense population. The algae in the water below the ice are located in an environment that is not much warmer. Similar observations have been made under Arctic and Antarctic ice.

SOIL ALGAE

Algae from a number of classes inhabit the soil, living both on and in the soil itself. Included in this group are the diatoms, green and red algae, blue-greens, euglenoids, xanthophytes, and perhaps an occasional member of another class. The study of soil agae is not a recent development, for there are excellent papers dealing with many unicellular and colonial forms dating back into the last century.

Many organisms may be found growing actively on the soil surface, or in the top millimeter. This is the most conspicuous flora, with organisms frequently producing a very visible bright green, blue-green, red, or even blackish film. These populations are as dense as can be collected in nature. Individual organisms are easily observed and identified under the microscope. On wooded trails in damp areas, many hikers have slipped on blue-green algae. Along the bank of a stream, especially in a protected area with a tree canopy, there is often a dense algal mat. In this environment the organisms do not receive the direct rays of the sun. Although protected from desiccation, there is a reduced light intensity and a filtering of wavelengths at both ends of the visible spectrum. Thus, growth is slow and there is selection for organisms able to survive in such an environment.

The algae growing in the soil are less conspicuous, or not visible, and may be missed by the collector. When examining a sample under the microscope, few organisms can be found, and it is often impossible to identify them. Cell shape can be distorted, an abundance of reserve products might completely obscure the plastid, and it might be impossible to determine where a motile stage can be produced by the organisms being observed. Most researchers in this area rely on extensive studies of these organisms in culture.

A small quantity of soil, usually but not always measured, is incubated in a medium and allowed to grow for several weeks. The culture is maintained under standard conditions, that is, only one defined medium is used, the temperature is held constant (at about 22°C), and both intensity and periodicity of light are controlled.

Few exhaustive studies of one soil have been carried out. Seldom has there been any attempt to determine the number of organisms present, or to gain knowledge of the complete flora. Such studies would be most time consuming. Nevertheless several hundred algal species have been isolated from soils throughout the world.

In their laboratories at the University of Texas, Professor Bold and his students have made excellent contributions, isolating into pure culture and describing many new genera and species from several classes of algae.

What is the size of the population of algae in the soil? How are organisms distributed horizontally and vertically? What is their role? In a few cases we know that there can be as many as 10^6 cells per gram of soil, and that soil algae occur to a depth of at least 1 meter. The numbers and distribution pattern depend on the type and composition of the soil. There are few investigations dealing with these questions.

What are these potentially photosynthetic organisms doing in the soil? How can they survive below the level of light penetration? Some are merely suriving for a period of time, having originated at the soil surface, or they may be growing heterotrophically. There are only limited data to support the concept of heterotrophic growth of soil algae, at least sufficient growth to enable the population to increase in size. With soils collected and dried for later study, or those taken from roots of vascular plants in herbaria, we know that algae can survive for several decades. We do not know how organisms in soil, subject to variations in both temperature and available soil moisture, are able to survive nor how many can be rejuvenated.

Most organisms that grow in the soil resist desiccation, and in the laboratory survive periods without much moisture. In the dry state they are not killed when heated to over 100°C for an hour; some can be grown after being heated to 130°C for an hour.

Many of the organisms isolated from soils are unicellular and packet-forming green algae (Chapter 4). In some investigations of soil algae these organisms are studied exclusively, not only because they are important in many soils, but because they are quite easy to grow, especially under the standard conditions usually employed. The isolation media would be considered quite concentrated when the levels of nutrients are compared with those normally found in the soil. In such media the green algae thrive! Are we getting a true picture of the soil flora using these growth conditions? Certainly the picture is not complete. In the future we may find that diatoms, or other organisms, are more important in soil than we were led to believe.

AIRBORNE ALGAE

For well over a century we have known that microorganisms are transported in the atmosphere. Proof did not come until the late 1800s, and much of the work at that time dealt with bacteria and fungi. It is only in the last 40 years that investigators have attempted to look at the algae themselves and to determine the numbers and kinds of organisms that can be found in the atmosphere.

Many of the forms probably originate from aquatic environments where the spray action at wave surfaces provides an aerosol, which is then easily transported into the atmosphere. Other forms can and do originate from soil surfaces. Organisms are plentiful in soils, at least in the upper millimeters. As the soil dries and becomes powdery, wind action can place small particles, with their accompanying microorganisms, into the atmosphere. Early studies with other microorganisms did not deal with algae, because they did not appear under the conditions used for culturing. With algal culture media and modern techniques, investigators have isolated members of several algal classes, including chrysophytes, cyanophytes, xanthophytes, diatoms and euglenoids. It would not be surprising to find members of other algal classes.

Airborne algae can be collected in a number of ways. The simplest technique is to expose a culture vessel to the atmosphere for a period of time. This would allow particles to settle into or on the substrate used for growth. Rods coated with silicon grease have been exposed to air. The silicon surface can then be streaked on agar (Chapter 21), or a portion of the rod is incubated in liquid medium. One way to obtain quantitative information involves pumping an air sample through a filter. The number of viable cells in a known volume of air can then be determined.

A simple technique has been devised for sampling airborne algae that have their origin in the soil. Petri dishes are exposed to the atmosphere during a period of high wind. To increase the possibilities of obtaining airborne algae, the exposed dishes are inoculated from the window of a moving vehicle. If these collections are obtained close to the ground, and within a short distance from the exposed soil, the procedure is probably just another method for collecting soil algae. However, dishes have been exposed on aircraft flying at known altitudes.

Most of the airborne forms are small unicells or colonies. With direct observations of freshly collected material under the microscope, it is frequently impossible to identify these algae. Thus the samples collected are incubated in culture media. After several days, or even weeks, there are populations of organisms, and with periodic observations most of the stages in the life history can be detected. Identification is then possible.

The airborne algae provide a mechanism for dispersal of both soil and aquatic organisms. A particular species recently introduced into an area can soon be transported to other habitats, at least in the direction of prevailing winds.

Some attention has been paid to the role of airborne algae in allergic reactions, especially respiratory problems. Although probably of minor significance when we consider the host of other possibilities, these microorganisms must not be overlooked in determining the cause of certain inhalant allergies.

SUBAERIAL ALGAE

Because they are present in the air, algae can be dispersed over a wide area. Some cells may then land on almost *any surface,* and if there are favorable conditions for a few days, especially a little moisture, growth can occur. In damp areas, such as near the ocean, in protected areas, or in the tropical rain forests, such algal populations are called subaerial algae. Extensive floras have been reported from the leaves, bark, and flowers of tropical trees.

FOSSIL ALGAE

The best known fossil algae are among the microalgae, including blue-green algae, diatoms, and coccolithophores. The blue-green algae are among the most ancient, known for at least 2 billion, and possibly even 3 billion, years. Coccoliths, scales from some haptophycean genera (Chapter 11), dinoflagellates, and diatoms are more recent, no older than 200 million years old. Included with these common forms is an occasional reproductive cell of a chrysophyte, or the skeleton of a silicoflagellate. Some fossils are remarkably similar to present-day genera and are often described as species in these extant genera (Table 15-2).

The best fossils of macroalgae are found in those groups of algae on which there are lime deposits. Fleshy genera lacking this firm outer coating, and microalgae lacking firm cell coverings, preserve poorly if at all. As a result they are almost entirely missing from the fossil record. A number of calcareous forms were once described as animal forms, for example, the nullipores, but are now known to be coralline red algae. Calcareous macroalgae are usually charophytes, coenocytic green algae, or red algae. Only occasionally will there be a calcareous macroalga in another class.

In most algal groups there are surprisingly few fossils known. When they are found, there are suddenly many genera of one class (Table 15-2). Preservation of earlier forms, which would aid in determining the

Table 15-2. DATES OF EARLY FOSSIL RECORDS IN VARIOUS ALGAL GROUPS

Algal Group	Geologic Time[a]
cyanophytes	Precambrian
chlorophytes	Ordovician, possibly Cambrian
rhodophytes	Cretaceous
phaeophytes	Silurian
diatoms	Jurassic
dinoflagellates	Jurassic
haptophytes	Jurassic

[a] In some cases there are earlier reports, but records are poor, and it is often difficult to place accurately some fossil material. Organisms are then tentatively placed in one class.

origin of a particular group, was apparently not good. After considering the facts that some genera are not at all unlike present-day genera, that reproductive parts often do not preserve, and that the record is so incomplete, it is obvious that the fossil record does not offer much assistance when we attempt to determine relationships among groups of algae. Considerable research, and new discoveries, might improve our present understanding of this subject.

REFERENCES

Ahmadjian, V. 1967. *The Lichen Symbiosis*. Blaisdell Publ. Co., Waltham, Mass. 152 p.

Apollonio, S. 1961. The chlorophyll content of arctic sea ice. *Arctic* 14: 197–9.

Barghoorn, E. and J. Schopf 1966. Microorganisms three billion years old from the Precambrian of South Africa. *Science* 152: 758–63.

Barghoorn, E. and S. Tyler 1965. Microorganisms from the Gunflint chert. *Science* 147: 563–77.

Bold, H. 1970. Some aspects of the taxonomy of soil algae. *Ann. N. Y. Acad. Sci.* 175: 601–16.

Brock, T. 1967. Life at high temperatures. *Science* 158: 1012–9.

Brook, A. 1968. The discoloration of roofs in the United States and Canada by algae. *J. Phycol.* 4: 250.

Brown, R., D. Larson, and H. Bold 1964. Airborne algae: their abundance and heterogeneity. *Science* 143: 583–5.

Bunt, J. and C. Lee 1970. Seasonal primary production in antarctic sea ice at McMurdo Sound in 1967. *J. Mar. Res.* 28: 304–20.

Cameron, R., H. Conrow, D. Gensel, G. Lacy, and F. Morelli 1971. Surface distribution of microorganisms in Antarctic Dry Valley soils: a martian analog. *Antarctic J. U. S.* 6: 211–13.

Engel, A., B. Nagy, L. Nagy, C. Engel, G. Kremp, and C. Drew 1968. Algal-like forms in Onverwacht series, South Africa: Oldest recognized lifelike forms on earth. *Science* 161: 1005–8.

Hoek, C. van den 1975. Phytogeographic provinces along the coasts of the northern Atlantic Ocean. *Phycologia* 14: 317–30.

Hoham, R. 1975. Optimum temperatures and temperature ranges for growth of snow algae. *Arct. Alp. Res.* 7: 13–24.

Jackson, J. and R. Castenholz. 1975. Fidelity of thermophilic blue green algae to hot spring habitats. *Limnol. Oceanogr.* 20: 305–22.

Mitchell, R. 1974. The evolution of thermophily in hot srpings. *Quart. Rev. Biol.* 49: 229–42.

Norris, R. 1967. Algal consortisms in marine plankton. In: V. Krishnamurthy, (Ed.) *Seminar on the Sea, Salt and Plants*. Central Salt and Marine Chemicals Res. Inst., India. pp. 178–89.

Schlichting, H. 1964. Meterological conditions affecting the dispersal of airborne algae and protozoa. *Lloydia* 27: 64–78.

Schlichting, H. 1975. Some subaerial algae from Ireland. *Br. Phycol. J.* 10: 257–61.

Schlichting, H. 1974. Ejection of microalgae into air via bursting bubbles. *J. Allergy Clin. Immunol.* 53: 185–8.

Stein, J. 1967. Studies on snow algae and fungi from the front range of Colorado. *Can J. Bot.* 45: 2003–45.

Taylor, D. 1973. Algal symbionts of invertebrates. *Ann. Rev. Microbiol.* 27: 171–87.

Classification and morphological variability

CLASSIFICATION

Thus far the emphasis has been on organisms as well as the classes in which they are placed. Another highly useful approach to the study of algae might include not only these categories but also a complete classification of algae discussed. In such a study one would list the divisions or phyla, orders, families, and species, as well as the characteristics of each of these categories. However, the beginning student might not have the background for a thorough appreciation and evaluation of these categories until she (he) is familiar with a large number of organisms. Comparisons are then more easily made.

Kingdoms Of Organisms

Some time ago *all* organisms were usually placed in either the plant or animal kingdom. With greater knowledge of the cell structure of simple forms, it became apparent that a third kingdom could be easily justified. This third kingdom would include the bacteria and blue-green algae—the prokaryotic forms. More recently, there have been suggestions that organisms are more naturally grouped into four, five or even eight kingdoms (Fig. 16-1). In the four-kingdom proposal, the Protista would be added. Plants would then be defined as those organisms containing the higher plant or chloroplast pigments, animals would include only those forms with blastula and gastrula stages, and thus many *simple* eukaryotic forms would be available for the kingdom Protista. In the five-kingdom proposal there would be a kingdom of fungi, and some additional modifications of the previous four kingdoms. Edwards has recently suggested that there are eight kingdoms: prokaryotes, three kingdoms of eukaryotic, photosynthetic plants, three of fungi, and a kingdom for animals. He emphasizes the evolution of pigments, a very conservative character.

Ways of Classifying Algae

In all these approaches the blue-green algae are not only separated from the remaining groups of algae, but are allied with the bacteria. Although the fossil record is poor for many algae, it now appears clear that the bacteria and blue-green algae were well established 10^9 years ago. Satterthwait also proposes that in the Precambrian the red algae, green algae, and brown algae were well established. With a poor fossil record, most of the theory regarding classification of the algae comes from knowledge of various structures and methods of reproduction in extant organisms. Some emphasis has recently been placed on certain ultrastructural detail.

The task of providing a unified picture of the algae is complicated by the fact that they are really a very diverse group, ranging from prokaryotic to eukaryotic forms, unicells of just a few microns to macroscipic forms over 50 m in length, as well as both flagellated organisms and attached forms. Although most phycologists would characterize true algae as photosynethetic, some variety in food uptake is also possible, either with or without photosynthesis.

Often included with the algae are some "plants" or "animals" that do not easily fit our rigid descriptions of just what plants and animals should be. These organisms would probably be called protists in one of the proposals in Fig. 16-1.

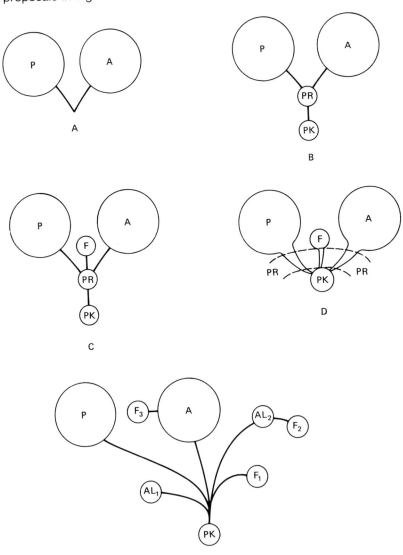

Two traditional approaches to the classification of any group could be called monophyletic or polyphyletic, or, in the loose sense, "lumpers or splitters" at the division or phylum level. The monophyletic approach holds that there was one independent origin of the algae, whereas a polyphyletic approach would consider few to many possible ancestors. A number of important points are to be considered in attempting to choose between these two approaches. In addition to concern about the evolutionary relationships of these organisms, an approach complicated by the paucity of the fossil record, algae are often classified merely for convenience. Thus the investigator might choose to assemble various categories so that an orderly study and cataloging of data could be achieved without concern about origins. Scientists might rely on quite different characteristics in these approaches to classification.

Characteristics Used at the Division or Class Level

For some time there has been general agreement that the major groups of algae could be separated by examining pigment composition, flagellated stage present, if any, and the chemistry of the food reserve. Certain nuclear details and wall structure are sometimes very important. Ultrastructural information can, on occasion, be essential.

In this book I have used this information to indicate the various classes, but have not yet indicated how these classes might be interrelated. The classes and common names applied to each are listed in Table 16-1. In the past these classes of algae were assembled into various proposed classifications. Some of these are presented in Table 16-2. As a student in an introductory course not all names would necessarily be familiar to you. Note that the list of classes, and to a lesser degree the list of divisions, tend to increase with time. This indicates not only a tendency toward polyphyletic approaches, but also

Figure 16-1. Evolution of the various kingdoms of organisms.
Key: P - Plant
    A - Animal
    PR - Protist
    PK - Prokaryotes
    F - Fungi, separated into three kingdoms in E.
    AL - Algae, separated into various kingdoms in E.
A. Two kingdoms.
B. Four kingdoms. The prokaryotic forms are recognized in this and the remaining diagrams.
C and D. Five kingdoms. In four the protists include organisms primitive to plants, fungi and animals.
E. An eight kingdom approach, recognizing three groups of fungi, and dividing the algae into categories based on their pigmentation. P would include the green algae as well as land plants.

**Table 16-1. THE CLASSES OF ALGAE AS TREATED IN THIS
             BOOK, AND THEIR COMMON NAMES**

Class	Common Name
Cyanophyceae	blue-green algae
Chlorophyceae	green algae[a]
Euglenophyceae	euglenoids
Rhodophyceae	red algae
Phaeophyceae	brown algae
Bacillariophyceae	diatoms
Xanthophyceae	yellow-green algae
Chrysophyceae	golden algae
Haptophyceae	haptophytes
Eustigmatophyceae	eustigmatophytes
Dinophyceae	dinoflagellates
Cryptophyceae	cryptomonads

[a] In the future the green algae will be separated into Chlorophyceae and Charophyceae [sensu Stewart and Mattox, Chapter 4]. The classes are placed in divisions in Table 16–3.

our increased knowledge of many organisms. The new class Eustigmatophyceae, based on a number of features seen with the electron miscroscope, was described after the schemes listed in Table 16-2 were proposed.

Discussion

Smith suggested a system that is now widely used, but for present times is perhaps too conservative. Chapman felt that the euglenoids were really not algae, but should be grouped with colorless flagellates. In addition, he consolidated several of Smith's divisions into one division, the Euphycophyta. He based this decision on biochemical and morphological grounds, including the absence of many simple forms in the red and brown algae. Thus the evolution from simple cell to filament would occur as in a simple chlorophytes, and the red, brown, and higher green algae, including stoneworts, would evolve from such a filamentous form. However, in the second edition (1973) authored by Chapman and Chapman the euglenoids are included with the algae. The classification employed by Chapman and Chapman is similar to that of Prescott (Table 16-2). The three classes within the Chrysophyta of Prescott are raised to the level of division.

Chapman and Chapman place the haptophytes with the new, more restricted, group of chrysophytes.

TABLE 16-2. THE CLASSIFICATION OF THE ALGAE BY VARIOUS AUTHORITIES

Smith (1950)	Chapman (1962)	Silva (1962)	Prescott (1964)	Scagel et al. (1965)	Christensen (1962)

Cyanophyta
  Myxophyceae
Chlorophyta
  Chlorophyceae
  Charophyceae
Phaeophyta
  Phaeophyceae
Rhodophyta
  Rhodophyceae
Euglenophyta
  Euglenophyceae
Pyrrophyta
  Desmokontae
  Dinophyceae
Chrysophyta
  Chrysophyceae
  Bacillariophyceae
  Xanthophyceae
	Myxophycophyta
	  Cyanophyceae
	Euphycophyta
	  Chlorophyceae
	  Charophyceae
	  Phaeophyceae
	  Rhodophyceae
	Pyrrophycophyta
	  Cryptophyceae
	  Dinophyceae
	Chrysophycophyta
	  Chrysophyceae
	  Xanthophyceae
	  Bacillariophyceae
		Cyanophyta
		  Cyanophyceae
		Chlorophyta
		  Chlorophyceae
		Phaeophyta
		  Phaeophyceae
		Rhodophyta
		  Bangiophyceae
		  Florideophyceae
		Cryptophyta
		  Cryptophyceae
		Pyrrophyta
		  Desmophyceae
		  Dinophyceae
		Euglenophyta
		  Euglenophyceae
		Chrysophyta
		  Chrysophyceae
		Xanthophyta
		  Xanthophyceae
		Bacillariophyta
		  Centra-
		  bacillariophyceae
		  Pennatibacillario-
		  phyceae
			Cyanophyta
			  Subphylum
			  Coccogoneae
			  Subphylum
			  Hormogoneae
			Chlorophyta
			  Chlorophyceae
			  Charophyceae
			Phaeophyta
			  Isogeneratae
			  Heterogeneratae
			  Cyclosporeae
			Rhodophyta
			  Subphylum
			  Bangioideae
			  Subphylum Florideae
			Euglenophyta
			  Euglenophyceae
			Pyrrhophyta
			  Desmophyceae
			  Dinophyceae
			Cryptophyta
			  Chloromonadophyta
			Chrysophyta
			  Chrysophyceae
			  Bacillariophyceae
			  Xanthophyceae
				Cyanophyta
				  Cyanophyceae
				Chlorophyta
				  Chlorophyceae
				  Charophyceae
				Phaeophyta
				  Phaeophyceae
				Rhodophyta
				  Rhodophyceae
				Euglenophyta
				  Euglenophyceae
				Chrysophyta
				  Chrysophyceae
				  Bacillariophyceae
				Xanthophyta
				  Xanthophyceae
				  Chloromonadophyceae
				Pyrrophyta
				  Dinophyceae
				  Cryptophyceae
					Cyanophyta
					  Cyanophyceae
					Rhodophyta
					  Rhodophyceae
					Chlorophyta
					  Chlorophyceae
					  Euglenophyceae
					  Prasinophyceae
					  Loxophyceae
					Chromophyta
					  Phaeophyceae
					  Dinophyceae
					  Cryptophyceae
					  Haptophyceae
					  Raphidophyceae
					  Chrysophyceae
					  Xanthophyceae
					  Bacillariophyceae

Silva presented his classification in a reference volume as a framework for discussion of the physiology and biochemistry of the algae. He acknowledged that it is incomplete, but sufficient knowledge of the Cryptophyta had accumulated, especially the presence of phycobilins, for Silva to suggest that division. In addition, no longer were the Chrysophyta of Smith separated into classes for the diatoms, yellow-green, and golden-brown algae. (Note that Chrysophyta is used in different ways by Smith and Silva. In addition, Chapman and Chapman, Scagel et al., and Christensen suggest changes in the old [sensu Smith] division Chrysophyta.)

The approach used by Christensen is quite interesting (Table 16-2). First he separated the blue-greens from the remaining algae because they are prokaryotic. In the remaining eukaryotic forms the investigator can treat the red algae individually because they do not have, perhaps never had, a flagellated stage. In the remaining groups of algae, many forms are flagellated, many have a flagellated reproductive cell, while some have lost all motion by flagella. By emphasizing pigmentation, the remaining algae can be placed in the Chlorophyta (possessing chlorophyll b), or the Chromophyta (lacking chlorophyll b). At the same time he recognized that there are problems with such groups as the yellow-green algae, the dinoflagellates, and the euglenoids. A look at other pigments, food reserve, as well as nuclear cytology makes the dichotomy into chlorophytes and chromophytes less clear.

In a weak moment one might be tempted to present yet another system for classifying the algae. Rather than really solve the real problems of algal classification, the situation might be made more confusing if a new system were proposed hastily. As Dodge noted (1973): "Almost every new algal text book that is published presents a revised system for the classification of these organisms. This has made for a situation, which must be highly confusing to the student and which makes teaching extremely difficult. In a work such as this it is essential to have some 'labels' at a higher level than that of the genus yet at the same time it is probably better to use a neutral classification scheme. Thus the largest units employed here are classes. . . which . . . are the only stable feature which has persisted through the numerous classification schemes of the past few years."

Lewin (1973) pointed out that "the field is in such a state of flux that even the major subdivisions are a source of debate and are subject to frequent rearrangements."

But is it not possible to tidy up one of the schemes already in existence? For the benefit of those who might wish to have division labels for organisms discussed in this book, I have indicated how the system of Christensen can be brought up to date. The new class Eustigmatophyceae would be placed in the Chromophyta. The cryptophytes might be considered in their own division (Chapters 1 and 14). With these changes, along with elimination of classes not discussed in this book, the modified Christensen scheme is presented in Table 16-3. Because there is such emphasis on the evolution of the phycobilins and the various chlorophyll pigments, which are thought to be phylogenetically very conservative characters, some problems remain with this classification of the algae. For example, in the Cryptophyta, separated from other divisions because the cryptophytes are the only flagellated organisms with phycobilins, there is also chlorophyll *c*. Accordingly, there would have to be two independent lines with the latter pigment. Including the cryptophytes in the Chromophyta raises the problem of explaining the origin of phycobilins in a very large group of

Table 16-3. A MODIFICATION OF THE CLASSIFICATION OF CHRISTENSEN

Divisions	Classes
Cyanophyta	
	Cyanophyceae
Rhodophyta	
	Rhodophyceae
Cryptophyta	
	Cryptophyceae
Chromophyta	
	Phaeophyceae
	Bacillariophyceae
	Xanthophyceae
	Chrysophyceae
	Haptophyceae
	Eustigmatophyceae
	Dinophyceae
Chlorophyta	
	Chlorophyceae
	Charophyceae
	Euglenophyceae

organisms that otherwise do not have these pigments. Perhaps these organisms could be included in the Chromophyta, if plastid origin is by way of invasion of eukaryotic cells by pigmented prokaryotes (see below). Another difficulty is the classification of chlorophytes and euglenophytes in the Chlorophyta. With the exception of pigment composition and plastid ultrastructure, these organisms have little in common, for example, different food reserves, cell coverings, and plant body types.

Endosymbionts

In the discussion of the cyanophytes (Chapter 3) the possible origin of plastids of all groups of algae through the invasion of a number of different eukaryotic host cells was proposed. The invaders would have been simple prokaryotic cells similar in structure to present-day blue-green algae, but with a variety of pigments, especially the chlorophylls. In a general way this theory could be dovetailed with the classification of Christensen, or even with recent modifications of that approach.

The most primitive cell type, a prokaryote, would have lacked pigments. One line could then become eukaryotic, giving rise to nonphotosynthetic organisms, including all animals, as well as the host cells for other algal groups. A second line, the first algae, would have been prokaryotic forms that had evolved chlorophyll *a*. The blue-green algae as we know them today would not have been the most primitive of the algae, but would have developed from the primitive algae. The important step would be the evolution of the phycobilins (Fig. 16-2). Free-living forms with the latter pigments would be present-day blue-

Figure 16-2. Diagrammatic presentation of the evolution of the divisions of algae. The primitive cell evolved along two major lines. In one case membranes were developed around portions of the cell, forming organelles; this line developed into eukaryotic organisms. Motility by means of flagella soon evolved in this group. Certain of these eukaryotic cells, both with and without motility, served as *host* cells for the invasion of prokaryotic cells (see below).

The second line that developed from the primitive prokaryote first developed chlorophyll *a*, and later other pigments. Those forms with the phycobilins invaded two different eukaryotic cells, one with and one lacking motility. These combinations gave us the cryptophytes and the rhodophytes. The primitive prokaryotes, which had evolved both chlorophyll *a*, the universal chlorophyll, and the phycobilins *but* remained free-living, are the blue green algae of today. Then the development of chlorophyll *c* by some primitive types produced a cell type that later invaded a motile eukaryote, forming the chromophytes.

The final invasion of a eukaryote by a cell with chlorophyll *b* resulted in the formation of the Chlorophyta. The higher plants evolved from the charophyte branch [sensu Stewart and Mattox] of the Chlorophyta, or from closely related forms.

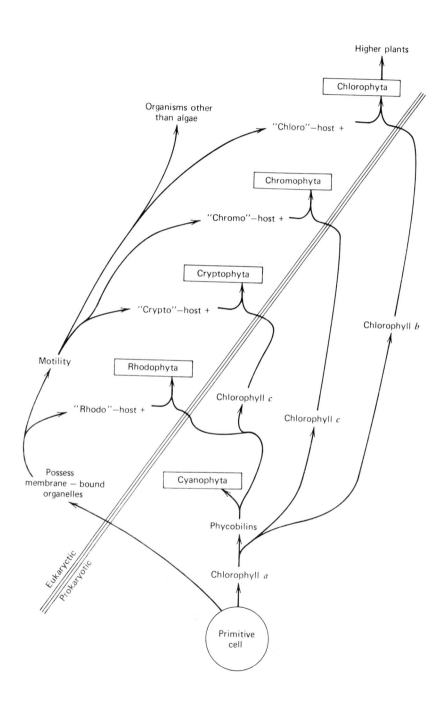

greens, but those that invaded a nonmotile eukaryotic cell would have produced the present-day rhodophytes.

A second invasion of another eukaryotic cell, this time a flagellated cell, would lead to the small group of cryptophytes. However, one must also have the invading prokaryote evolve chlorophyll *c,* for the cryptophytes are the only algal group with *both* phycobilins *and* an accessory chlorophyll c.

From the first algae, or those with chlorophyll *a* but lacking phycobilins, one line would have evolved chlorophyll *b,* and another group would separately evolve chlorophyll *c.* If these prokaryotes independently invaded flagellated eukaryotic cells (Fig. 16-2), both the Chlorophyta and the Chromophyta could have developed. Once the plastid type was formed with the appropriate pigment composition and thylakoid ultrastructure, then the various classes within each division evolved.

If in the scheme proposed in Fig. 16-2 the development of chlorophyll *c* in the two separate lines proves troublesome, then an *expanded* division Chromophyta might be proposed. This division would have the cryptophyte line possessing phycobilins and another line, one including most chromophytes. The latter group would have lost the phycobilins very soon after establishing a permanent relationship between the host cell and symbiont (Fig. 16-2). In this approach it might also be easier to propose that the chlorophyte ancestors at one time had phycobilins, and that no organisms with those pigments and chlorophylls *a* and *b* are presently free-living. The recent discovery by Lewin of a free-living "cyanophyte" lacking phycobilins, but presumably with an additional chlorophyll, perhaps chlorophyll *b,* would not support this approach. Would then one want to have loss of the phycobilins only in one line of evolution in the algae? And thus is the fun of developing phylogenetic schemes.

Ultrastructure and Classification

As noted in Chapter 12, many cytological features seen only with the electron microscope have been used to erect the class Eustigmatophyceae. In addition, we have accumulated sufficient structural detail in some classes that one might begin to use this EM information to show some phylogenetic relationships. Dodge has made some such suggestions which he feels are certainly preliminary at this point. Initially, he proposed that the cyanophytes be removed from the rest of the algae because they lack membrane-bound organelles. The remaining algae are eukaryotic. Among the diverse assemblage that comprises the eukaryotic algae, Dodge could not use any *combination* of ultrastructural features to place any one class of the *most primitive.* However, if some emphasis is placed on the mesokaryotic nucleus, a nucleus in which chromosomes are condensed during most of the nuclear cycle, the Dinophyceae could be considered in that position.

Dodge then looked at the ultrastructural information on eyespots and chloroplasts, as well as that available on flagella cytology and morphology. At the same time he compared data on flagella, cell covering, nuclear structure, and ejectile organisms (i.e., trichocysts) in an attempt to detect relationships among the various classes. He could make the following generalizations:

1. Most of the classes placed in the Chromophyta appear to cluster. Diatoms, with few flagellated cells and glass walls, are often isolated from other chromophytes.
2. The dinoflagellates, cryptophytes, eustigmatophytes, and euglenoids are difficult to place.
3. The rhodophytes and the chlorophytes, in this sense including the charophytes, are independently isolated from other algal classes. The isolation of the Chlorophyceae from Euglenophyceae is apparent in Dodge's schemes.
4. It is difficult to place the Haptophyceae, perhaps indicating that we need more investigatons of species found in that class.

In summary Dodge shows that certain ultrastructural information can be used to support a classification based on other factors, but at the same time the positioning of certain classes as shown in Table 16-3 cannot be based solely on the available ultrastructural information. We can certainly use more data not only from EM studies, but also using physiological, biochemical, and morphological characteristics.

Cell Division Phenomena.   As pointed out with the filamentous green algae (Chapter 4), ultrastructural investigations of the cytological features of both cell division and cytoskeletons of zoospores have shown us that there are really two major groups of green algae. The most exciting information has come from observations of nuclear and cytoplasmic division, including the movement of nuclei after division, presence of microtubules at the time of cytokinesis, and cell wall formation (Chapter 4).

Thus many primitive, green algae, including *Volvox* and relatives, as well as many relatives of *Ulothrix,* are not on the line of evolution toward higher plants. *Stichococcus, Klebsormidium, Spirogyra,* and allies, and the stoneworts would be closer to the possible ancestors of higher plants, if one were to emphasize nuclear and cytokinetic phenomena. Recent investigations on the presence of glycolate oxidase (in *Stichococcus* and allies) or its absence (from the *Volvox* group) add further support to such a theory. Such cytological information found at both the light and electron miscroscope levels *could* be important only in the green algae, but this is unlikely. What will similar intensive investigations in other algal classes find?

Biochemical Taxonomy.   Thus far in this chapter the emphasis has been placed on the higher levels of classification. When we examined the various algal classes in earlier chapters, three of the five characteristics normally used were at least partially biochemical features. These included pigmentation, food reserve, and wall composition. We have abundant information on these features in most classes, with several recent review articles on these subjects.

Throughout this book I have indicated how the various genera may be distinguished. Almost without exception the features used deal with morphology or reproductive phenomena. Is there another way, or are there features that might be used in combination with morphology? For certain small groups of organisms physiological or biochemical features have been found useful. For example, Shihira and Krauss found the responses of various strains of *Chlorella* to organic substrates, along with other data, sufficient to distinguish several species and strains. A combination of biochemical and morphological features have been used for laboratory cultures of other chlorophytes and cyanophytes.

Recently Lewin reviewed many of the accomplishments and problems in biochemical taxonomy. In addition to features already covered, he discussed the literature dealing with carbohydrates, lipids, proteins, and nucleic acids, pointing out the numerous uncertainties at the level of both genus and class. There is considerable promise to research in such a broad and fundamental area as biochemical taxonomy.

MORPHOLOGICAL VARIABILITY

In our treatment of the algae it is obvious that a great deal of emphasis is placed on what we see when we encounter an organism. We thus rely heavily on the morphology of both the microalgae and the macroalgae when we classify them, even though there have been recent advances in the use of physiological or biochemical attributes with microalgal taxonomy. In order to have a morphological system work, it is essential that there be stability of form, or at least little form variation. Although it appears that this is true of most organisms, there are now sufficient well-documented cases of polymorphism, or morphological variability, in the algae, that careful consideration must be given to the use of any character in the classification of the algae.

Related to this area is the study of abnormal form, or teratology. At times teratological studies are intermingled with good examples of morphological variability. For example, under stress a number of organisms will alter their cell structure or their gross form. Unicells can become abnormally large, or develop irregular shapes (Fig. 16-3), and macroscopic forms can become so warped and entangled that it is difficult, if not impossible, to make out the gross form. At the turn of the century, workers who were learning how to grow algae subjected some of the microalgae to the stress of excessive concentrations of inorganic nutrients, or even supplied sugars at levels of 10 to 20 g l^{-1}! More recently there have been papers dealing with diatoms showing pennate forms with a dent in the side (Fig. 16-3), or with the raphe system and ornamentation distorted (Fig. 16-3). These are teratological forms, and the study of this area, although of some interest in itself, will not be given further consideration here.

We will be concerned with variability in form, but of quite a different nature. We presently classify some organisms on the basis of the wall structure, and assume that organisms will always form that type of wall structure. For example, a diatom will be classified as a given species because it has a certain arrangement puncta (X). A related "species" will have a different pattern (Y). Traditionally we have considered them to be two species. When we encounter morphological variability, *one organism* is capable of forming pattern X under some conditions, and pattern Y at another time. The distinction here between this morphological variability and teratology is that there is nothing abnormal about the two forms. Both diatom forms were previously recognized as distinct species, and now we know that there is just one genetic entity, which is capable of producing two different patterns on the wall.

With the macroalgae a number of examples of morphological variability already have been presented. When we find that organisms formerly called *Ralfsia* or *Conchocelis* may actually be stages in the life histories of *Scytosiphon* and *Porphyra,* respectively, then we are certainly dealing with morphological plasticity. Inasmuch as we have given considerable attention to these benthic marine forms throughout the book, here we will emphasize the microalgae.

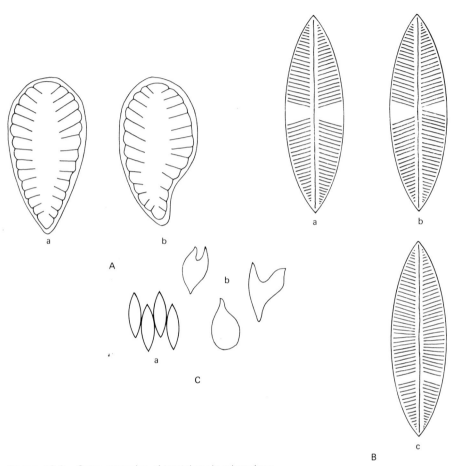

Figure 16-3.   Some examples of teratology in microalgae.
A. *Surirella* (a diatom) colony with its normal shape (a), and a colony with a dent in the side (b).
B. *Stauroneis* colony (a) with the normal shape. The central area is either broken by an additional stria or two (b), or it is displaced (c).
C. Typical colony of *Scenedesmus obliquus* (a). Abnormal cells are shown in b.

History

Although morphological variability in algae was reported more than a century ago, the controversial works of Chodat with unicellular and colonial algae provided the basis for most discussions of this subject in the early 1900s. At that time few investigators used unialgal cultures and those that were free of bacteria were even more rare. Certainly, laboratory media were imperfect by present standards, and in order to support luxuriant growth high concentrations of some chemicals were employed.

Because of the culture conditions he employed, a great deal of Chodat's work on morphological variability was held suspect. It was argued that the various forms produced were probably abnormal and that there was little or no consideration of the conditions under which organisms grew in nature. One might also argue, and in some cases justifiably, that some examples of morphological variation were the result of poor culture technique, for example, the use of bialgal cultures. If one is attempting to prove for the first time that an organism that looks like a *Franceia* (Fig. 16-4) can reproduce and then resemble a *Scendesmus* (Fig. 16-4), it is essential that one work with unialgal cultures or a clonal culture.

In the early 1900s Smith from the United States and Chodat in Switzerland had independently monographed the genus *Scenedesmus.* Chodat found numerous examples of polymorphism, with changes in shape, reduction in spine number, and formation of unicellular stages. On the other hand, Smith reported little variation in his cultures and in several attempts could not demonstrate a filamentous stage, similar to a *Dactylococcus,* in cultures of *S. obliquus.* Chodat and others had reported these chains of cells in several strains. Smith and Chodat were strongly critical of each other in several publications.

During the early part of this century the outstanding phycologist from Great Britain, Fritsch, offered no support for polymorphism. In his work on the green microalgae he discussed diversity of shape of organisms, but felt that the unicellular stages of *Coelastrum* and *Scenedusmus* were abnormal forms. In addition, he thought that pure culture studies were quite valuable in that one could examine cultures repeatedly, but were unreliable for the study of normal form variation of the algae. In addition, he thought that they might even give a distorted view of the life history of the organism. He also expressed some doubt that morphological variability could exist in nature. Later he became more tolerant of the use of cultures, but his many early publications had a marked influence on developing phycological thought.

However, during the same period, Pringsheim emphasized the importance of cultures in dealing with morphological studies. In his valuable work, *Pure Cultures of Algae,* he pointed out that we know far less about many common species than appears from a casual look at algal floras. The range of variability possible for an organism, due to

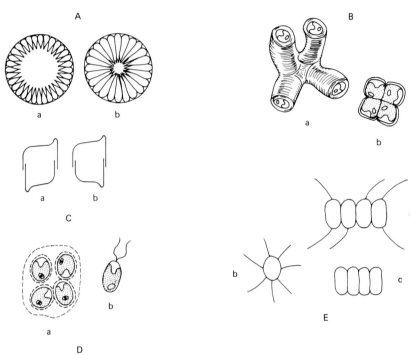

Figure 16-4. Some examples of morphological variability.
A. One clonal culture can produce individuals of *Cyclotella* that resemble form a or form b.
B. A culture of *Chlorosarcinopsis* can produce packets that resemble typical members of the genus (b), or *Hormotila*-like forms with abundant mucilage (a).
C. These are diagrammatic cross sections of a pennate diatom. The individual was observed dividing and some daughter cells had the keels opposite (a) and thus would be called *Nitzschia,* while others had keels on the same side (b) and thus resembled *Hantzschia.*
D. *Chlamydomonas* cells (b) have been produced as palmelloid colonies (a).
E. *Scenedesmus* culture 16 can form colonies with spines (a) or without spines (c), or unicells which resemble *Franceia* (b).

environmental influence, was certainly not known. That range would vary from organism to organism and could be determined in culture.

In a classic 1967 paper Pringsheim regretted the fact that there were so many misunderstandings between laboratory and field investigators. This problem was one of the main obstacles to an adequate understanding of polymorphism. In addition, he supported some early work on variability by stating that Chodat ''was impelled to apply bacteriological methods to small Chlorophyceae by the difficulty of distinguishing between species similar in appearance. His brave attempt has been much underestimated because his media were inadequate, often causing deformities. I think, like that eminent botanist, that cultures must supplement inspection of natural material for a true picture of the properties of the organisms.''

Laboratory and field workers do follow different paths and therefore have had difficulty in recognizing the extent and implications of morphological variability. Field workers often examine organisms at one point in time, especially when material is fixed in the field for later microscopic examination. If the same site is sampled a few days later and some ''different'' organisms are present, one can list the organisms as ''different'' rather than consider that the changing environment has brought about a morphological change. On the other hand, laboratory investigators can be criticized for certain of their methods of culture, such as use of unrealistically high concentrations of chemicals in media, and production of some truly teratological forms. The gap between laboratory and field studies then widens.

For 30 years prior to the 1960s morphological variability was not a popular subject for research. In the eyes of some, Chodat's work could be dismissed because of his culture methods and techniques. Fritsch and Smith had spoken out strongly against polymorphism, and field and laboratory phycologists continued to be critical of each other, instead of entertaining the possibility of morphological variability in nature.

With many organisms, a lack of understanding of generic and specific limits is now a severe handicap. Hasty abandonment of either the genus *Scenedesmus* or the name *Franceia* because both forms can be produced by one organism should be avoided. For we really do not know the limits of variation of most of these organisms and should eventually describe the *complete* organism. In addition, some *Scenedesmus* might never form a spiny unicellular stage. And do we really know that all *Franceias* have a colonial stage? For the present, use of old, established names indicates only morphological forms and not genetic entities. As long as it is so stated in a published article, investigators could continue to use established names in that way.

Current Studies

Recently there have been increased numbers of well-documented reports of morphological variability. In laboratory studies some organisms do not conform to established limits for genera, and older concepts are being altered. Bold discussed attempts to classify palmelloid green algae and some close relatives in which the occurrence of motility and the formation of gelatinous layers are used in diagnoses (Fig. 16-4). He stated: "The extent to which the external environment is responsible for evoking these diagnostic criteria has never been adequately investigated." Braarud felt that much could be learned from detailed studies of the effect of environmental factors on form. Since form variation could occur in many species of marine organisms, he deplored the absence of information on the actual effect of environmental factors on structure.

Green algae.   In the green algae there are numerous examples of polymorphism (Table 16-4; Fig. 16-4). We have this wealth of information probably because of ease of culturing, as well as the frequent field observations. Control factors are better documented with this group.

Table 16-4.   EXAMPLES OF MORPHOLOGICAL VARIABILITY IN CERTAIN UNICELLULAR, COLONIAL, AND FILAMENTOUS GREEN ALGAE[a]

| Organism Studied | Organism May Resemble | | Other Possibilities |
	Other Genera	Other Species in the Genus	
Ankistrodesmus	X		
Coelastrum		X	colony formation can be inhibited
Chlamydomonas	X		
Haematococcus	X		
Gonium	X		colony formation can be inhibited
Golenkinia			spine formation can be inhibited
Spongiochloris			zoospore formation can be inhibited
Chlorosarcinopsis	X		
Stichococcus	X		helical plastid may be formed
Stigeoclonium	X		
Draparnaldia	X		
Micrasterias	X	X	
Scenedesmus	X	X	cell shape can be altered
Zygnema		X	cell width unstable
Pleurotaenium		X	viable zygotes when "species" were crossed

[a] When populations of an organism are examined critically, one may see forms which resemble other "species" or "genera" or changes in a typical life history. Most studies were conducted in the laboratory. In certain cases precise environmental control of the various morphological types has been achieved.

Some organisms can produce several morphological forms, for example, a *Haematococcus* may be in a *Chlamydomonas* stage and a *Hormotila* phase of a *Chlorosarcinopsis* is possible. In soil extract some *Scenedesmus* are not in a colonial form, but resemble *Franceia* or *Chodatella,* and spines no longer form on certain strains (Fig. 16-3), if the iron level is not adequate. *Stichococcus,* defined as having a parietal banded plastid, may develop a helical plastid. *Schizomeris* may be a stage in the development of a *Stigeoclonium.* With the desmids, variability at the subspecies level is well known.

Blue-green algae.   As pointed out by Pearson and Kingsbury, variation in the blue-green algae has long been recognized but has frequently been ignored when classification is considered. They showed variability in several morphological characters in the *Lyngbya* and *Calothrix* species studied, including abundance of heterocysts, types of constriction at the cross wall, changes of color, and twisting filaments. Perhaps the most far-reaching examples of morphological variability in the blue-green algae would be the reports of Drouet and Daily, already mentioned in the chapter on blue-green algae. In these cases, marked alterations in classification are suggested; valuable information on the subject of morphological variability is presented along with these findings. What we had formerly thought were distinct species are now considered merely ecophenes or ecological variants. The work with *Microcoleus vaginatus* growing under natural conditions shows that cell size and shape, as well as condition of the sheath, vary considerably. Thus the organisms can grow in forms that resemble organisms previously described from four other genera, including nine different species.

The type and extent of branching have been used in the Cyanophyceae to delimit genera. However, for almost 100 years we have known that single and double false branching can occur on one filament. A *Nostoc* mutant formed true branches in a typically unbranched genus! In another study the thickness of the sheath of an *Oscillatoria* varied directly with the level of calcium supplied. Also *Oscillatoria* produced shorter filaments when nutrients were limiting, and *Calothrix* hairs formed when iron or phosphate concentrations were low.

Diatoms.   One might expect that the last group in which we would find variation would be the diatoms. But this expectation is probably due to the prejudice we have inherited from a history of dealing with rigid glass walls in ashed or acid-cleaned material. Inasmuch as several cases

Table 16-5.   EXAMPLES OF POLYMORPHISM IN PENNATE AND CENTRIC DIATOMS

Organism	Phenomenon
Coscinodiscus-Asteromphalus	Each valve resembles one genus
Coscinodiscus-Actinocyclus	Each valve resembles one genus
Coscinodiscus	3 "species" from one clone
Chaetoceros	2 "species" on one chain
Cyclotella	2 "species" from one clone
Nitzschia	Sometimes can form *Hantzschia*-like valves
Mastogloia	2 "species" from one clone
Diatoma	Seasonal variation
Phaeodactylum	Ovate cells in low calcium or on solid media. Fusiform cells with more than 15 mg Ca l^{-1}, or in liquid culture
Thalassiosira	Forms *Coscinodiscus*-like cells.

of morphological variability were reported in the last few years (Table 16-5), it is time to challenge a system of classification that fails to allow for plasticity.

Dinoflagellates and silicoflagellates.   Although most dinoflagellates have exhibited little varability, a *Gymnodinium* in culture could develop characteristics of a related genus. A nonmotile form may have a swarmer that resembles a second genus. With the silicoflagellate *Dictyocha* in culture there was sufficient variability in skeletal morphology to assign more than one generic name to the clone, and thus skeletons are not sufficient for diagnosing species.

Other marine algae.   The filamentous alga *Asterocystis* forms unicells resembling *Chroothece* when grown in 25% sea water. Na^+ and Mg^+ are needed for the filamentous form. Among the larger algae, especially members of the red and brown algae, the gametophyte and sporophyte may be markedly dissimilar. As pointed out when these groups were discussed in detail, the *n* and *2n* stages have been classified as different genera. The development of a particular stage may be influenced by day length and involve phytochrome, a pigment associated with day length phenomena in vascular plants.

General Discussion

Clearly there is a strong case for morphological variability with evidence from many diverse groups, as well as from both field and laboratory. It is true that in some cases a phenomenon has been reported only once, or with only one species in a genus. Demonstration of a phenomenon in a laboratory does not indicate that it occurs in nature. Laboratory media are sometimes much too concentrated and thus make comparisons with nature very difficult. But a wealth of information on morphological variability has now been accumulated, and we must entertain the possibility that, if an organism has the genetic potential to alter its form in culture, some similar alteration *might* occur in nature. Has a *Stichococcus* evolved a mechanism for developing a helical pastid only in laboratory culture?

Frequently we have accepted or rejected certain laboratory reports, and some field investigations, because the work was not performed with "normal" material. We might dismiss some work in which media were very concentrated with inorganic or organic compounds, but again we may have several prejudices built up when considering what is now normal. What bodies of water could be considered normal habitats? What is normal for an organism? If subtle changes in the level of a cation determine whether an organism possesses appendages, and whether an organism occurs as a unicellular or colonial flagellate, which form is normal? Is the normal form that which was originally collected? Are all laboratory media suspect unless data obtained confirm concepts based solely on field observations? Soil water cultures are considered "safe." However, with the origin of the soil, the proportions of soil to water, and the method of sterilization varying from one laboratory to another, soil extracts might not only be much too variable, but also as prepared by some, much more concentrated than natural waters. Over 500 mg l^{-1} total solids are common in soil extracts, but most natural waters have less than one-third that level.

We now know that we cannot rely exclusively on some charactieristics formerly found useful, such as skeletons, wall structure, and sheaths. C. Sorokin's recent study with *Stichococcus,* showing that plastid form is not stable, is of great importance, for in many of the green algae plastid form is a generic characteristic. Inasmuch as the alternate forms of polymorphic organisms may be in different genera, families, or even orders, as we currently define these categories, changes in classification could be far-reaching. In the long run, polymorphism could be an asset in taxonomy. For, eventually one could describe an organism as

possessing a range of possible morphological forms, expressed in
different environments.

Distribution

For the most part we have overlooked the ecological implications of
variability. What effect does change in form have on distribution?
Certainly the thickness of the sheath might affect the resistance of some
blue-green algae to desiccation. At one time we thought that a very
different sheath meant that we were observing a different blue-green
genus in a particular environment. But by modifying its structure the
blue-green alga might be merely reacting to stimuli of a different
environment. Some methods of branching exhibited by filamentous algae
could ensure an increase in surface area.

Certainly with some unicellular and colonial forms, alterations in shape
could affect flotation. A diatom with a change in the number of costae
per unit of wall length might alter its buoyancy. Unicellular stages of
some colonial forms not only have more surface area, allowing an
increase in rate of nutrient exchange, but also have an increased
frictional resistance to settling. Modification in number and position of
types of appendages would directly affect the flotation of the organisms.
Form also affects orientation of microalgae as they respond to
turbulence.

Since laboratory experimentation shows that slight changes in the
levels of some elements or compounds have pronounced effects on
structure, we might reconsider the possibility of indicator organisms.
With a new interpretation, the presence of a *Scenedesmus* unicell might
indicate a certain chemistry of the water, whereas the colonial stage of
the same organism found at some other time could show us that the
conditions in the water had been altered.

Why should organisms maintain a static structure when they are being
subjected to a constantly changing environment? Short-term changes
might be quite subtle, and laboratory studies have indicated that very
modest changes in the chemistry of the medium can alter structure. The
evolving organism has the genetic potential for several types of
structures. It remains then for the organism to be located in the
environment which will trigger one of the possible structures. In time,
selection might work against some structural types, but we might also
consider that two or more forms could be possible for a species. When
optimal environmental conditions exist, one structure allows distribution
of the organisms in the plankton where rapid growth can occur. But with

a nutritional environment in which rapid growth could not occur, and damage might take place in bright light, an alternate structural form would result in settling of the organism. Thus variability could mean survival.

In this area the future is filled with optimism. Many difficult problems can be solved by combining the advantages of field and laboratory studies through a team approach. Laboratory conditions will never be exactly like nature, but the gap can be closed. In the revised classification of many organisms new descriptions must present organisms as possessing a range of possible forms. Let us accept the challenge.

REFERENCES

Bailey, D. and G. Samsel. 1971. Bibliography on the ecology and taxonomy of freshwater algae. In: J. Rosowski and B. Parker (Eds.) *Selected Papers in Phycology.* Dept. Botany, Univ. Nebraska, Lincoln. pp. 5–11.

Chapman, V. 1962. *The Algae.* Macmillan and Co., London. 472 p.

Chapman, V. and D. Chapman. 1973. *The Algae.* Macmillan and Co., London. 497 p.

Chodat, R. 1926. Etude de genetique, de systematique experimentale et d'hydrobiologie. *Rev. Hydrobiol.* 3: 71–258.

Christensen, T. 1962. Alger. In: T. Böcher, M. Lange, and T. Sorensen (Eds.) *Botanik.* Systematisk Botanik. Vol. 2, No. 2. Munksgaark. Copenhagen. pp. 128–46.

Dixon, P. 1970. A critique of the taxonomy of marine algae. *Ann. N.Y. Acad. Sci.* 175: 617–22.

Dodge, J. 1973. *The Fine Structure of Algal Cells.* Academic Press, New York 261 p.

Edwards, P. 1976. A classification of plants into higher categories based on cytological and biochemical criteria. *Taxon* 25: 529–42.

Leedale, G. 1974. How many are the kingdoms of organisms? *Taxon:* 22: 261–70.

Lewin, R. 1974. Biochemical taxonomy. In: W. Stewart (Ed.) *Algal Physiology and Biochemistry.* Univ. Calif. Press, Berkeley. pp. 1–39.

Margulis, L. 1970. *Origin of Eukaryotic Cells.* Yale Univ. Press, New Haven. 349 p.

Pickett-Heaps, J. 1975. *Green Algae. Structure, Reproduction, and Evolution in Selected Genera.* Sinauer Associates, Inc. Sunderland, Mass. 606 p.

Scagel, R., R. Bandoni, G. Rouse, W. Schofield, J. Stein, and T. Taylor. 1965. *An Evolutionary Survey of the Plant Kingdom.* Wadsworth Publ. Co., Belmont, California. 658 p.

Schopf, J. and D. Oehler. 1976. How old are the eukaryotes? *Science* 193: 47–9.

Silva, P. 1962. Classification of algae. In R. Lewin (Ed.) *Physiology and Biochemistry of Algae.* Academic Press, New York. Pp. 827–37.

Smith, F. 1950. *The Freshwater Algae of the United States.* McGraw-Hill Book Co., Inc., New York. 719 p.

Stewart, K. and K. Mattox. 1975. Comparative cytology, evolution and classification of the green algae with some consideration of the origin of other organisms with chlorophylls *a* and *b*. *Bot. Rev.* 41: 104–35.

Trainor, F. 1970. Algal morphogenesis: nutritional factors. *Proc. N.Y. Acad. Sci.* 175: 749–56.

Benthic organisms

INTRODUCTION

When we visit the seashore, the macroscopic algae attached to rocks and waving in the tides, the benthic forms, attract our attention (Fig. 17-1). Although these coastal waters are frequently highly productive (Table 17-3), we know far more about the production by the microscopic forms found floating in these waters, or in the open ocean, than we do about the attached forms. There are published figures on production from many phytoplankton studies, but the data on production in coastal waters are limited. In fresh waters there is even less information on benthic organisms. There is, in addition, not the variety of benthic macroalgae found along coasts of freshwater lakes that we find along the seashore.

The composition of the attached marine floras, which are found along most coasts, depends on the interactions of myriad factors, such as physical, chemical, biological, and dynamic considerations. On a worldwide basis, several specific factors are involved in the distribution of various marine floral components. These would include tidal amplitude, upwellings and currents, sea ice, temperature, and, in warmer waters, calcification. The marine algal communities found in different parts of the globe vary considerably, while the freshwater macroscopic communities in any area are less complex. Throughout the world there is quite a similarity in the freshwater benthic communities. As one closely examines *any* benthic flora *in situ,* it becomes apparent that there is a very definite zonation, that is, organisms are found horizontally stratified (Fig. 17-2).

Figure 17-1.   Algae of the intertidal zone on the New England coast.

Figure 17-2. Zonation. The black zone. (a) Just below the latter is *Enteromorpha* (b) and then barnacles and fucoids (c).

There are many more algal species in the Pacific than the Atlantic Ocean, and, as is true for other organisms, more tropical species than temperate or Arctic forms. There are areas of especially large numbers of algal species, for example, Japan, southeast Australia, western New Zealand, southwest Africa, west central Pacific Ocean. About one-third of the known types of kelps are found in Australia. Certain areas also have large masses of algae, for example, the western Pacific, western New Zealand, southwestern Austrialia, Japan, among others.

In this chapter, I will first outline factors that must be considered in understanding the distribution of organisms in either marine or fresh waters. Certain factors, such as salinity, can be more important in the marine environment. However, in salt lakes high concentrations of various elements affect algal distribution. The subject can become quite complex, especially when the interrelationships of factors are discussed. An outline is presented in Table 17-1.

Table 17-1. THE VARIOUS PHYSICAL, CHEMICAL, BIOLOGICAL, AND DYNAMIC FACTORS THAT AFFECT THE DISTRIBUTION OF BENTHIC ALGAE

I. PHYSICAL FACTORS
  A. Light
    1. quality, affected by latitude, clouds, turbidity, tides, etc.
    2. quantity, affected by turbidity, depth, tides, etc.
    3. periodicity, both daily and seasonal, even yearly
  B. Substrate
    1. chemical composition
    2. texture; e.g., density, porosity, solubility
    3. water-holding capacity
    4. location
      a. high or low in the tidal region, and thus with more or less contact with water
      b. covered, as in a tide pool, in the intertidal area
      c. sprayed or vigorously splashed
    5. available space or substratum
  C. Temperature
    1. water temperature
      a. daily variation
        rate of change
        maximum
        minimum
      b. annual variation
      c. variation with depth
      d. marked increase in tidal pools
    2. air temperature, when exposed
      a. daily or seasonal variation
      b. maximum
      c. minimum
  D. Rain
    1. annual maxima
    2. seasonal aspects, relationship to tides
    3. duration
  E. Pressure
    1. little known direct effect
  F. Relative humidity
    1. for exposed algae, relation to desiccation
    2. microniche in dense rockweeds

Table 17–1. THE VARIOUS PHYSICAL, CHEMICAL, BIOLOGICAL, AND DYNAMIC FACTORS THAT AFFECT THE DISTRIBUTION OF BENTHIC ALGAE *(Continued)*

II. CHEMICAL FACTORS
    A. Nutrient levels
        1. major elements, including carbon dioxide
        2. minor or trace elements
        3. growth requirements, e.g., vitamins
    B. Nutrient increase, resulting in eutrophication
    C. Extracellular products
    D. Pollution
    E. Concentration of nutrients by evaporation
    F. Dilution of nutrients by rain
    G. Salinity
    H. Buffering and pH values
        1. more variability than in the open ocean
III. BIOLOGICAL FACTORS
    A. Competition from other organisms as
        1. grazers
        2. symbionts
        3. epiphytes
        4. endophytes
        5. parasites
    B. Effect of these organisms on available light, nutrients, space, etc.
        1. harmful effects, e.g., reduced light penetration
        2. beneficial effects, e.g. resistance to desiccation
    C. Diseases
        1. control
IV. DYNAMIC FACTORS
    A. Tides, including surf action
        1. periodicity
        2. amplitude
        3. seasonal
    B. Currents
    C. Upwellings
    D. Winds
    E. Storms
    F. Ice action

The coastal marine environment offers the littoral alga a unique habitat, one that greatly differs from either a freshwater or a terrestrial habitat because it is often characterized by extremes in temperature and salinity. Although rainfall can affect the organism exposed at low tides, or dilute the seawaters in small areas, both rainfall and humidity are of less importance than for land organisms. Dilution of seawater can occur in an estuary or near the mouth of a river, but the decrease in salinity which might affect some organisms can be balanced by an increase in major elements, which could stimulate the growth of others. However, in general, the chemical composition of the sea is far less variable than that of fresh water.

With periodic tidal exposure (Fig. 17-1), the macroalgae along a rocky coast certainly have an extremely harsh environment. Consider a *Fucus* plant, which is bathed with a fresh rain, and splashed by many waves. Or alternately, it is subjected to the temperature of a cold and windy winter day, probably around −20°C, and then seawater at 0°C. On the other hand, consider the temperature of the thallus, which is exposed to the hot summer sun with temperatures up to 40°C, and then bathed in seawater at 20°C. Certain of these factors and the way organisms cope with them, will be discussed below. Some organisms are obviously not able to cope with such extremes in salinity and temperatures, while others can with adaptive mechanisms evolved over long periods.

The harshest environment for algae would be in the spray zone, frequently called the black zone (Fig. 17-1). Some researchers consider this to be the upper fringe of the littoral zone (Fig. 17-3). The composition of this zone depends in part on the composition of the rocky substrate, with more kinds of organisms found on the granular substrates such as sandstone. The lichens *Verrucaria* and *Lichinia* and blue-green algae *Nostoc, Gloeocapsa, Lyngbya, Calothrix,* and *Microcoleus* may be found, along with some green algae. One may even find zonation among the blue-green algae.

Recall that when we consider the coastal marine algae we are thinking about organisms that inhabit about 1 to 2% of the marine environment. For it is the macroalgae that inhabit the coasts, while the planktonic forms, usually microscopic, populate the open oceans.

PHYSICAL FACTORS

Among the physical factors to consider in a discussion of marine algal distribution are the substrate, air and water temperatures, and illumination.

The Physical and Chemical Nature of the Substrate

The texture of the substrate, or the degree of hardness, is important to algal growth. This can be seen on a rocky coast where some materials

Figure 17-3. The black zone. Note how closely the microscopic organisms, many blue-green algae, adhere to each contour of the substrate.

are covered by organisms, whereas other substrates do not provide the physical environment for attachment. The chemical nature of the substrate, except for forms epiphytic on other living things, is not as important for marine as for land plants. Land forms get their nutrients from the soil, whereas the marine algae are actually bathed in their mineral solution. There are not the elaborate roots, which function in part in absorption, found in higher plants. The holdfasts function in anchorage.

The substrates that may be colonized include rocks, other algae, animals, shells, wood, or rope. Certain hosts, *Zostera,* for example, may have several kinds of algae growing on them at one time. It is also not uncommon to see epiphytes growing on epiphytes. Some of the algae grow on the surface of the substrate, whereas others, such as the *Conchocelis* stage of *Porphyra,* actually digest away some of the material, such as a mollusc shell. The chemical nature of a rock is for the most part not as important as its texture, but the chemical character of *Ascophyllum* might be important to *Polysiphonia lanosa,* an epiphyte growing on *A. nodosum.* At first glance the *Polysiphonia* appears to derive something stimulatory from *Ascophyllum,* perhaps some amino acids. However, some phycologists disagree with this theory, pointing out that studies with labeled carbon reveal that *Ascophyllum* releases only about 1% of its fixed carbon, hardly enough to support a host plant. However, it is possible that a germinating *Polysiphonia* spore could develop *only* in the presence of this limited amount of organic matter released from *Ascophyllum*. Or the surface characteristics of the host might be the important factor in the determination of this obligate relationship.

Temperature as a Factor

The yearly temperature variation must be considered in the analysis of algal zonation or geographic distribution. In warmer areas this might be just a few degrees in sublittoral waters, for example, 8°C at Miami, Florida, but as much as 20°C in northern waters. At Port Aransas, Texas there is a seasonal change of 17°C! But the diurnal variation can be as important as the seasonal, especially when one considers that for the littoral organism there is also exposure to air. Recall the previous discussion of a possible 20°C temperature change with each tide.

The littoral organism is subjected to the greatest diurnal variation, while the submerged, especially the deeply submerged, sublittoral forms are subjected to the least seasonal variation. On the east coast of North America there are fewer species and numbers than on the west coast. In part this is due to the more limited yearly temperature range on the west coast. In the Arctic, genera survive in relatively cold waters by growing at twice the depth at which they are found in more southerly waters. At that depth in the Arctic there is a more stable temperature.

Some algae can survive −8°C, in solid ice, for at least the length of one tide, but this cold tolerance is found to vary seasonally in some organisms. On the other hand, summer plants of *Fucus* and *Ascophyllum* survived temperatures of −20°C in the laboratory. Most littoral algae are not as tolerant and they cannot survive freezing. In the upper temperature extremes *Bangia* can tolerate 40°C exposed on dry rocks, but when it is *in* seawater, it will not survive at 35°C. We know little of the environmental requirements of different stages in life histories of many organisms.

Illumination

At first, when discussing illumination, emphasis was placed on both the quality and quantity of light. Now we know that some algae, certain red algae, for example, will express a certain stage in the life history only in response to length of day. Thus with *Bangia-Conchocelis* under short days the *Bangia* stage will be expressed. When there are more than 12 hours of light, spores will germinate to form the *Conchocelis* stage.

The depth of the photic zone, that area in which there is visible light penetration, varies with the locality, the season of the year, amount of cloud cover, wave action, and so forth. But in northern waters the depth of this photic zone is usually no more than 50 m, while in the Arctic and tropics it may be 100 m. Deeper light penetration has been recorded in the Mediterranean, down to 130 m, with the highest recorded reading at 200 m.

In a broad way there is a pigmentation-illumination-distribution story. In spite of the exceptions, for example, green algae living at great depths in the Arctic, red algae growing in the upper littoral, and gametophytes and young sporophytes of browns living in deeper waters, the following story can be told.

Green algae are found in greatest numbers near the surface and in the upper few meters of water, usually disappearing at 30 m. They use red light quite efficiently. As the light penetrates the water there is progressive and selective absorption. The longer wavelengths of light, the red, and those of the orange and yellow, are absorbed first. Note the connection with green algae distribution. Many, but not all, brown algae are found in intermediate depths. The red algae absorb in the blue and blue-green parts of the visible spectrum. These shorter wavelengths of light penetrate to deeper waters, where red algae can be found. Naturally there is also a decrease in the available intensity, but the red algae, as a group, show optimum photosynthetic activity at low light intensities. Not only do sublittoral algae thrive at low light intensities, but also some are so adapted that they can be killed by exposure to the light intensity in the littoral zone for one hour. There is thus a correlation between pigments, light intensity and depth, but there is no absolute correlation between the algal color and the depth at which it is found. For example, some of the red algae (*Chondrus,* certain *Polysiphonia* spp., *Gracilaria* and *Rhodymenia*) can grow in strong light and are common in littoral waters.

The light intensity that reaches an intertidal organism can vary a great deal, with full sunlight when it is exposed, but with considerably less intensity when the organism is covered with water laden with particulate matter.

CHEMICAL FACTORS

Salinity
The salinity of the open ocean is about 35 parts per thousand, with the Na and Cl ions making up the bulk of the figure. In coastal waters, in estuaries, and after heavy rains this salinity can be lower, but the proportions of the major elements remain more constant than they are in the freshwater environment. Some organisms, such as *Enteromorpha*, survive the extremes in salinity from a few parts per thousand to full sea water salinity. When there is a heavy rain and the salinity in an estuary drops to near a freshwater level, or when ships pass from the ocean into rivers, organisms in the estuary or growing on the ship must be able to withstand this osmotic shock. On the other hand, in evaporating tide pools organisms can be subjected to salinity three to four times that of seawater. *Dunaliella* is a common flagellate that survives at high salinity, but *Prasiola* and *Enteromorpha*, both macroalgae, are also found under such conditions. Sublittoral forms, generally "more delicate" in all respects, can survive 0.5 to 1.4 times normal seawater concentrations, whereas littoral forms withstand salinity changes from 0.2 to 3.0 times normal seawater concentrations.

As one proceeds from the marine environment into an estuary and up a river, the red algae are the first to disappear. Most are not at all tolerant of reduction in salinity; *Caloglossa* and *Bostrychia* are exceptions. Then the brown algae and many diatoms can no longer survive. The green algae are most tolerant, and in the Firth of Clyde in Scotland *Enteromopha* can grow, even in fresh water.

pH
The pH of the sea water is alkaline, usually 8.1 to 8.2, but it can be lower where there is freshwater intrusion, and higher in tide pools. Some marine algae can live in tide pools at pH 10. *Capsosiphon*, *Percursaria*, *Enteromorpha*, and *Ulva* have been reported in such an environment. Diurnal variations in pH have been reported in such environments, with the availability of carbon dioxide important in the changes taking place. (See the discussion on this subject in the freshwater environment. Page 398).

Oxygen
The amount of oxygen dissolved in water (DO) varies with temperature, with lower values in warmer waters. It is close to saturation near the surface, with wave action, as well as photosynthetic activity, responsible

for the oxygen. Unlike freshwater environments few areas are "dead," that is, show zero DO. Possible exceptions might be in the immediate vicinity of solid waste which is deposited in the ocean along our coasts, or the very low diurnal DO levels which are sometimes found in estuaries.

Nutrients

There are only a few examples of the effect of a specific nutrient on the distribution of a benthic marine form but this subject is under investigation at several laboratories. Benthic marine algae have the benefit of the nutrients that come from the land, either directly or through our rivers. The nutrients are quickly mixed and brought in contact with the marine vegetation. Nitrate and phosphate are typically at low levels, especially in the summer, but sufficient to support growth. Ryther has reported that nitrogen is the nutrient that normally limits growth along the continental shelf. Phosphorus is the prime limiting nutrient in fresh waters.

An organism such as *Prasiola* thrives where there is a high nitrogen level and thus is found on rocks or in pools where there are considerable bird excreta. *Ulva* and *Enteromorpha* species can also occur there.

Unfortunately, we now have a number of examples of the adverse effect of sewage effluent, or specific pollutants from industry, on the distribution of some of the marine forms. There is evidence that some kelp populations have been reduced from previous levels because of the effluent from California towns. Sea urchin populations have also increased and these feed on the kelp. Thus the kelp population is being reduced by pollutants, sea urchins, and industry harvesting for alginic acid.

BIOLOGICAL FACTORS

We will not treat biological factors in detail because many of these considerations have been covered elsewhere. The primary producers are constantly subject to grazing by herbivores (Fig. 17–4). This includes not only the macroalgae, which are most visible, but also the local plankton and epiphytes on macroalgae. Control of some local populations has been maintained by microbial activity. The loss of *Zostera* because of a fungal parasite is an extreme example of such control. For many years in the 1940s and 1950s little of this eel grass could be found in the

Figure 17-4. Grazing by *Littorina*. Above attached filamentous algae have been grazed clean when the animals and algae were submerged by the tide.

northeast. Now it is coming back and dense *Zostera* beds can be found in many areas.

As one collects along the shore, it becomes apparent that some algae are epiphytic on others, and at times this is an obligate response. Thus substrate can be a biological factor. *Polysiphonia lanosa* is closely associated with *Ascophyllum,* for the former is always found attached to the latter. Investigators have studied this host-epiphyte interaction, using tracer materials. These associations are found not only with the macroalgae, but the smaller epiphytic forms, such as diatoms or small red and green algae, may also have specific hosts. *Codium,* the nuisance alga of oyster beds, attaches to the shells. The large fronds interfere with water circulation and, because of the buoyance provided by the increased surface, disturb oysters from their habitat.

Some littoral forms provide protection for the epiphytes associated with them. When a *Fucus* bed is exposed at low tide (Fig. 17-5), the thicker, heavier plant body provides some protection for smaller attached forms. In addition, when submerged, the larger algae filter some of the bright surface sunlight.

Algae of the littoral zone can lose considerable water and still resist desiccation. The higher *Fucus* lives in the intertidal zone, the thicker are the walls, and the more fucoidin is present. Such organisms apparently

Figure 17-5. Fucoids exposed with the lowered tide.

can store large quantities of water in the walls, sufficient to enable them to survive until the next tide. *Fucus* plants can lose 70% of their water content and yet survive. Thus the effects of desiccation hardly reach the protoplasm. In extreme cases *Bangia* and *Porphyra* can survive even after 2 to 3 weeks desiccation in air. Certain algae of the littoral zone will die if they are always immersed, for instance, *Fucus.* Some sublittoral algae will not survive unless always submerged. Exposure to air, too, could be lethal because resistance to desiccation is inadequate. The probability of death through exposure is greater for sublittoral forms than littoral.

Contrary to first impressions, the main differences in resistance to desiccation lie in the protoplasm and not in overall morphology or wall structure. In one study, the rate of drying of both littoral and sublittoral forms (of similar morphologies) was almost identical. After death of the sublittoral alga the protoplasts of the littoral form were still alive.

In general, most species of seaweeds in the littoral zone are annuals, and perennial forms are located in the sublittoral. In the littoral zone plants apparently grow well when all factors are favorable. They can survive until the next growing season through sexual reproduction or spore production.

Perennial algae may die back to a prostrate basal portion yearly. *Fucus,* often dominant in the littoral, is not an annual and is an obvious exception to this generalization.

DYNAMIC FACTORS

Tidal Activity and Wave Action

Many of us are accustomed to two tides per day, but there are regions where there might be just one, depending on the latitude, deep ocean currents, and bottom contour. On the east coast of the United States there are generally two regular tides each day. On the west coast the two tides are irregular; that is one high tide is higher than the other. One tide each day is the rule on the Texas Gulf coast. The extremes of high and low water each month occur when the sun, moon, and earth are approximately in line.

Wave action, especially during storms, tears algae from the substrate and deposits them on the shore (Fig. 17-6).

If there is a great deal of wave action, the turbulence generally prevents any large changes in salt concentrations or variations in temperature. But as mentioned above, the organisms can be subjected to these variations when the tide is out. There might be considerable desiccation or ice action in the winter, which could denude the area by scouring the rocks of attached vegetation (Fig. 17-7).

In addition to these considerations, tidal activity must be discussed when thinking of availability of nutrients. Local upwellings introduce

Figure 17-6. Sand dune (upper) and two layers of algae (and eel grass) washed ashore after spring storms.

Figure 17-7. The intertidal of the Arctic. Ice action scours attached algae from most surfaces, except in cracks or depressions.

nutrients from deeper waters, and the coastal water movement provides good mixing. Some organisms have reproductive cycles corresponding to tidal rhythms, thus favorable conditions for reproduction and dispersal of the products are present. *Ulva* releases both zoospores and gametes at times of favorable tides. Organisms such as *Dictyota* or *Zonaria* discharge gametes, but not necessarily spores, in rhythm with tides. Photosynthesis in *Porphyra* and *Fucus* is maximum when exposed to air during low tide. In contrast the rate in *Ulva* was higher when submerged.

THE FLORAS

We now know of benthic annuals such as *Cladophora, Enteromorpha, Porphyra,* and *Polysiphonia,* and benthic perennials such as *Fucus, Sargassum, Laminaria, Hildenbrandia, Chondrus,* and the tropical *Acetabularia.*

Some benthic organisms such as certain species of *Ulva, Ectocarpus, Sphacelaria, Gracilaria,* and *Hypnea* are cosmopolitan, but most marine forms have a definite, restricted distribution. This could be a few kilometers or an entire continent. The most prolific floras are found on rocky coasts, but there are salt marsh and flat floras. Because of the mass movements of water by currents and storm activity, it is difficult to

determine that a species is endemic. Spores and juvenile stages of
marine algae may be attached to ships and transported thousands of
miles to suitable habitats. How long have humans, sailing the oceans,
aided in the transport of algal spores?

Along rocky coasts the dominants vary in different parts of the world.
On the African coast the red algae are the littoral dominants, in New
Zealand there are larger brown algae. *Fucus* and *Ascophyllum* species,
the rockweeds, are common in New England. The polar regions have a
permanent flora reduced to about 30 species of small stature, unicells,
crusts, and filaments, because the ice scours the rocks. Larger forms
can survive only in cracks, but even then are subject to considerable
scour (Fig. 17-7).

In previous years as one collected along the north Atlantic coast, it
appeared that there were distinct floras in the following areas. An Arctic
flora existed with few components reaching New England. There would
be a northern New England flora and a southern New England flora with
a distinct break at Cape Cod. Few algae were noted along the mid-
Atlantic because of the paucity of suitable substrates, with the exception
of mudflats and marshes. There was also a tropical flora. Now, with
information from more studies, including those conducted by SCUBA as
well as year-round investigations, we realize that there are not the sharp
breaks once believed. Thus some colder water forms from higher
latitudes are present in deeper waters at latitudes where they were not
previously expected. Wilce and colleagues now consider floral breaks
south of Belle Isle, Cape Cod, and Cape Hatteras. Recently, van den
Hoek described five phytogeographic regions for the entire north
Atlantic. These include tropical west Atlantic, Carolina, warm temperate
Mediterranean-Atlantic, cold temperate Atlantic-Boreal, and Arctic floras.

In Table 17-2 we can examine one example of vertical zonation of
marine algae along a rocky coast. There is considerable local variation,
but some aspects of this zonation pattern can be applied to other areas.

On the Pacific coast of the United States a northeast boreal flora
extends from the state of Washington to Pt. Conception, California.
Further south there is a divergence of the California current, resulting in
a warm water flora. The latter is characterized by the extensive kelp beds
and considerable algal diversity around San Diego and Los Angeles.

PRIMARY PRODUCTION BY BENTHIC MACROALGAE

For many years we did not pay too much attention to marine macroalgae
production. Early emphasis had been placed on the phytoplankton, for

Table 17-2. MARINE ALGAL ZONATION ALONG A ROCKY CONNECTICUT COAST, A VERY BRIEF ACCOUNT

Zone	Organism
Spray	Black zone with *Calothrix*
	Nostoc
	Lyngbya
	Gloeocapsa
	Small green algae, such as coccoid greens and *Pseudendoclonium*
Littoral	Lichens such as *Caloplaca, Verrucaria* and *Lichinia*
	Enteromorpha
	Bangia
	Cladophora
	Porphyra
	Fucus, Ascophyllum
	Ectocarpus, Elachistea
	Ralfsia
	Chondrus
	Hildenbrandia
	Codium
	Ulva
	Scytosiphon, sometimes in tide pools.
Sublittoral	*Polysiphonia*
	Ceramium
	Chondrus
	Scytosiphon
	Laminaria
	Corallina
	Champia
	Codium

these organisms are found in all bodies of water, are much easier to collect, and at least some were easily placed in culture some time ago. Thus, with these organisms, field observations could be carefully duplicated in laboratory investigations. Wherever vessels could sail it was possible to collect phytoplankton samples for measurements, even in winter months.

On the other hand, collection of marine benthic algae, especially those in deeper waters, was not easy. This would be especially true in cold weather, or during stormy periods. Only intensive studies for many months would yield necessary information on the standing crop and the time needed for turnover. However, with the use of SCUBA and the more active interest in the problem of productivity, several interesting studies have been conducted in the last decade. Mann, from Nova Scotia, recently summarized much of the Canadian work on production in kelp beds. Laminarians have proved especially useful for these investigations because of their pattern of growth. Plants were identified and numbered, and holes were placed in the blade. The first hole was about 10 cm from the juncture of the blade and the stipe. Because the blades of these kelps grow from the meristem activity at the tip of the stipe, there is a continuous movement of tissue from that region to the tip of the plant itself. The measurement of the movement of the hole is a good indication of the increase in blade biomass. However, the blade also increases in thickness and width, and these factors must also be considered.

Growth in the kelp beds has been found to increase in the fall months, despite the concurrent decrease in temperature and available light (daily). Thus these laminarians have evolved a growth pattern enabling them to utilize available nutrients at that time. Both inorganic and organic compounds come directly from land runoff, rivers, estuaries, and upwellings. Wave action keeps the water well mixed, and thus the organisms are always in contact with a more than adequate nutrient supply. They maintain their active growth until late winter and early spring.

In the Canadian studies the data indicated that the increase in a new tissue was balanced by the erosion of the blade at the tip. There was complete blade renewal at least once, but as much as five times each year. Primary production is the weight of new organic material produced by these organisms over a period of time. This figure is conveniently presented as grams of carbon in a unit of volume for a day, season, or year. Frequently, the data are presented as grams of carbon produced in a square meter for a year (Table 17-3). Net primary production in the Canadian studies reached 1700 g C m^{-2} yr^{-1} (Table 17-3). In other studies with littoral algae, *Fucus* and *Ascophyllum* species, the wet weight of developing organisms could be doubled within five to ten days. The plants do not grow at that pace very long. Productivity of littoral algae has been measured at about half that of kelp beds (Table 17-3).

In the studies with littoral zone algae there was a correlation between morphology and primary production. The best figures were for organisms that were freely branched or those that were in flat sheets. The poorest net production was from prostrate and encrusting algae. In the temperate

Table 17-3. NET PRIMARY PRODUCTION BY ALGAE IN THE MARINE ENVIRONMENT, INCLUDING SOME ESTIMATES[a]

	$g\ C\ m^{-2}\ yr^{-1}$
Phytoplankton	
oceanic	50–100
coastal	100–200
Littoral zone algae	600–1000
Sublittoral zone algae from,	
Nova Scotia	1700
California	400–800
Indian Ocean	possibly up to 2000
Spartina	200–1500
Forests	500–700
Mature rain forests	1300
Managed crops, e.g., alfalfa	1500

[a] For comparison, figures from some other habitats are presented. After Odum, Mann.

zone net primary production decreases where there is poor light penetration or in dense algae forests where there is mutual shading. With a steep slope to the continental shelf in some regions there is a limited area for growing algae.

The bulk of the production from benthic macroalgae, or as much as 90% of that from the littoral zone, eventually finds its way to dissolved or particulate organic matter. The organisms continually release some organic matter which goes into solution. Particulate matter is found as organisms erode away. Many littoral brown algae are not heavily grazed. It is the smaller organisms, as well as the filter feeders, that benefit directly from the extensive growth of these algae by using dissolved organic or particulate matter.

Thus the fringe areas between the land and the sea, especially in those regions with a gradual slope to the continental shelf, can be excellent habitats for algae. Figures for net primary productivity can match those for highly productive land habitats.

REFERENCES

Abbott, I. and G. Hollenberg. 1976. *Marine Algae of California.* Stanford Univ. Press, Stanford. 827 p.

Dawson, E. 1966. *Marine Botany.* Holt, Rinehart and Winston, Inc. New York. 371 p.

Edwards, P. 1969. Field and cultural studies on the seasonal periodicity of growth and reproduction of selected Texas benthic marine algae. *Contrib. Mar. Sci.* 14: 49–144.

Hoek, C. van den. 1975. Phytogeographic provinces along the coast of the northern Atlantic Ocean. *Phycologia* 14: 317–30.

Holdren, J. and Ehrlich P. 1974. Human population and the global environment. *Amer. Sci.* 62: 282–92.

Lee, T. 1977. *The Seaweed Handbook. An Illustrated Guide to Seaweeds from North Carolina to the Arctic.* The Mariners Press, Boston. 217 p.

Littler, M. and S. Murray 1974. The primary productivity of marine macrophytes from a rocky intertidal community. *Mar. Biol.* 27: 131–5.

Mann, K. 1973. Seaweeds. Their productivity and strategy for growth. *Science* 182: 975–81.

Neushul, M. and D. Coon. 1971. Bibliography on the ecology and taxonomy of marine algae. In J. Rosowski and B. Parker (Eds.) *Selected Papers in Phycology*. Dept. Botany, Univ. Nebraska, Lincoln. pp. 12–17.

Odum, E. 1971. *Fundamentals of Ecology*. W. Saunders Co., Philadelphia. 574 p.

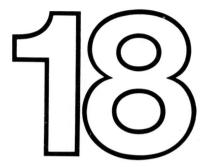

The plankton and their environment

INTRODUCTION

The study of the plankton and their distribution is a difficult one because of the many interrelationships that must be considered. For convenience I will begin with a discussion of the water itself.

For many years we have been accustomed to having clean water piped to our homes, and we learned to expect a high standard of taste and clarity. Potable water comes either from wells or from local reservoirs, and we rarely give it much consideration, unless a pipe breaks or an animal falls into the well!

More than 97% of the water on earth is in the oceans and is thus not useful in the home and in much of industry. About 2% is in ice at the poles. Thus less than 1% of all available water serves as the habitat for all fresh water organisms, as water for industry, recreation, and rainfall, as well as for our personal use around the home. This 1% represents 1400 cubic km of water in liquid form and 54 cubic km in the atmosphere at any one time.

Because half of the U.S. population dwells on about 2% of the land, for example, the dense metropolitan strips from Boston to Washington, Chicago to Pittsburgh and San Francisco to San Diego, sometimes referred to as Boswash, Chipitts, and Sag, the available water from a considerably larger area has to be collected and piped to these areas. After use, the water is discharged, into the immediate surroundings and not always returned to its source. The discharged water is deposited into the ocean, or bays and sounds, as well as in rivers and lakes. Urbanization thus is the cause of pollution in the Mississippi River as well as in Lake Erie and Lake Washington. We will later discuss the effect of this pollution, in part a problem of distribution of water, on the phytoplankton suspended in the water column.

Rainfall in the United States averages 76 cm per year. About 53 cm go directly into the soil and vegetation, most of which is returned by evaporation and plant transpiration to the atmosphere. Of the remaining 23 cm that might be available to us, we use only 7.5 cm. More than 5 cm are used as water for industry, for cooling, or washing of industrial products. It is the use of this 7.5 cm which adds so many nutrients, increases the temperature, changes the turbidity, or cuts down on the light, when it is sent back into our rivers or lakes. Sometimes the water will be used many times over before a river empties into the ocean. All of this then has a direct effect on plankton distribution.

Even though most of our water-related problems are found in fresh waters, and thus studies of plankton in these waters have accelerated, we must also examine marine distribution. Close to 70% of the earth's surface is covered by the seas.

Earlier estimates extrapolated from a few areas of high production held that from 66 to 90% of global photosynthesis took place in the ocean. Ryther now estimates that about one-third of global photosynthesis occurs in the oceans. In contrast the benthic algae are said to contribute from 1 to 10% of oceanic production. Recent data on production in kelp beds indicate that the higher number is more accurate. Actually very little of the ocean area provides a suitable substrate for benthic organisms. Phytoplankton primary production accounts for about 5×10^{13} kg of dry matter per year.

ORGANISMS CALLED PLANKTON

At least two dozen prefixes have been commonly used with the term plankton, but we will attempt to concentrate on the term phytoplankton, the plants found floating or weakly swimming in bodies of water. In this discussion I will not differentiate between euplankton and other organisms found suspended in the water. Such organisms might have only recently been scoured from the bottom, or in ponds, water currents might temporarily suspend bottom organisms or remove periphyton from aquatic vegetation. Perhaps some of the data included in the discussion that follows deal with organisms that are not true plankton. I will attempt to discuss some of the interesting data reported in the literature and avoid a judgment.

Planktonic organisms contribute to the food web directly as organisms, both when living (when grazed) and dead. In addition, living organisms can release extracellular products, including growth stimulating and inhibiting compounds. As they decompose they become part of a large carbon pool called particulate and soluble organic matter.

Distribution

Although there frequently is considerable mixing and then uniformity in distribution of phytoplankton, it might also be discontinuous and patchy, both horizontally and vertically. Sometimes this can be seen from the surface, but it is also noticeable while diving. The layers might be just 1 cm thick and from 1 to 10 m broad. In the ocean, elliptical patches can be 60 to 240 km in size.

We also have to consider the seasonal distribution (Table 18-1) and how this varies with geography. In both polar and temperate waters there is no appreciable growth in the winter, either in marine or

Table 18-1. THEORETICAL SEASONAL SUCCESSION OF
PHYTOPLANKTON IN BOTH MARINE AND
FRESHWATER ENVIRONMENTS

	Freshwater	Marine
Winter & Spring	Diatoms	Diatoms
Late Spring	Green algae	Diatoms
Summer	Blue-greens	Dinoflagellates
Autumn	Blue-greens	Diatoms

freshwater environments. The numbers of phytoplankton increase in the
spring. Factors involved in the production of a particular bloom include
temperature, increases in nutrient levels, and an increase in solar
radiation. Some phycologists claim that there is a built-in metabolic
cycle, a biological rhythm. There is often great emphasis placed on a
nutritional trigger for the bloom. But in many areas there are sufficient
nutrients even before the pulse, with a marked decrease during, or just
after, maximum growth.

Peak spring growth is frequently followed by a rather steep decline,
with decidedly lower numbers of phytoplankton in the summer.
Sometimes, as in the plankton in coastal waters, there is a second pulse
in the autumn. In the Arctic and Antarctic there is a single peak, and this
is later in the season, in the summer. There can be a thousand-fold
increase over numbers found earlier. On the other hand, in the tropics
the maximum growth is frequently only five times the typical standing
crop, and this peak growth can occur in the winter. The region is
characterized by some growth year round, and highly productive lakes.
Typical eutrophic lakes, when compared to those that are oligotrophic,
have more nutrients and little algal diversity.

We are still quite ignorant of myriad factors controlling plankton
distribution. Because of the simplicity of many laboratory systems, many
researchers look to information from cultures in an attempt to apply
these data to field studies. But much laboratory work is with *Chlorella,
Chlamydomonas,* and *Scenedesmus,* which really are not important
organisms in plankton, and little progress is made. In addition, until the
1960s progress was slow in culturing the imporant plankton organisms.
Until we recognized the need for vitamin B_{12} with the majority of
planktonic algae, culture was impossible. Why not grow the organism in
natural waters, at least for initial isolation? For a variety of reasons which
will be discussed in the section on nutrition, these waters are poor

laboratory solutions, if used in the same way one uses typical laboratory media.

We should examine some of the following data with caution. Most of the early information, even into the 1950s, dealt with organisms larger than 3 mm (macroplankton) or in the 0.06 to 3 mm range (microplankton). It did not include consideration of nannoplankton (10 to 60 μm in size), and the ultraplankton (0.5 to 10 μm size). As we begin to gather more information on ultraplankton we now realize that they can be members of many classes, including chlorophytes, chrysophytes, cyanophytes, haptophytes, and diatoms. Much of the data obtained are true for only certain species, in one place at any one time. We often do not have sufficient information to make broader generalizations. Finally, although there are the same considerations whether dealing with marine or freshwater environments, there are a few distinct differences.

Sampling

In a flowing stream it is obvious that samples collected at a single station may vary from day to day. But it should also be clear that the same can be true for samples taken from almost any aquatic environment. Water moves in many ways, so that one does not always get uniform distribution. Plankton may settle from suspension when there is no water movement, can float at the surface, or pile up on the shore away from the wind. The best data are thus obtained from thorough and repeated sampling.

Until the late 1950s number 20 or 25 plankton nets were widely used. These were dragged through the water, or placed in a stand so that buckets of water could be passed through them. The sample is called net plankton. But these nets, even though prepared from fine bolting silk, have 40 to 50 μm holes (number 20 net) or 60 to 70 μm spaces (number 25 net). As they began to get occluded, they retained smaller and smaller organisms, but much of the plankton from the first portion of the sample passed through them (Fig. 18-1). Verduin measured the photosynthesis from the filtrate that passed through a 64 μm net. About 65% of the activity of his total collection was in the filtrate. In his studies of Scandinavian lakes Rhode found that the photosynthesis was correlated with an abundant nannoplankton, organisms smaller than 60 μm, rather than with organisms trapped in plankton nets. In the Coral Sea 90% of the producers are in the nannoplankton, while in the south Atlantic the coccolithophorids account for 45% of the plankton, and many of these are nannoplankton organisms. In a recent two-year study

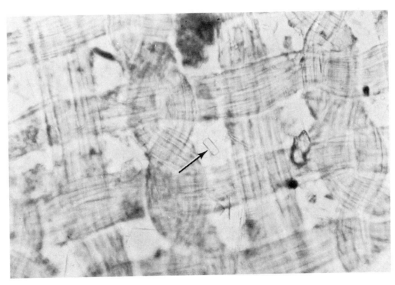

Figure 18-1. A number 20 plankton net with some diatoms. Note the complex weaving pattern in the net. With an average net opening of 40–50 μm a *Melosira* (center) would easily pass through when the net is towed. 325 X.

in Chesapeake Bay, about 90% of the plankton passed through a net with 35 μm holes.

In the North Sea, larger algae, or net plankton, are the dominant forms near the British coast, but in most other areas ultraplankton can account for 90% of the productivity.

In several English lakes, Lund estimated that nannoplankton and ultraplankton accounted for less than 10% of the algal biomass. Nannoplankton and ultraplankton organisms are more abundant in less fertile lakes. Desmids, larger organisms, or microplankton, are well known in these English lakes. (It should be pointed out that 10% of the mass could account for more than 10% of the photosynthesis, because of the efficiency of the nannoplankton and ultraplankton.)

One then has to use net sampling with caution, but the same is true of some of the other standard methods. The procedure used is determined in part by the study to be conducted. A water sample can be collected and then concentrated by sedimentation, but this is more efficient when the collection is fixed. There are disadvantages in that some of the organisms smaller than 10 μm will not settle out from a water column even after several days. Various membrane filters offer excellent possibilities, but cells can rupture because of pressure, and specimens can temporarily dry resulting in cell destruction. Centrifugation is a quick and simple procedure for small samples, but use of large quantities of material, or the partial destruction of the collection as a result of the high speed, makes the procedure less desirable.

Various devices have been used to gather surface water or water at various depths. These include Van Dorn water bottles in which the sample can be kept free of air-borne organisms, or simple plastic bottles, with a device for opening the bottle at the appropriate depth. Plankton nets can be lowered or dipped into the water. In shallow waters Fogg has suggested inserting a pipe or tube, and thus collecting a "core" of water.

Patrick and co-workers have suggested use of artificial substrates, such as glass slides, to obtain a more accurate picture of the phytoplankton. Surprisingly, even though organisms must land on, and sometimes attach to, the artifical substrates, this approach gives an accurate picture of the organisms present and the species diversity. In comparative studies, various artificial substrates can be employed.

For a number of reasons some phycologists find all of these sampling procedures inadequate for their investigation. In a number of laboratories artificial streams have been constructed, simulating the environment under study. These models have the advantage of allowing the investigator to use different approaches in duplicate streams or in another run. It is then only another step to rely completely on laboratory cultures. With culture studies and use of artificial streams it is argued that it is difficult to relate findings directly to nature.

PRIMARY PRODUCTION BY PHYTOPLANKTON

Some consideration should be given to the role of the phytoplankton in the environment. These organisms are the primary producers, the base of the food chain, over much of the surface area of the globe. By primary production we mean the weight of new organic material produced by these organisms over a period of time. Frequently the data are presented as grams of carbon produced in a square meter for a year (Table 18-2). The square meter takes into consideration the production in the water column under the surface area.

Investigators have used many methods to obtain accurate figures for productivity. One method commonly used involves identification and counting of organisms, and a later determination of the actual volume of cellular material. In addition to having information for determining increase in biomass, and thus production, this method provides other

Table 18-2. THE PRIMARY PRODUCTION IN VARIOUS FRESHWATER ECOSYSTEMS, WITH A COMPARATIVE FIGURE FOR SUBLITTORAL MARINE ALGAE[a]

	$gC\ m^{-2}\ yr^{-1}$
Phytoplankton, marine	
Oceanic	50–100
Coastal	100–200
Upwellings	300
Phytoplankton, fresh water	
Lakes	2–400[+]
Lake production	
Temperate	2–900
Tropical	30–2000
Oligotrophic	2–180
Mesotrophic	75–300
Eutrophic	300–2000
River production	
Temperate	$<$1–600
Tropical	1–possibly 1000
Sublittoral Macroalgae, Marine	400–1700

[a] After Odum, Wetzel, Likens.

useful data about organisms and populations. However, it is among the most tedious of methods. Other methods measure the weight of the phytoplankton. However, both wet and dry weight determinations can be quite inaccurate. Dry weight has been measured by drying organisms at 105°C to a constant weight. In a second approach, after obtaining such a dry weight, the organic matter is released by incineration at a temperature of 550°C. The mineral composition is weighed. The weight of organic matter is then determined by difference. The latter approach is preferred because the nonashed dry weight of some organisms would include inorganic walls or skeletons, and we would get a false picture of the amount of carbon fixed by photosynthesis.

Other indirect methods for measuring productivity include the rate of oxygen evolution, pigment content, especially the chlorophylls, levels of ATP, and the changes of levels of both carbon dioxide and oxygen. Uptake of labeled carbon dioxide from matching light and dark bottles is the basis of a widely used technique suggested by Steeman-Nielsen.

According to Wetzel, the oxidation of organic material to carbon dioxide provides one of the most accurate measurements.

Data on production in freshwater ecosystems are presented in Table 18-2. Freshwater environments can be very productive ecosystems, but there is a wide range of figures. Figures for lake production are higher than those for phytoplankton production alone because of the input from periphyton and vascular plants. In Lawrence Lake, Michigan, phytoplankton accounted for 25% of the production, macrophytes produced twice that amount of carbon, and periphyton, attached algae, among others, accounted for the remainder.

FACTORS INVOLVED IN PHYTOPLANKTON GROWTH

Light
The most important role played by illumination is in the photosynthetic reaction, but it might also be considered as a factor involved in aggregation of some organisms, phototactic responses of some flagellates, spore germination, and triggering of some cyclic phenomena, for example, rhythmic migrations of diatoms or other organisms. The upper layer, where there is adequate light penetration, is called the photic zone (Fig. 18-2). Where there is some light, but the net gain from

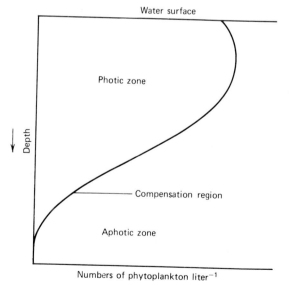

Figure 18-2. Diagrammatic presentation of the effect of light penetration on the distribution of phytoplankton. Although fewer organisms are found at the surface than several cm below, in general cell number decreases with depth.

photosynthesis equals that lost by respiration, there is an area called the region of compensation. Light can penetrate a little deeper, but soon a sinking organism reaches the region called the aphotic zone where there is no sunlight.

The compensation region depth can be more than 100 meters in the Caribbean or only 10 meters at Georges Bank off the northeast coast of North America. In Long Island Sound (New York) it occurs at 2.5 m in the winter, but increases six-fold in the summer. The deepest part of Long Island Sound is approximately 100 meters at the eastern end, the Race. In freshwater the photic zone is usually within 5 m, but it can be even closer to the surface in lakes or ponds with humic substances in solution.

Up to 25% of the sunlight may be lost by reflection from choppy waters, but the loss decreases in calmer waters. Then there is loss by light scatter and absorption by particulate matter. Hard water lakes usually have less depth of light penetration than soft water lakes, in part because of their water chemistry.

Chlorophylls absorb both red and blue light in the upper waters, but as the red light is lost in the upper 10 m, they can still function by absorption of blue light down to 50 m. The carotenoids remain as functional pigments below 50 m. At least some algae outwardly appear to have pigment adaptations to utilize efficiently these available wavelengths at various depths. Thus you must consider not only the maximum depth to which light can penetrate, but also the quality of light in that region.

Light penetration is affected by factors such as the season of the year and the latitude. Some distance from the equator days are longer during the warmer seasons. In addition to shorter winter days, the angle of the sun cuts down on light penetration. Ice cover provides another layer that absorbs light. Obviously, cloud cover, in some areas more typical in winter months, and particulate material in the atmosphere must also be considered.

The biota itself will absorb and scatter light, but the phytoplankters do not proliferate in most environments to the point where self-shading would markedly decrease photosynthesis. With laboratory cultures, self-shading must always be considered, for cell numbers are higher. In nature, mats of floating algae and dense growths in eutrophic lakes would shade organisms below. This can account for low levels of dissolved oxygen (DO) in deep water.

In examining the vertical distribution of algae in the ocean, the maximum number of organisms is frequently found a few centimeters below the surface layer. The intense light at the surface, over 1,000,000

Table 18-3. FIGURES FOR LIGHT SATURATION OF PHOTOSYNTHESIS IN THREE GROUPS OF MARINE MICROALGAE[a]

Organism	Light Saturation in Lux
Green algae	8000 or less
Diatoms	C. 15000
Dinoflagellates	30000

[a] Data after Ryther.

lux on a bright summer day, can be detrimental to photosynthesis. In one study Ryther reports that light saturation for three groups of organisms is in the 8,000 to 30,000 lux range (Table 18-3). An additional 10,000 lux produced some inhibitory effects. When the green algae, with light saturation below 8,000 lux, were exposed to full intensity sunlight, photosynthesis was markedly reduced to about 5 to 10% of the rate at saturation. In some cases the intense light at the water surface can kill plankton, unless the cells develop the protection of additional pigments or opaque wall material.

As light penetrates bodies of water and is absorbed, the intensity falls off rapidly (Fig. 18-2), especially in turbid waters. In some lakes and reservoirs in the plains states the water is commonly turbid, and lack of sufficient light limits phytoplankton growth. Light quality also changes. Thus light can be a limiting factor not only in such situations, and at the water surface, but also under ice (and snow), in the ocean at the high latitudes, and on cloudy days, especially in the winter.

Temperature

In the temperate zone most lakes of sufficient depth (over 10 m) show stratification, with an upper warm layer called the epilimnion and a lower hypolimnion. In the summer at any one time the temperature in the epilimnion is quite uniform. The summer heat will cause a temperature rise, and any wind will cause circulation in the upper layers. (There is much less circulation of waters in the hypolimnion.) In the region between the two there are sharp temperature changes, about 1°C each meter; this is the thermocline. The temperature in the hypolimnion would be the lowest in the lake, and rather uniform throughout. In the epilimnion, which is the photic zone, there should be a high reading of dissolved oxygen (DO), good light for photosynthesis, and uniform chemical and physical factors. The organic matter in the system is generated in this layer. In the hypolimnion the DO is lower, sometimes at zero level, and there is little if any light for photosynthesis. This then is a poor environment for growth, even though there may be an adequate supply of nutrients (Fig. 18-3).

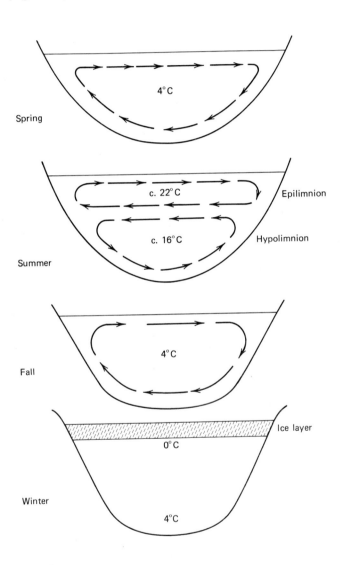

In the summer, organisms are circulating in the epilimnion and thus are rather uniformly distributed. In the hypolimnion, although there is some circulation, the number of cells settling frequently exceeds the increase due to multiplication. This settling is attributable to various factors limiting growth.

With stratification, the nutrients tend to increase in the hypolimnion. Many of these nutrients can be recycled during turnover, but others can get trapped in the sediment. If there is sufficient organic matter, active heterotrophic growth can eliminate the DO. When there is little DO, inorganic nutrients can be more easily released from the sediments.

Overturn occurs twice a year. In the fall as the upper waters cool, they sink. This is at first only to the level of the thermocline, but soon there is good mixing throughout. At this time there can be a significant increase in growth, as well as a different distribution of the organisms, because of the changes in nutrient composition at the surface. In the hypolimnion the hydrogen sulfide produced by bacterial growth can be flushed from the system, and the DO can be brought back to saturation (Fig. 18-3).

Inasmuch as water has its greatest density at $4°C$, the deep water of a north temperate lake is at that temperature in the winter. There is a drop in temperature as one proceeds up to the ice cover. Since there is little effect from wind, there is a minimum of mixing. Because of the ice cover, snow cover, and short days, there is little available light. Winter growth is controlled by light and not temperature. Some researchers have suggested that organic utilization must be considered when trying to explain phytoplankton growth under these conditions. But there is some doubt that the level of organics is sufficient to allow for growth, when there is bacterial competition for the same molecules. Under ice and snow there is usually a decrease in numbers of phytoplankton species, but at times there can be sufficient numbers to designate the population a bloom.

After the ice melts in the spring, the winter temperature throughout will be low; mixing can occur because of wind action. If the water is driven to one side of the lake, upwellings on the opposite shore will account for the mixing. The DO is high and the nutrient level is sufficient for the spring or early summer pulse or bloom.

In shallow lakes, even if there is stratification, there can be mixing other than during the spring and fall overturn. There can be bottom movements, however slight, called seiches. Wind action or the flow of water through the body of water can also account for some mixing. Whenever the conditions at the sediment are anaerobic, nutrient release can occur, and mixing will supply adequate nutrients throughout. It should also be remembered that decomposition of organic matter occurs throughout the water column, so that complete mixing is not essential for a recycling of some nutrients. As we will see in the section on flotation, organisms can sink to a level where there is a nutrient supply, and grow more rapidly at that level.

Stratification might not be explained so simply in some lakes. Because of topographical features, prevailing winds, depth, or seiche activity in the water itself, there might be no stratification or its duration could be shortened. In the ocean there may be little mixing of the water column for several months, as in the Sargasso Sea. Stratification may occur only in the summer as in temperate waters, or at any time throughout the year. The severity of storms, which provide various degrees of mixing, prior to the growing season will affect the primary production for that year.

Stratification has to be considered in supplying water to municipalities from reservoirs. In the summer one wants to avoid the epilimnion, where the abundant algae could cause taste and odor problems, or clog filters. In addition, if there is an anaerobic hypolimnion, this region is avoided because of the production of the gas, hydrogen sulfide, or formation of reduced heavy metals. The thermocline is the proper location for the intake to the water supply.

In the Atlantic the increase in numbers of phytoplankton in the water column at different latitudes has received a great deal of attention. For example, it was thought that there was more production in the colder, northern waters. However, the figures were based on spring or summer collections, and had little to do with annual production. On a yearly basis, growth is more nearly the same at all latitudes. Thus there is not the temperature effect that seemed apparent. In the Pacific, studies confirmed the above results, and showed that the areas near the continents, or in upwellings, were the most productive.

There are optimum temperatures for growth for individual species and there is selection for certain types in different habitats. Often the time when the water temperature is highest, favoring certain types of organisms, does not coincide with the weeks when there would be the most hours of available light. In addition, there can be marked differences between field and laboratory findings. For example, a *Chlorella* isolated from a Lappland lake in which the temperature never was above 7°C, had a laboratory optimum for growth of 20°C. For *Skeletonema* various strains are reported to have a field temperature optimum from 12 to 20°C, but in the laboratory the range is 20 to 30°C. It appears that other factors, such as level of nutrients, also have an effect.

In our brief account we can only mention a few other parameters. For example, the optimum temperature for growth will differ from that for reproduction, or even for photosynthesis. The annual *range* of temperature must be considered, especially in colder waters, for this can be from 0 to more than 30°C. At the lower temperature there are effects of ice on organisms, such as injury to cells, or nutrient concentration such that there could be osmotic problems.

Inorganic Nutrients

Before discussing specific nutrients for marine forms, the first consideration is the salinity. It is normally at 35 parts per thousand (ppt). The figure may go over 40 ppt as in some parts of the Red Sea, in certain salt lakes, and especially tide pools that are not flushed with every tidal cycle. Typically the proportion of the major elements remains constant, and, because of this, the salinity can be measured directly from a chlorinity determination.

Along with adequate light and temperature, one must consider the availability of nutrients as one of the most important factors regulating plankton growth. It is clear that the absence of one essential element, or its presence in the least favorable concentration, will limit growth. As a result there has been much effort toward identifying the nutrient or nutrients present when maximum growth is triggered, as well as to determine which element eventually limits growth. The level of a nutrient present at a given time represents a steady state between uptake and a variety of processes whereby an organism loses the nutrient to the environment. At one time or another N, P, S, Fe, Co, Mo, Mg, Mn, Si, and Zn have been shown to be limiting in fresh waters. In the marine environment, N, Fe, Si and perhaps one or two other elements have been implicated. In Antarctic lakes the limiting nutrient is known to be ammonium, whereas neither N nor P were found to be limiting in Arctic lakes. In the latter a combination of several minor elements stimulated growth. The problem is most complex, but now in most situations we usually look to nitrogen and phosphorus as the possible limiting, and thus significant, elements.

Phosphorus may be available in microgram per ml quantities in nature and frequently is not detected after a bloom. One can follow the spring increase in plankton populations and show that when maximum growth is achieved, the levels of nitrogen and phosphorus are reduced. Pearsall, over 40 years ago, was one of the first to demonstrate the increase in both N and P during the winter, with a marked decrease after a bloom. The numbers of phytoplankton then drop off. Since humans greatly increased phosphorus concentrations through use of detergents, fertilizer, and some industrial processes, we have seen much research and discussion in this area. And rightly so! However, the pendulum does swing back, and we now know that nitrogen is the limiting factor for phytoplankton growth along the continental shelf. Nitrogen, readily available as a gas to only a few organisms, is plentiful in that form. There are 700 g cm^{-2} of earth's surface. It is frequently in short supply in fresh waters, but seldom is it limiting. It appears that we have paid little attention to carbon availability, and most researchers assumed that it was never limiting. No doubt in some selected cases, especially in midafternoon in some lakes, carbon can be in short supply.

As one might expect, this is a difficult area of research, and precise measurements are not always easy. Phosphorus, for example, can be recycled two times a week, or the mean life in the water column can be only minutes. If you measure the concentration of the element or compound in the water, you are probably not taking into consideration the possibility of luxury consumption. For example, *Asterionella* can concentrate phosphorus 150-fold.

In English lakes, such as Lake Windermere, there may be only a microgram or two of phosphorus per liter in solution. However, a release of phosphate from the sediments in one English lake would increase the phosphorus concentration one thousand times. With one microgram of phosphorus certain diatoms, and perhaps many kinds of microalgae, will form a million cells, whereas *Asterionella,* with a low phosphorus threshold, will produce 16 times as many cells.

Increased phosphate levels have been one of the main causes of excessive algal growth in many streams and lakes throughout the world. A good example is Lake Washington, near Seattle. Now algal growth has been reduced by sewage diversion; phosphate concentrations are decreasing. Some link the spring algal bloom in small lakes to a phosphorus increase from melting ice. It would be difficult to separate this phenomenon from turnover of nutrients from deeper waters.

In fresh water, an *Asterionella* bloom terminated when the level of silica dropped below 0.5 mg l^{-1}. Fresh waters throughout the world frequently have 10 to 50 times that level. A figure of 0.5 mg silica per liter as a limiting factor is in close agreement with some laboratory findings. However, this amount is not always limiting, for when lake water is examined in the laboratory, these same levels do not limit growth if phosphate is added. Growth of *Asterionella,* in the presence of the additional phosphate, continues until the silica concentration reaches zero. In other experiments Fogg found that *Asterionella* required 2 μg l^{-1} phosphorus in natural waters from the lake, but to get the same cell numbers in culture 40 μg l^{-1} were needed. Some unknown interactions account for these different figures.

Skeletonema can grow even when the level of silica in the marine environment is greatly reduced. However, a relatively thin wall is formed. Various species of diatoms might be able to dominate at different levels of silica. In the ocean levels as high as 2.8, or even up to 5 mg l^{-1}, have been reported. In Long Island Sound, New York, the water leaving the Sound has 0.08 mg l^{-1}, but 0.16 mg l^{-1} is coming into the Sound with the ocean water. Thus silica storage in the Sound amounts to 1.2 kg per second.

In an Alaskan lake, magnesium was the limiting factor, while molybdenum was responsible for low productivity in a California lake. In the latter case the addition of 0.1 mg l^{-1} stimulated an increase in photosynthesis. Patrick has shown that low temperature and high Mn levels (40 to 300 mg l^{-1}) stimulate diatom growth. Blue-green algae dominate when there is low Mn and high temperature.

We should perhaps emphasize nitrogen and phosphorus as elements that control the amount of growth, but should not overlook the possibility of any elements, at least in isolated cases.

Nor should we overlook the fact that several nutrients can be limiting simultaneously. This was pointed out by Smayda in recent bioassay studies in coastal waters. In addition to questioning the view that growth of marine phytoplankton is limited by one specific limiting nutrient, he showed that the combinations of nutrients involved varied seasonally. The natural water was conditioned by growth of endemic phytoplankton. He had to add EDTA, along with other nutrients, to stimulate growth of the assay organisms.

Lange has shown in culture that blue-green algae can competitively exclude other forms. A *Phormidium* would grow best in a combined culture with *Microcystis* and *Nostoc* because the former could use iron and other trace elements more efficiently.

In this brief survey I have only touched on a few elements of the whole inorganic nutrient story. Consult references and recent publications for detail.

Carbon Dioxide

Because of the reserve of bicarbonates in the sea, low levels of carbon dioxide are not found there, at least levels that will limit growth. However, in freshwater lakes there are wide variations in carbon dioxide content, depending on region, time of year, and time of day. In acid lakes the free carbon dioxide may be close to 200 ppm. In lakes in which the pH is near neutral the free carbon dioxide will be about 30 ppm. Alkaline or hard water lakes contain little if any free carbon dioxide, but organisms will use available bicarbonates. In lakes with pH values over 8.5 there may be between 30 and 200 ppm carbonates. Precipitated carbonate forms, with the result that these are called marl lakes.

Levels of CO_2 will be low in colder lakes under ice, as well as in the afternoon of a bright sunny day. In the latter case, photosynthesis causes this depletion, along with a pH rise. Blue-green algae bloom under conditions of low CO_2. In some cases, as in soft waters, there is the hypothesis that bacteria associated with the blue-greens can supply the carbon dioxide.

Carbon dioxide can be, at least temporarily, a limiting factor in fresh waters. In addition, the form of inorganic carbon available must be considered, for some *Chlorella* species will not utilize bicarbonate. Thus, in culture, they exhibit maximum growth when the pH is low, and free carbon dioxide is available. How many organisms are selective in their utilization of inorganic carbon?

Oxygen

In the ocean, both at the surface and in deep waters, the dissolved oxygen (DO) is at acceptable levels because of surface mixing and deep water circulation. In lakes the DO varies with depth and season. Lakes may be supersaturated or, in bottom waters, have no DO. Only in the latter case would the oxygen level affect phytoplankton. Excess algal growth at the surface can lower the DO in a lake. This is accomplished because of shading and a loss of photosynthesis in deeper water. Decaying algal blooms utilize oxygen.

Organic Nutrients

At present the effect of organic nutrients is a fertile area of research but we still lack complete information. However, the B vitamins have been examined in some detail. Thiamine, biotin, and B_{12} are the vitamin growth factors that could limit growth of phytoplankton. Many truly planktonic organisms require one or more of these vitamins and, as the work of

Provasoli and others has demonstrated, it appears quite reasonable that one or more would be limiting in some areas. Seasonal studies have shown that the level of vitamin B_{12} is lowest in the summer because it has apparently been used up in the spring pulse.

Many organisms require other organic compounds for growth, but there is an even larger number of algae that utilize organic compounds, while photosynthesizing, if the compounds are available. If organisms have the potential for organic assimilation, is it not reasonable to assume that, at some time, the absence of the compound would affect distribution of the alga? Let us examine this further. Uptake of organic carbon is impossible for a number of organisms, but others, in laboratory experiments, have been shown to utilize one or more organic compounds. As one might expect, there is a wide range of possibilities. Certain organisms can utilize only one compound, others absorb many different compounds. For some, utilization may take place only in the light, but other organisms can metabolize organic compounds for growth in complete darkness. The possible specialization is indicated by a *Cyclotella,* which can metabolize only glucose. However, utilization does not take place when there are long days, but rather only under light conditions occurring in winter months.

Because considerable laboratory work has been performed with organic concentrations much higher than levels occurring in nature, there is some question as to how organisms operate in nature. Bacterial uptake systems are able to remove organics at microgram per liter levels. Few algae possess similar mechanisms, and at such low substrate concentrations diffusion into algal cells cannot occur. Some blue-green algae can take up organics at $\mu g \, l^{-1}$ levels. It has been suggested that they use the growth achieved through organic utilization to maintain all the pigments and enzymes necessary for immediate active growth. When the blue-greens do receive sufficient light for active photosynthesis, they can immediately trap the first light. The organism in the sediment without available light would thus not become moribund, but would be prepared for relocation to a more favorable environment.

Some compounds that affect phytoplankton growth are algal extracellular products. There is some excellent information available on the effect of extracellular products of one organism on growth of another organism, both in nature and the laboratory. Pratt found that a diatom, *Skeletonema,* and a flagellate, *Olisthodiscus,* never bloomed together, *Skeletonema* is highly successful because it has a better growth rate than *Olisthodiscus.* However, *Olisthodiscus* has some control over the

Skeletonema bloom, because of its production of an extracellular product, which stimulates growth at low concentrations but inhibits development at high concentration. Among the blue-green algae a bloom of *Microcystis* can be followed by one of *Aphanizomenon*. These genera are seldom found together. Perhaps this is another example of organic inhibition. The "bloom" and "crash" can occur within a few days, followed by a different bloom. However, in several eutrophic lakes in India *Microcystis* can be dominant for months. Bloom organisms from Linsley Pond, Connecticut, have been observed in nature and grown in culture. Each dominant bloom organism exhibited either an enhancing or neutral effect on successor organisms, but either inhibitory or neutral effects on their predecessors. For example, a stimulation of the growth of organism B grow in the filtrate of organism A (Table 18-4) was obtained both with filtrate from cultures of organism A and water from a Linsley Pond bloom of organism A. Other laboratory studies have shown antagonism between two green flagellates in culture because of fatty acids released. It has been proposed that certain phytoplankters can inhibit growth of some zooplankters.

Vitamins required by many phytoplankters are released into the environment by other microorganisms. A few algae are known to release vitamins required by other algae in the same habitat (Fig. 18-4). In culture studies *Coccolithus* gives off vitamin B_{12}, which is necessary for

Table 18-4.   THE EFFECT OF CERTAIN BLOOM ORGANISMS FROM LINSLEY POND, CONNECTICUT ON THE GROWTH OF OTHER ORGANISMS IN THE POND[a]

Filtrate of	Organism			
	A	B	C	D
A		+		
B	−		+	
C		−		+
D			−	

[a] Plus indicates a stimulatory effect, minus an inhibitory effect. Neutral effects are not listed. Data after Keating.

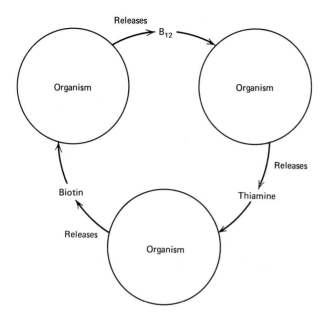

Figure 18-4. Theoretical cycling of vitamins in inland marine waters. Algal species are known that release B vitamins into the environment. Some of these algae, or other forms, have requirements for these same B vitamins.

growth of *Skeletonema*. The latter can then grow and form the thiamine needed by *Coccolithus*.

Several of the compounds we have discussed come from algae or bacteria in the plankton, but they, or other compounds, could originate from soil or from bacterial and fungal decomposition, especially decomposition of the remains of a bloom. The activity of all compounds can be modified by other factors, including pH, temperature, other chemicals, and so forth.

After the early work with *Chlorella* in which the organisms released chlorellin, a compound that inhibited further growth, emphasis was placed on autoinhibition as a controlling factor. But there is no indication that this occurs normally in nature. The report on chlorellin was obtained from laboratory cultures in which the cell densities were excessively high, when compared to nature. We now know that not all *Chlorella* cultures will produce the autoinhibitor, even with very dense cultures.

Recently there has been research in progress to determine if inhibitors produced by a colonial green flagellate can be used to control growth of troublesome organisms in nature (Chapter 20).

Floating or Sinking

A planktonic organism must have some special adaptation to remain in the photic zone, for protoplasm, skeletons, and walls are heavier than water. Often cells do this through active growth, for these organisms often are readily suspended in the water column regardless of their morphology. As organisms age, or become nutrient deficient, they may sink at a more rapid rate; dead cells are not good floaters.

Certain organisms, not only the diatoms but also some other nannoplankton forms, can reduce size through many divisions, and thus have an increased surface area to volume ratio. This ratio increases frictional resistance to settling. More often there are also additional adaptations to retard settling, such as the spines of *Chaetoceros*, bristles, or an elongate, needlelike shape, as in *Synedra*. Many organisms form chains of cells or ribbonlike growths. Certain chains of cells lose little surface area through the joining cells. For example, there is considerable exposed surface area in adjacent *Skeletonema* cells or in some *Tabellaria* colonies.

Vacuoles are formed by some organisms and the ions held within the vacuole might make such an organism lighter than its vacuole-less counterpart. The pseudovacuoles of blue-green algae, as in *Microcystis* and *Aphanizomenon*, probably increase the alga's buoyancy. In a Minnesota lake an *Oscillatoria* was found at about 5 m. When placed either above or below that depth, the organism regulated its density, perhaps involving gas vacuoles. It soon became distributed at the 5 m level once more. This response was influenced by availability of nitrogen, and by temperature and light conditions. Larger organisms, such as the floating *Sargassum*, have conspicuous air bladders, and close relatives attached at the shore also appear to display photosynthetic portions more readily with placement of air bladders along the plant body. These bladders also could aid in keeping organisms from becoming entangled.

The accumulation of fatty food reserves or oil could also provide some buoyancy, but another factor must also be considered. For example, a diatom might store oil, and at the same time produce a much thicker silica wall. The end result would be a heavier cell and rapid settling.

Some organisms appear to have a shape that would provide maximum surface area, but under certain conditions accumulate abundant mucilage. This might be detected only after examination in India ink. The mucilage could be quite light, but there is a decrease in the ratio of the surface area to cell volume.

There have been experiments dealing with the selective accumulation of ions by organisms as a mechanism for regulating buoyancy. Although there are many such examples, Smayda recently stated that no one has proved that this system applies to photosynthetic organisms in nature.

There appears to be good evidence for such a system in the colorless *Noctiluca*.

Regardless of the adaptations that might occur, there is always the factor of turbulence. Since most protoplasm and all walls and skeletons are heavier than seawater, turbulence must be considered. At the same time, the orientation of cells and colonies moving through a mass of water is affected by ornamentation. Some organisms might be using their ornamentation to regulate orientation toward favorable light intensities.

Using settling chambers and an inverted microscope Smayda has examined the sinking rates of several cultures of marine diatoms (Table 18-5). Organisms were layered on the top surface of the settling chamber and then the time required for touch down on the bottom was recorded. For *Thalassiosira* and *Nitzschia* it would take between 2 and 2000 days for the cells to sink 100 meters, and between 5 and 2000 days for *Skeletonema* and *Rhizosolenia* to settle the same distance. From these data it is apparent that with a little turbulence some cells in each genus will remain in the photic zone indefinitely.

There have been reports of morphological differences in species found in distinct environments. *Skeletonema costatum* found in Florida is adapted to the less dense water by a size reduction. Cells are smaller than in Puget Sound, Washington. Similar information was obtained with a *Chaetoceros*. When *Ceratium*, the dinoflagellate, is found in warmer water, there is an increase in spine or horn length; there might also be an extra horn in warmer waters. However, a *Rhizosolenia* appears to be larger in warmer waters. Perhaps the wall is more lightly silicified in the larger form, giving the *Rhizosolenia* more time in the plankton.

The shape of the cell and the presence of appendages, known to be of importance in providing resistance to settling, should be considered when discussing grazing by zooplankters and mechanisms for increasing

Table 18-5. MEAN SINKING RATES OF DIATOMS AS
DETERMINED FROM LABORATORY STUDIES[a]

		Meters day $^{-1}$		
	Minimum	Young Cells	Old Cells	Maximum
Thalassiosira	0.05	0.1	0.28	45
Nitzschia	0.2	0.26	0.5	3
Skeletonema	0.2	0.3	1.4	7
Rhizosolenia	0.05	0.2	1.8	23

[a] After Smayda.

nutrient uptake. Also the increased surface area, at least of the plasmalemma, no doubt affects nutrient uptake. Appendages can provide mechanisms for keeping organisms somewhat separated and thus avoid mutual shading.

When the nutrient levels are low, some settling could be advantageous. Detrimental effects of high light intensities on cells that are showing signs of nutrient limitation would then be avoided. The organisms could recover in water with a richer nutrient supply and later be moved back to an area where maximum photosynthesis could take place.

Grazing

When considering the growth of phytoplankton, one must also look at the grazers, organisms at the next energy level, especially if they occur in relatively large numbers. Also the rate of growth of these herbivores is of importance. Even a continuous grazing of a small portion of the phytoplankton has a marked effect on the algal population. For example, if 100 organisms were to divide six times during a period of a few days, 6400 phytoplankters would result. However, if there were 10% grazing at each step, after each division there would be an end result of 3410 cells, with only 413 grazed. With 20% grazing, there would be just 1692 cells of phytoplankton after six divisions.

Organisms can regulate the amount of grazing, to some extent, by their position in the plankton, release of extracellular products, and the formation of certain structures, such as spines, trichocysts (formed by some dinoflagellates), or walls and shells, which offer protection. After reproduction or sporulation certain cell types may be produced which will enable the organism to survive intensive grazing. The reproductive stage will merely pass through the gut undigested. In one recent study large forms, or filaments, were seldom found in the gut of zooplankters. Encased gelatinous organisms were present, but they were not digested. Many small unicells and colonies were suppressed by grazing. Thus grazing pressure can affect the relative proportions of algal species. In some cases the smaller flagellates are removed and blue-green algae become dominant in warmer months.

BLOOMS—MAXIMUM GROWTH OF ALGAE

It has been known and recorded for centuries that algae can grow in bodies of water to sufficient numbers to change the water color. In the Old Testament there is an account of a river changing to a blood red color; the Romans recorded similar phenomena. Throughout the centuries there have been records of algal blooms in Europe.

Determination of the cause of the spring pulse, a rapid increase in algal numbers as a result of exponential growth, is a complex process. Rarely are these blooms the result of only one factor, but one can be the dominant factor. In various situations the factors differ, but the temperature increase, the increase in radiant energy with longer days, and nutrient availability are all involved. In the ocean, the maximum nutrient levels can be achieved in late fall, but the pulse of growth occurs 2 to 3 months later. But in fresh waters there is a spring nutrient increase due to overturn. The nutrients could have been released several weeks earlier from sediments because of anaerobic conditions at the bottom, but are not available to the epilimnion until overturn. Spring run-off could supply increased nutrients to coastal waters and also decrease the salinity.

A population decrease, and even a leveling off of numbers, are caused by the exhaustion of at least some nutrients. Also involved in at least some environments are extracellular products, inhibitors, death of cells (even involving parasites), competition for light, loss in outflow (as in a lake), sedimentation and grazing.

In temperate marine and fresh waters, a spring and fall pulse can often be found (Fig. 18-5). As the water becomes more nutrient-rich, the decrease in cell numbers after the spring bloom becomes less noticeable, or even disappears (Fig. 18-5). Polar waters are characterized by a single pulse later in the season, occurring in the summer. Higher standing crops are found in tropical waters but the seasonal increase in cell number is not as marked. However, yearly figures for production are equal to, or better than, those from other areas.

What is the source of organisms in a bloom? The inoculum could be always present, but in low cell numbers. This is probably true for the freshwater diatom, *Asterionella*. Other bloom organisms could be in a resting stage during winter months. In other cases there could be a periodic reintroduction from adjacent waters. This is probably true of oceanic blooms, and might be the source of the red tide organisms.

As we have discussed previously, some organisms are not simultaneously present. For example, *Olisthodiscus* and *Skeletonema* are not found together in Narragansett Bay, and diatoms and dinoflagellates are not observed together in large numbers at one time. On the other hand, *Oscillatoria rubescens* is frequently present in lakes that have

Figure 18-5. The seasonal increase in numbers of phytoplankton in the temperate zone, eutrophic lakes in the temperate zone, in the arctic, and in the tropics. There is a spring increase and a fall pulse in temperate lakes, with but one pulse in Arctic lakes. In eutrophied lakes three curves are presented, with the highest curve indicating the greatest nutrient supply.

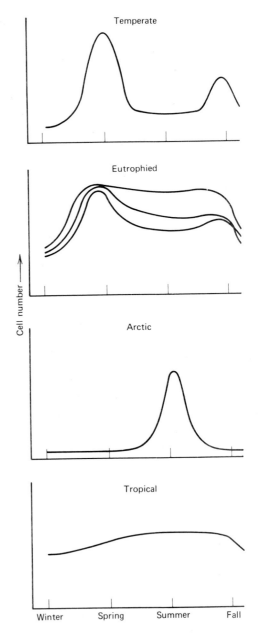

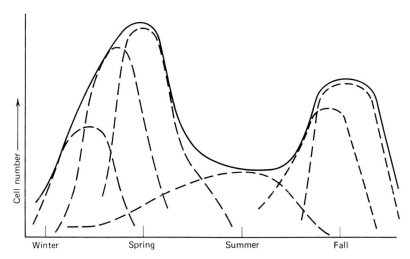

Figure 18-6. The composition of the spring and fall pulses in some temperate waters. Each pulse may be actually made up of marked increases in numbers of several different species (individual broken lines).

received a high pollution load. The growth of a single organism could be merely by chance, but the yearly occurrence of a particular bloom species, or some other troublesome organisms, might indicate the unique suitability of that precise environment.

In fresh waters one can find diatoms in the spring followed by green algae whereas by late summer blue-green algae may be dominant (Table 18-1). The organisms in each of these groups could be represented by 10 to 20 species at one time, or there could be a dominant species for one short period followed by other dominants in quick succession (Fig. 18-6). In the latter cases release of extracellular products could be involved in succession. Among the blue-green algae, *Microcystis* can be one of the short-term dominants, or it may bloom for a couple of months as in Lake Waubesa, a Wisconsin Lake that has had considerable pollution. It is generally true that blue-green algae dominate where there has been a high discharge from sewage treatment plants.

Red Tide; A Single Species Bloom

The dinoflagellates *Gymnodinium* and *Gonyaulax* are usually organisms responsible for the large marine fish kills or paralytic shellfish poisoning. When present in large numbers, as up to 6×10^7 cells per liter, they may color water red, or yellow to brown. They may account for more than 99% of the organisms present at that time. It is now thought that increased nutrient levels along with a drop in coastal salinity, both the result of run-off from the land, can provide an environment favorable to red tide development. Also heavy rain, calm, sunny days, and the meeting of dissimilar waters all cause an increase in vertical stability, which some feel is responsible for blooms. These tides are found off the Gulf Coast of Florida, in California, and in New England. They have been reported in India, South Africa, Peru, and Japan and appear to be concentrated at midlatitudes. Some phycologists have suggested that such single species blooms must result from a release of substances that either are self-stimulatory, or are toxic to other phytoplankton. Additional information is presented in Chapter 20.

OTHER ENVIRONMENTS

Little has been said about other organisms that are related to these planktonic forms or about the environments in which they may be found. Similar forms can be found in soil, streams, artificial ponds, tide pools, the air, and rivers. Each of these provides a unique environment for the alga. An introduction to many of these habitats is presented in Chapter 15. In addition the student might wish to check the abundant literature on this subject.

REFERENCES

Alexander, M. 1971. *Microbial Ecology.* John Wiley and Sons, Inc. New York. 511 p.

Boney, A. 1975. *Phytoplankton.* Crane, Russak and Co., New York. 116 p.

Cairns, J. 1971. *The Structure and Function of Fresh-Water Microbial Communities.* Res. Div. Monograph 3, Virginia Polytechnic Inst. and State Univ., Blacksburg. 301 p.

Davis, C. 1955. *The Marine and Fresh-Water Plankton.* Michigan State Univ. Press. 562 p.

Duddington, C. 1967. *Flora of the Sea.* Crowell Co., New York. 207 p.

Ferguson Wood, E. 1965. *Marine Microbial Ecology.* Reinhold Publ. Corp., New York. 243 p.

Fogg, G. 1975. *Algal Cultures and Phytoplankton Ecology.* Univ. Wisconsin, Madison. 2nd ed. 175 p.

Goldman, C. 1966. *Primary Productivity in Aquatic Environments.* Univ. Calif. Press, Berkeley 464 p.

Gorham, E., J. Lund, J. Sanger, and W. Dean. 1974. Some relationships between algal standing crop, water chemistry and sediment chemistry in English Lakes. *Limnol. Oceanogr.* 19: 601–17.

Hickman, M. and D. Klarer. 1975. The effect of discharge of thermal effluent from a power station on the primary productivity of an epiphytic algal community. *Br. Phycol. J.* 10: 81–91.

Maeda, O. and S. Ichimura. 1973. On the high density of a phytoplankton population found in a lake under ice. *Int. Rev. Ges. Hydrobiol.* 58 : 673–85.

Pringsheim, E. 1967. Phycology in the field and in the laboratory. *J. Phycol.* 3: 93–5.

Reid, G. 1961. *Ecology of Inland Waters and Estuaries.* Van Nostrand Reinhold Co., New York. 485 p.

Smayda, T. 1974. Some experiments on the sinking characteristics of two freshwater diatoms. *Limnol. Oceanogr.* 19: 628–35.

Titman, D. 1975. A fluorometric technique for measuring sinking rates of freshwater phytoplankton. *Limnol. Oceanogr.* 20: 869–75.

Vallentyne, J. 1974. *The Algal Bowl.* Dept. Environment, Fisheries and Marine Service, Ottawa. 186 p.

Algal control

METHODS OF CONTROL

Limit Nutrient Input
Chemical Compounds
Application of Chemicals

CONTROL INVOLVING LIVING
ORGANISMS

Grazing
Phycoviruses

MECHANICAL REMOVAL

REFERENCES

During the last decade or two more people have become interested in algal control not only because they might have seen algae growing in their swimming pools or local ponds, but also because algae are responsible for an unpleasant taste to drinking water. Extensive coverage has been given to regional problems on TV and in the press. Millions of people have heard of the pollution in Lake Erie, Lake Washington, and the Mississippi and Hudson Rivers. Only occasionally must we concern ourselves with control of algae in nonaquatic environments (Fig. 19-1).

Figure 19-1. Filamentous green algae growing on the side of an abandoned house along the California coast. This is not an uncommon sight in areas with high humidity.

Fortunately, we have some previous history of the successful use of algicides. (Some compounds do not kill algae, but instead temporarily inhibit growth; they are algistatic.) Information about the effectiveness of algicides was obtained from localities where algal control has been a long-standing problem. The lakes in and adjacent to Madison, Wisconsin, have been treated with copper sulphate for several decades. From 1950 to 1954 10^5 kg (125 tons) of copper sulphate were used in Lake Waubesa alone. The offensive algal blooms occurred because of spreading urbanization and increased use of the lakes, even for discharge from a sewage treatment plant. The problem was eventually alleviated because of diversion of the sewage effluent and the upgrading of both domestic and sewage treatment plant facilities. Prior to diversion algal control was achieved yearly with algicides.

The number of algal nuisance problems is on the increase in all developed countries. In general, the trend parallels the increase in the level of nutrients that we expect rivers, lakes, and coastal marine waters to carry. The most effective form of algal control, surely not most convenient nor the least costly, is to stop using bodies of water for nutrient disposal. We should develop procedures to recover nutrients before discharging any effluent.

However, in the real world we must often settle for something less than ideal, at least temporarily, or for a degree of control. The problem of control of excess algal growth must be attacked differently in various systems. First consider the water use. In reservoirs we do not want to allow much growth of any alga that would impart even a slight taste or odor, nor to use an algicide that would provide an objectionable taste or health problem. In ponds used for fish culture, the owner must have a selective algicide, one effective against the objectionable alga, but not harmful to other algae, nor to organisms that are constituents higher in the food web. An algicide that is not effective on contact might be quickly removed from the system. In swimming pools, the bulk of several algicides can be lost on the first complete cycling of water through filters, which could occur in as short a time as 6 hours.

Cooling towers are frequently associated with power plants. They can be used to lower the temperature of a thermal discharge. In such towers it is essential to keep surface areas clean and avoid clogging of tubing. A particular type or concentration of algicide could be found necessary, especially if attached forms become dominant.

In swimming pools water maintained near pH 7.2 will usually not support much algal growth unless there is a poor water source or detritus is allowed to accumulate in the pool. Additives of chlorine, for example, as trichloro-S-triazinetrione, are useful not only in bacterial control but also in limiting algal growth. The pool filtering system, using diatomaceous earth, will remove both algae and particulate matter.

METHODS OF CONTROL

Limit Nutrient Input

In most instances the algal problem was caused by a previous activity of humans. Too many summer cottages were built in the watershed of a recreational lake. A drain field failure released nutrients in a small pond in the back yard. A farmer used too much fertilizer. Industry was careless in the discharge of chemicals into waste water.

Even though one of the compounds described below would solve the immediate problem, a permanent solution requires that the cause of the problem be located. Both Lake Washington, which for many years received effluent from waste treatment facilities in the greater Seattle area, and the Madison, Wisconsin, lakes serve as examples of control by limiting nutrient input. In Lake Washington the figures for phosphate concentration, presence of the blue-green *Oscillatoria* and objectionable growth of algae all are in agreement. With a metro sewer system, and diversion, there is less phosphate and fewer algae. Recovery of the lake is taking place.

Chemical Compounds

Copper Sulfate.   This algicide is used more often than all other compounds combined. It is applied at the concentration of approximately 1g in 4 l in many water purification plants in the summer. Copper sulfate contains about 40% copper. Table 19-1 provides comparative figures so that the reader can see how much of this compound is needed on the job. The rate of application is often determined by using acre foot measurements. One acre foot is a surface area of one acre, one foot deep. Frequently an algal problem is restricted to surface waters, and the copper sulfate is then sprayed on the lake or reservoir. Calculations of the amount of algicide needed can then be made for the *surface* acre foot. When there is excess growth of plankton, or when bottom-dwelling organisms are to be killed, the volume of the body of water is calculated, and application is made. Determine the surface area and the average depth, and then convert to acre feet. Often estimates are made when bottom contours are unknown. In the laboratory only a few micrograms of copper are needed for optimum growth, and as little as 5 μg Cu l^{-1} can decrease the growth rate (Table 19-2). Sensitivity, however, is related to the binding of the chelator. In some culture studies 5 mg Cu l^{-1}, which was strongly chelated, proved stimulatory.

Other Chemicals.   Many of the compounds used for algal control are identical to or similar to bactericides or fungicides. There has been a long history of use of these compounds, for plant pathologists have long been interested in preventing crop loss caused by bacteria or fungi.

Table 19-1. THE LEVELS OF COPPER SULFATE, IN MILLIGRAMS PER LITER, USED FOR ALGAL CONTROL IN VARIOUS WATERS

	CuSO$_4$ in mg l^{-1}
Water near neutrality	0.2
	(or 2 lb in 10^6 gal,
	or 10 oz acre-foot^{-1})
Water with low alkalinity	0.1
Water with high alkalinity	up to 2.0

Table 19-2. IN NATURE AND IN LABORATORY EXPERIMENTS COPPER IS FOUND, OR IS USED, IN MICROGRAM PER LITER AMOUNTS

	Cu in $\mu g \, l^{-1}$
Fresh water	15
Marine waters	2
Culture media	4

Those interested in obtaining new algicides would screen compounds already in use as bactericides or fungicides.

Halogenic Elements.   These elements are successful when used as disinfectants, but the effect is too short-lived for usefulness as an algicide. In addition, the use in alkaline waters, or where there is organic enrichment, as near sewage outflows, is quite restricted. They are not as effective in these conditions. Recently, the addition of chlorine to drinking water has been seriously questioned. Chlorinated organics formed from such usage may be carcinogenic.

Silver Compounds.   In addition to the expense, these compounds would not be used where they could be filtered from the system. They have been used in small ponds, and even in swimming pools, but most of the activity can be removed with the first filtering. The water is normally filtered about once every 5 to 6 hours in a swimming pool. There is also precipitation and loss if the chlorine level is about 12 mg l^{-1}. The recommended concentration in indoor pools is 80 mg l^{-1} and three times that amount out of doors. Thus it is not realistic to use silver compounds under these conditions.

Potassium Permanganate.   Although this compound is corrosive and can form a dangerous precipitate, it is sometimes used on a limited scale. Under certain conditions there appears to be some value to potassium permanganate, but research in this area is not intensive. Potassium permanganate kills a wide spectrum of algae, as well as other organisms. It is sometimes used with other compounds as a mixture.

Sodium Arsenite   is an excellent algicide for filamentous algae and for larger aquatic weeds. It is used in southern waterways for control of water hyacinth, but its use has to be carefully controlled, since most forms of arsenic are highly toxic to humans.

Ammonium Sulfate.   In Israel in brackish water fish tanks, there may be excessive growth of *Prymnesium,* a flagellate that secretes a toxin. The control of the protist is achieved by use of ammonium sulfate. The ammonium and the alkaline pH affect the motility of *Prymnesium* first, and populations diminish because the organism can no longer feed.

Mercury Compounds.   At one time algimycin 200 and algimycin 300 were being suggested as algicides for use in swimming pools and had received some acceptance. Algimycin MT–4 was suggested for application in cooling tower problems. The two algimycins were banned from use in 1971 because they contain some mercury. A number of control chemicals popular at that time, especially fungicides, contained mercury.

Strong Oxidizing Agents   are now thought to be too dangerous to be used, except in small areas where there can be some control. They affect the complete biota and thus can not be strictly called algicides.

Sodium Pentrachlorophenate.   This compound is noncorrosive, readily soluble and quite stable for a period of time long enough for algicidal action. It has been used in cooling towers, but it is harmful to the eyes of those who apply it. Anyone swimming in waters where the cooling tower water is discharged may experience some discomfort in the eyes. This could be a most serious problem when the cooling tower is cleaned, and the water is discharged. This material is also toxic to fish and has a persistent taste. Although it appears that it could be useful in closed systems, the compound cannot be disposed of when its algicidal action is completed.

Dichlone.   This compound, 2, 3 dichloronapthoquinone, is known by its shorter trade name, Dichlone. It is selective for blue-green algae and can be used in μg l^{-1} quantities. However, the effect is of short duration, and the repeated application might be dangerous. More research is needed.

Chloranil.   Another quinone, chloranil, has been effectively used in cooling towers, but it is used in mg l^{-1} quantities, a far higher concentration than is needed for Dichlone.

Monuron or Chlorophenyldimethylurea   is useful for aquatic weeds and filamentous algae, but it is most effective when applied before there is considerable growth. Thus it is best as a preventive compound. However, it absorbs on the sediments, and will build up in the ecosystem.

2, 5 Dichloro-3, 4-dinitrothiophene   is very effective against both *Microcystis* and *Anabaena*. Less than 1 mg l^{-1} is needed. The cost is half that of copper sulfate, and it is safe to handle. The compound is not yet widely used.

Ethionine.   This compound, studied in the laboratory, has been found to be a selective algicide. At low concentrations it inhibits blue-green algae, which are 100 times more sensitive to the compound than *Ochromonas* and *Euglena*. Of all microorganisms tested, a protozoan was the least affected. However, we also know that ethionine is carcinogenic. Perhaps similar compounds could be used, if it can be demonstrated that they are not carcinogenic.

Mixtures.   Much of the early work with mixtures was an attempt to keep copper sulfate in solution or to arrive at some combination that would have a synergistic effect, that is, a combination that would be more effective than the sum of the individual responses. Citrate and citric acid were applied along with the copper sulfate. Later, chelators such as EDTA, ethylenediamine-tetraacetic acid, or sequestrene salts with a similar chemistry were used.

In Czechoslovakia researchers have used mixtures which they call CP 310 or CP 350. Mixture CP 310 contains 300 μg l^{-1} copper sulfate and 10 μg l^{-1} sodium pentachlorophenate; CP 350 has 300 μg l^{-1} copper sulfate and 50 μg l^{-1} silver nitrate. With mixture CP 350 there is a definite synergistic effect.

In the summer of 1971 mixture 350 was used effectively to control algal blooms in reservoirs. Blue-green algae were particularly numerous that summer and after application of the algicide mixture there was almost immediate control.

When there is mass algal mortality, either because of the use of an algicide or a natural population crash, a rapid decomposition of the organic material usually follows. When this takes place, there may be a drastic lowering of the dissolved oxygen (DO). (At times an excessive blue-green growth has already shaded much of the water column, and has already reduced DO. Decomposition would thus keep it low for a longer period.) Researchers from Czechoslovakia report no such adverse effects on DO when mixture CP 350 is used. Apparently there is still sufficient photosynthesis to retain adequate levels of oxygen.

Coagulants.   In some cases algae have been chemically precipitated from solution. There is then the problem of removing the algae from the system. Alum, AlK $(SO_4)_2$ is applied, causing a neutralization of charge, and precipitation results. In Australia the alum has been used to precipitate phosphate from an impoundment. A recurring bloom did not form during the year that alum was applied.

Research on Algicides.   An ideal algicide has yet to be found, but there is considerable experimentation with natural products. There is interest in experiments dealing with two membered cultures, or of growth of organism A in the filtrate of organism B, with the hope that one organism will produce a natural inhibitor. This compound could be a very specific algicide or might be effective on a broad spectrum of organisms. Knowledge of the succession of certain phytoplankton species leads us to believe that it is not always the physical and chemical environments that have marked effects on phytoplankton succession. Perhaps we can learn to manipulate some of these natural products.

Extensive screening experiments have been carried out with actinomycetes, but there has been little success. Research with *Pseudomonas* has shown that some strains release compounds that are highly toxic to some algae. Two compounds that have now been identified are phenazine-1-carboxylic acid, and hydroxyphenazine-1-carboxylic acid. This line of research seems to be most promising.

Application of Chemicals

The method of applying algicides depends on the season of the year, the temperature (especially if there is ice), the type of organism, and the presence of other biota. Large filamentous algae are treated a little differently than are microscopic forms. In addition, the type of body of water, the water chemistry, and the use of the water must be considered. When applied from a boat, the algicide can be sprayed, dragged behind the boat, as with large crystalline types, or trickled from a supply tank as the boat travels. Helicopters and airplanes can be used, and algicides can be supplied in the inlet waters at a concentration that will be effective in the lake or reservoir itself.

The response achieved by spraying usually occurs right after application. After the algicide mixes with deep waters the concentration is inadequate for any further effect. What is not taken up by the biota would then be tied up on the sediments, or flushed from the system.

Any algae killed may accumulate on the shore. These organisms should be removed from the water, as well as the immediate area.

CONTROL INVOLVING LIVING ORGANISMS

As stated previously some organisms produce compounds that affect other algae. However, other organisms have more direct and immediate effects.

Grazing

Intensive grazing of troublesome forms would introduce a natural and very effective control. However, we know little about producing a controlled grazing pressure, or even how to keep grazing in balance. Recently *Ochromonas* has been found to graze on the troublesome colonial form *Microcystis*. The latter species, toxic to domestic animals, did not inhibit the grazer.

Phycoviruses

For the blue-green algae, it has been suggested that we could use viral control if we could find a virus specific for a troublesome algal species. Most viruses now known have hosts that are not troublesome in nature. However, in 1972 Granhall reported that cyanaphages actually regulate the termination of the *Aphanizomenon* bloom in Lake Erken (Sweden). The virus is found associated with the vegetative cells. When the bloom was regulated by the virus, the heterocysts and akinetes were not destroyed. A distinct problem could materialize if resistant algal strains developed from repeated use of viruses.

MECHANICAL REMOVAL

Once confronted with excessive growth of algae on the water surface, or washed up on shore, people can remove the growth from the water. In addition, in order to eliminate possible recycling of the nutrients found in cell constituents, the algae should be removed from the watershed. The organisms can provide good material for compost or fertilizer and have been used on a small scale for years.

In a small lake there has been the successful use of an artificial control, provided by mixing the water column with compressed air. A 5 cm pipe at the deepest part of the lake supplied air through a ceramic diffuser. With the bubbling action, troublesome floating algae disappeared and clarity, or Secchi disc readings, increased. The algae that grew were then used at the next trophic level. The report termed the budget for the operation "modest."

REFERENCES

Anon., 1976. *How to Identify and Control Water Weeds and Algae.* Applied Biochemists, Inc., Mequon, Wisconsin. 64 p.

Fitzgerald, G. 1971. *Algicides.* Univ. Wisconsin Water Resources Center. 50 p.

Harris, D. and C. Caldwell. 1974. Possible mode of action of a photosynthetic inhibitor produced by *Pandorina morum. Arch. Microbiol.* 95: 193–204.

Lueschow, L. 1972. Biology and control of selected aquatic nuisances in recreational waters. *Tech. Bull. 57,* Dept. Natural Resources, Madison. 35 p.

Mulligan, H. 1969. Management of aquatic vascular plants and algae. In: Anon. (Ed.) *Eutrophication: Causes, Consequences, Correctives.* Nat. Acad. Sci./ Nat. Res. Council. Publ. 1700. pp. 464–82.

Prows, B. and W. McIlhenny. 1973. *Development of a Selective Algaecide to Control Nuisance Algal Growth.* Office of R. and D. USEPA, Washington. 126 p.

Sladeckova, A. and V. Sladeck. 1968. Algicides—friends or foes. In D. Jackson, (Ed.) *Algae, Man and the Environment.* Syracuse Univ. Press. pp. 441–58.

Algae can be both useful and harmful

Introduction and History   **421**

WASTE TREATMENT

AQUACULTURE

ARE THESE STEPS ECONOMICALLY FEASIBLE? IS THERE A MARKET FOR ALGAE?

REFERENCES

INTRODUCTION AND HISTORY

We know from early publications and records that humans have gathered both marine and freshwater algae for more than 4000 years. Early Chinese and Japanese writings refer to the use of these organisms for medicinal purposes, for food, and to extract products from them.

Although algae have been used as food in Asia for centuries, writings from the years before Christ indicate that the Romans were not fond of algae as food. But the Romans did extract some products such as for use in cosmetics. Considerably later, in the 1000s, the 1700s and 1800s, algae were used in the Orient and in Europe as fodder, fertilizers, for isolation of specific nutrients, for example, iodine or potash. In the twentieth century there are even industrial applications, including production of both gelling agents and food. Historically, almost any country with a coastal region has had at least some interest in their local marine algae.

In the past, organisms were often collected by hand, or rakes were used for harvesting local kelp populations. A small boat was convenient for gathering them from the intertidal region. Often nature assisted in the collection by providing a wash of seaweed after a storm (Fig. 20-1). (Now in the twentieth century kelp populations are harvested using large ships with cutting devices below water level and conveyor belts for moving the harvest into the hold of the ship (Fig. 20-2).) Harvesting could take place at almost any time of the year, but once there was some demand for a specific organism or growth form, collecting seaweeds became much more precise.

Organisms were washed, at least with seawater to remove debris, but sometimes with fresh water. They were allowed to dry on shore, on

Figure 20-1. Eel grass and algae washed ashore in Massachusetts. Such algae were once collected by colonists for use as mulch and fertilizer.

rocks, on special platforms, or even on the road. During these drying stages it was important to avoid contamination and to keep predators to a minimum. After drying, considerable loss of product could occur during a sudden shower because of the leaching of some materials.

COMMERCIAL PRODUCTS

The algae can be considered as useful material themselves, or as organisms from which a profitable product can be obtained. Only four major algal products have true economic value, and these are all algal wall components. Three of them, agar, alginic acid, and carrageenan, are extracted from walls of seaweeds, and the fourth, diatomaceous earth, consists of millions upon millions of diatom glass walls deposited over time in either fresh water or the ocean. Simply put, agar and carrageenan are polymers of galactose, or galactose-containing compounds, with sulphate groups. Some of the latter, the sulphate groups, are involved in the bonds between individual components. Agar and carrageenan are both called sulphated galactans, with more sulfur in carrageenans. On the other hand, alginic acid is composed of uronic acid units.

All three of these compounds extracted from marine algal walls are used either for their ability to make a solution viscous, or for their gelling properties. Agar, when the buyer gets the very best product, as from *Gelidium,* will gel with less than 10 parts per thousand (ppt). Carrageenan from *Chondrus* does not produce a good gel until there are 30 ppt, whereas 60 ppt of a *Eucheuma* product are needed for such a gel. A gel of phycocolloids now has to be a consistently uniform product, or it has little commercial value. The material is insoluble in cold water, and a 1.5% hot water solution must be clear. Then, when cooled to a little below 40°C, it will form a gel, and the gel will not dissolve below 8°C.

Carrageenan

This term is spelled in several ways, probably different interpretations of a Gaelic spelling. Carrageen is the raw material from which carrageenan, an extract from the walls of several red algae, is obtained. Species of *Chondrus, Gigartina,* and *Eucheuma* (Fig. 20-3) are most frequently used. It is one of the three most valuable commercial products obtained from marine macroalgae. In small amounts, carrageenan will make a

Figure 20-2. A kelp harvesting ship, the *Kelsol,* on the California coast. Courtesy of the Kelco Company.

Figure 20-3. *Eucheuma*, the red alga used for carrageenan culture in tropical and subtropical waters. 1⅝ X. Courtesy of C. Dawes.

solution viscous, and with further additions one can obtain a gel. It has been used as a stabilizer or emulsifier in foods such as ice cream, other milk products, and relishes. It is used as a binder in toothpaste or in pharmaceutical products, as well as an agent in ulcer therapy. Carrageenan has been found useful as a finishing compound in the textile and paper industries, as a thickening agent in shaving creams and lotions, and in the soap industry. Where one can gather the material and use it soon after collection, it is used to thicken sauces and soups, or fresh in salads.

Carrageenan is a water-soluble compound which we formerly thought was composed of just two fractions, kappa and lambda carrageenan. Kappa, the branched form, was precipitated by the addition of potassium chloride, was soluble in hot water, and was found mostly in haploid plants. Lambda carrageenan, the unbranched form, was soluble in cold water and found in abundance in diploid plants. The more kappa, the better the gel strength.

Now we realize that we do not have a complete picture of the structure of carrageenan (Fig. 20-4). It really includes a family or group of structural types, with D-galactose-3, 6-anhydro-D-galactose as the basic unit. The differences among the various components of carrageenan depend in part on the number and location of the sulphate groups. Carrageenan has a higher sulphate fraction than agar.

There are large amounts of carrageenan in *Chondrus, Gigartina,* and *Eucheuma,* even as much as 80% of the salt free dry weight. The level can vary with the species, time of the year, and perhaps stage in the life cycle. With *Chondrus* it has now been shown that seasonal factors are not important in determining the proportions of the various fractions that a plant will produce.

Mathieson and co-workers have demonstrated that the level of carrageenan in individual plants of *Chondrus* in New Hampshire can vary. When there is sufficient nitrogen available, and organisms are growing actively, there is more protein production, and less carrageenan. The organism produces higher concentrations of carrageenan in the sublittoral along the open coast, rather than in an estuary, and in late fall and early winter. The levels are lowest in the spring and early summer.

Along the coasts of the North Atlantic, in the United States, Canada, and several European countries. *Chondrus* and *Gigartina* have been

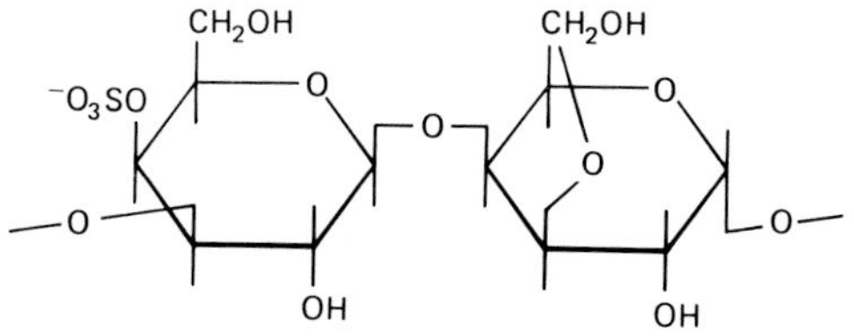

Kappa Repeating Structure

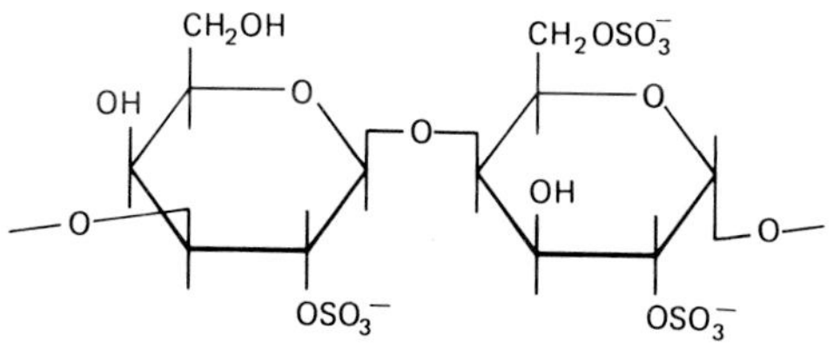

Lambda Repeating Structure

Figure 20-4. Certain of the components of carrageenan. See text for details concerning the complex structure.

collected for years. In the tropical Americas, but especially in the Pacific, *Eucheuma* has been a part of the diet for several centuries. In New England and eastern Canada, where the populations of *Chondrus* are quite dense, small factories are in operation to process carrageenan.

Properties.   About 3 to 6% of the dry material added to a solution will produce a gel with low gel strength. It will go into solution at the relatively low temperatures of between 27 and 40°C, depending on the source and purity of the material.

Preparation and Extraction.   First, the organisms are washed with fresh water to remove most of the foreign material. Then a mixture of one part of the raw material to 50 parts of water is boiled in order to put into solution the maximum amount of carrageenan. In order to remove debris, the solution is centrifuged and filtered, followed by evaporation and a thorough drying. Material with less purity is prepared by omitting some steps. However, the rapid freeze-out method is not used with carrageenan. If material of higher gel strength is desired, the kappa fraction, the form insoluble in the presence of K^+ ions, is selectively precipitated.

Culturing Carrageenan-Producing Algae.   Because of the demand for carrageenan, there is a very real problem of depleting the natural populations of *Chondrus* or *Eucheuma*. With this in mind, plus the greater costs of harvesting wild populations, there has been some interest in growing these two organisms commercially (Fig. 20-5). The life histories of at least some species have been worked out in culture, allowing production of material from a favorable clone. One can then get both haploid and diploid populations for chemical analysis.

In the Philippines where *Eucheuma* has been harvested by hand from small boats, dried and bleached in the sun, we have obtained about 20% of the world's supply of carrageenan. Now there are attempts to farm the seaweed, that is, to put it under cultivation (Fig. 20-5). In that way, with strain selection and proper management, productivity should increase. In shallow water 2.5 × 5 m nets are strung out on mangrove posts or stakes. Plant fragments of about 200 g size are tied on the nets. About 100,000 plants can be attached to nets on a one hectare farm, or community farm.

The farmer and his family remove damaged plants, weed out all epiphytes, and repair the nets. Sea urchins have been the natural

Figure 20-5. Tanks used for *Eucheuma* culture, Florida keys. Courtesy of C. Dawes and Marine Colloids.

predator. With the nets suspended above the bottom, predation is reduced. In about 60 days the *Eucheuma* plant has increased four-fold. Harvesting is by pruning the plant back to the inoculation size. On the small scale, with farms of about 0.25 hectare size, it has been demonstrated that *Eucheuma* can be cultivated profitably in the Philippines.

Dawes and co-workers on the Gulf coast of Florida have proposed farming different species in shallow waters. One can rely on growth in shallow bays, or work with tank culture. They have demonstrated that at that latitude good growth, allowing a doubling of the weight every 35 days, is attainable, and that *Eucheuma* could be cultivated at a profit.

Agar

Agar is used as a solidifying agent in the preparation of microbiological media. This material, obtained from red algae, is one of the best-known algal products. The word itself is of Malayan origin, but the material was first used in China in the seventeenth century. Soon agar was introduced to Japan where now it is produced and used extensively. Among the red algae that produce a useful product are *Gelidium,* extensively used for a good product for many years, *Gracilaria, Gigartina, Pterocladia, Ceramium, Suhria, Ahnfeltia,* and *Acanthopeltis.*

Gelling properties vary with the species from which the agar is obtained, but it will set at from 35 to 50°C. The laboratory material commonly used will set at 42°C. Various types of agar will go into solution at from 85 to 100°C.

The major component of agar is called agarose (Fig. 20-6), with the unit structures 1,4 linked 3,6 anhydroalpha-L-galactopyranose and a 1,3 linked beta-D-galactopyranose. The agarose is combined with agaropectin, which contains 4,6-carboxyethylidene-D-galactose, uronic acids, D-galactose, 3,6-anhydro-1-galactose and sulphates. Agar is unlike carrageenan in that it has a very low sulphate content. Now we know that this structural information is an oversimplification! The molecular structure is described as a continuous distribution of similar, but variously substituted compounds, including the above.

The amount of agar that one can extract from a kilogram of dry weight can be as much as 30% of the algal dry weight. With any one organism the percentage of agar extracted will vary with the season. As one might expect, many organisms are capable of producing agar, but the amount produced is quite low for some species. In quality control in the preparation of an agar product there is concern about the gel strength, color, and ash content as well as melting temperature.

The material is prepared by washing (Fig. 20-7), sorting (Fig. 20-8), and then either boiling or autoclaving in order to get the agar into solution. It is decolorized, freed of odor, and a filtering and evaporation process follow. The material, while still in a gel state, is frozen. Upon thawing, the water with impurities can be easily separated from the purified agar.

Today most agar comes from Japan, Korea, Australia, New Zealand, and Morocco, but many countries have in the past produced their own agar. These would include France, Italy, Spain, Egypt, Ceylon, India,

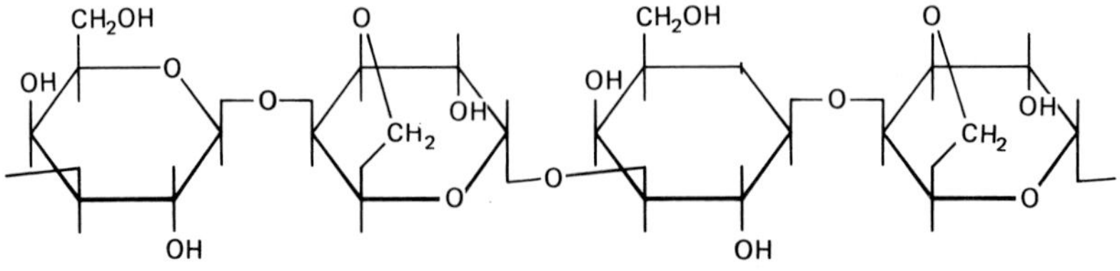

Figure 20-6. The structure of agarose, one of the repeating units in the structure of agar.

Figure 20-7.   Washing a collection of *Gelidium* in a pool in Shimada, Japan. Courtesy of W. Johanson.

Figure 20-8.   Sorting *Gelidium,* used for agar manufacture, in Japan. Courtesy of W. Johanson.

Mexico, some countries in South America, and also the United States. Many of us are aware of the use of agar in microbiological work, which was initiated in the 1880s. However, it is important in food preparation, in the transport of fresh fish to prevent damage and blackening of some species, and in the canning of other foods. It is valuable in the manufacture of processed cheese, mayonnaise, puddings, jellies, in baking, and in ice cream (but here alginates or carrageenan are favored). We frequently think of it as a stabilizer or emulsifier. It has some use in other industries as an emulsifier and also in forming molds, as well as in cosmetics. In the pharmaceutical industry it can be a carrier for a drug, especially when one wants slow release. Lotions and ointments can contain some agar.

With rapid increases in the price of agar in the 1970s there has been an interest in recycling material used in large microbiological laboratories. The used agar is autoclaved and solidified. The debris at the surface and precipitate at the bottom are discarded. After being cut in cubes and washed in cold water for 24 hr, it may be dried and reused.

Alginates

These compounds were first discovered in the nineteenth century. Alginic acid and its salts are obtained from the walls of brown algae (Figs. 20-9 and 20-10) where they may be as much as 25% of the dry

Figure 20-9. Bringing kelp to market in Chile. Courtesy of R. Wilce.

Figure 20-10. A bed of *Laminaria* on the Massachusetts coast. Courtesy of R. Wilce.

weight. The former is not water soluble, but salts of alginic acid are. The term algin refers specifically to sodium alginate. Until recently we believed that only brown algae, including species of *Macrocystis, Agarum, Laminaria, Alaria, Ecklonia, Eusenia, Durvillea, Cytoseira, Lessonia, Fucus,* and *Ascophyllum,* could produce alginic acid. Most of today's product comes from *Macrocystis*. Now we know that some bacteria can synthesize the compound. It occurs in both the walls and the intercellular spaces, and is sometimes thought of as an extracellular product.

The compound is structurally very complex. It is unlike the previously described red algal wall components in that an alginate is made entirely of uronic acid units. Originally we thought that the structure was a chain of beta-D-mannuronic acid units. Now L-guluronic acid is known to be present and both units are 1, 4 linked (Fig. 20-11). Perhaps the alginic acids rich in mannuronic acid are found in young cell walls or intercellular spaces, whereas those rich in guluronic acid are found only in older cell walls.

The properties of the gel formed depend on the interaction of the structure with cations. The guluronic acid residues have a strong affinity for divalent cations, and this parallels information we have of the great strength of gels richer in the guluronic acid fraction.

Figure 20-11. Structure of the two repeating units in the molecules of alginic acid.

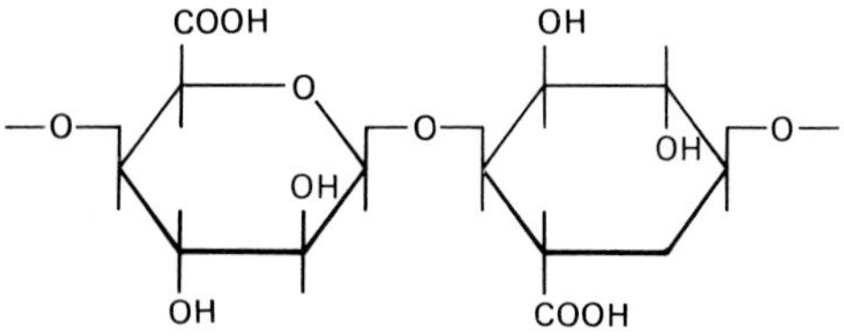

Chain of Mannuronic Acid Molecules

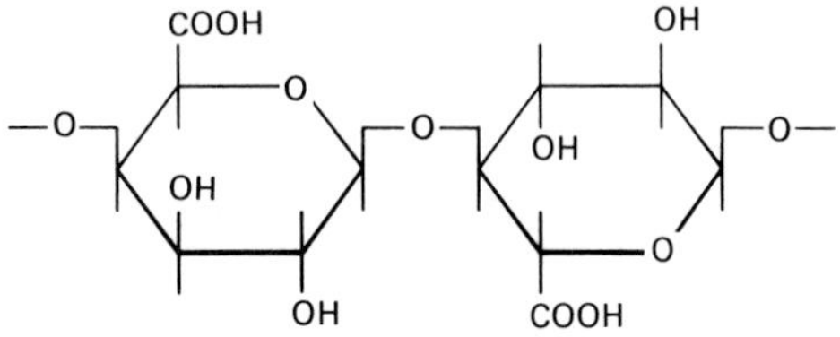

Chain of Guluronic Acid Molecules

Approximately 50% of the ice cream made in the United States contains alginates, which provide a smooth consistency and eliminate ice crystal formation, even in less expensive products. They are incorporated into cheeses and bakery products, especially frostings. Other industrial uses include paper products, paper adhesives, or the printing of fabrics. The pharmaceutical industry uses them as a filler, paint manufacturers as a thickener, and dentists to provide form and consistency in making dental impressions. About 50% of the world product is purchased by U.S. industry.

On the Pacific coast of North America the kelp industry harvests *Macrocystis* beds for their source of alginates. Large, specially designed ships (Fig. 20-2) and barges are used for cutting and collecting the harvest from kelp beds. Other countries involved in the production of alginates include Canada, Japan, Australia, and New Zealand, and some countries in Africa, South America, and Europe (Fig. 20-9).

Recently a sterile hybrid has been produced between *Macrocystis* and *Pelagophycus,* another useful kelp.

Silicified Algal Walls. Diatomaceous Earth
Diatom walls, with high percentage of silicon dioxide, are constantly being deposited in freshwater and marine sediments. In some areas, because of bottom contours, currents, and other factors, large concentrations of diatoms were deposited over millions of years. With later movements of the earth's crust, some of these deposits were

elevated and now are seen on the land. Much of the organic matter in and with the diatoms had decomposed, and with rain filtering through the deposits for many years, the deposits were thoroughly washed. The result is that some deposits are composed of about 95% silicon dioxide. When incinerated they lose less than 4% of their weight.

One of the best-known sites of diatom deposits, and apparently the largest, is in Lompoc, California. The deposits are about 225 m thick! Other deposits are found in Oregon, Wyoming, Florida, Montana, and New Hampshire in the United States, and in Tripoli, South Africa, and Great Britain. They can be either marine or freshwater deposits and can date from the Jurassic, Cretaceous, Tertiary, or Quarternary. Apparently the marine deposits are of greater depth than those formed in freshwater environments.

Although the Spaniards first saw the white outcroppings at Lompoc in the 1760s, the potential value was not realized until 1890, when it was recognized that they were identical to kieselguhr imported from Europe. Soon the material was used for thermal insulation and in 1916 a company was formed to mine the product.

Diatomaceous earth is used primarily for filters or filter aids. The material is especially useful when a product of the highest clarity is desired. Diatomaceous earth is especially satisfactory because it does not react chemically, is not readily compacted or compressed during use, and is available in many grades. The material is so finely divided that a gram has 120 square meters of surface area, and yet in use there is up to 90% open space in the filter cake.

Other uses of diatomaceous earth include polishing and burnishing, as in silver and auto polishes, insulation, in construction, in paper products, and as a filler. Certain grades of diatomaceous earth are especially useful for polishing delicate surfaces because the diatom walls are so lightly silicified that they collapse under mild pressure and do not damage the surface. When used as a filler in paints, they make a more useful product with little added cost. However, there are records of use as a filler in foods in times of famine. This practice, carried out by the food distributor, obviously adds no nutritional value to the food.

SOME TROUBLESOME ALGAE

In many ways we know far less about the algae than other groups of microorganisms. This is perhaps because historically the algae have been responsible for few problems directly affecting humans. On the other hand, bacterial and fungal infections and diseases cause considerable harm to humans, as well as to commercially important plants and animals. Of necessity, causitive organisms for wheat rusts, the potato blight, tree diseases, staph infections or tuberculosis, typhoid, and cholera have received a great deal of attention. Algae are currently considered problem organisms in euthrophic waters, and in a number of *isolated* areas the toxic algae have for centuries caused death to domestic animals.

Toxicity in Fresh Waters

In freshwater systems the toxic algae are small cyanophytes and include the genera *Microcystis, Anabaena,* and *Aphanizomenon,* and infrequently, *Nodularia, Coelosphaerium, Rivularia,* and *Gloeotrichia.* In the marine environment the dinoflagellates are the troublesome organisms responsible for red tide and shellfish poisoning.

The first confirmation that an alga, in this case a *Nodularia,* was responsible for the death of domestic animals was in 1878, when an Australian scientist reported a blue-green problem. Sheep, horses, pigs, and dogs died within hours after drinking from Lake Alexandrina, which had excessive growth of the *Nodularia.* In some cases the bloom of algae, blown ashore by the wind, was 5 to 15 cm thick, and had a pastelike consistency.

A few years later, in 1884, a serious toxic bloom was reported in Minnesota. Within 30 min of drinking water containing a bloom of *Rivularia* or *Gloeotrichia* cattle died. This particular geographical area, including southern Canada, Wisconsin, Iowa, Montana, and the Dakotas is the locality of frequent outbreaks. Around the fortieth latitude in both hemispheres there are repeated toxic blooms, for example, in Russia, Africa, North and South America, and Australia. Environmental conditions evidently play a role in the distribution of the particular blue-green algae. During the 1920s, 1930s, and 1940s, thousands of cattle died from *Microcystis* poisoning in South Africa. This organism is common in shallow freshwater lakes. Animals receiving large doses of this alga suffer from paralysis or convulsions, followed by death. Smaller amounts cause constipation, weakness, a drop in milk yield, as well as burning and sensitivity of the skin to sunlight. Repeated small doses can cause liver injury and jaundice. In the South African cases there have been some successes with treatments by enemata or heart stimulants. Cattle respond better if treated and left in the shade.

During this century in the north central states there have been periodic problems from early June through October, caused by *Microcystis, Anabaena, Aphanizomenon,* and *Coelosphaerium.* Cattle, chickens, sheep, hogs, turkeys, ducks, and geese have all been affected. In Iowa, the death of 700 game birds was attributed to the blue-green algae.

Not all blue-green algae can form toxic compounds. Prescott suggested that organisms that possess pseudovacuoles are the problem algae. However, there appears to be no direct connection between toxicity and these organelles. The pseudovacuole aids in keeping the organism in the upper limits of the photic zone, where large doses of the toxic algae can be obtained by the domestic animals.

The Schwimmer brothers have kept records of the numerous cases of toxicity to animals. At times it is extremely difficult to pinpoint the alga as the causitive organism, for there are other organisms present, as well as the possibility of chemical toxicity, not related to microorganisms. The animals which are discussed include livestock, domestic animals, birds, and wildlife which can show gastrointestinal, cardiovascular, hepatic and respiratory complications, often resulting in death. In many cases the problem organism is a blue-green, frequently *Microcystis*. In laboratory experiments with naturally occurring blooms the material has been given to laboratory organisms by inoculation (subcutaneous or interperitoneal), in the food, or contact has been made by immersion. The organisms are collected from a bloom and used immediately. Quick frozen material is usually most powerful, probably because the endotoxin is released from cells. Dried or refrigerated preparations are not as useful, or have lost some potency, probably because the toxin is degraded during the storage or preparation. In general, the material is more toxic when living cells are involved.

In problems reported in nature, as well as in laboratory studies, a distinction should be made between the *primary* effect, caused by the toxin itself, and a *secondary* effect. The secondary effect is frequently death because of oxygen deficiency, brought about by bacterial action on a bloom or by large masses of algae utilizing more oxygen than they produce. This may occur if light cannot penetrate the dense mass of algae, or if a great deal of oxygen is used during decomposition of the bloom organisms.

Laboratory Experimentation

Various approaches have been taken in the laboratory with cultures of toxic blue-greens, as well as with weedy greens such as *Chlorella* or *Scenedesmus*. In certain experiments, organisms, or extracts of organisms, have been given in several ways—orally, intravenously, or interperitoneally. After interperitoneal injection of selected strains of mice, weight loss, damage to vital organs or death can be demonstrated. In attempts to distinguish various grades of toxins, a variety of laboratory or domestic animals have been used. There are reports of:

1. a very fast death factor (VFDF), causing death within 15 minutes.
2. a fast death factor (FDF), usually taking 1 to 2 hours.
3. a slow death factor (SDF), usually associated with the bacteria found in combination with the blue-green algae. Death occurs within 2 days.

Other experiments have dealt with isolation and identification of the toxic element with *Microcystis* populations. Gorham has studied the problem in Canada and has examined material from mass cultures of *Microcystis*. In order to get sufficient material for analysis many 9 liter containers are used (40 to 50 day^{-1}). During these studies it was not possible to obtain an axenic culture, for bacteria continued to be associated with the blue-green. Without the bacteria the organism grew slowly for a while, but soon died. A symbiotic relationship between *Microcystis* and bacteria, or a single bacterium, might someday be discovered, but the *Microcystis*-bacterium combination is not responsible for the toxic properties. Toxicity came directly from the *Microcystis;* the associated bacteria could not produce the toxin. Each bacterium was isolated independently, studied in pure culture, and attempts were made to isolate toxins. When bacteria from toxic strains were associated with a nontoxic *Microcystis,* the *Microcystis* remained nontoxic. This experimentation was carefully carried out over many years, with at least five different strains of *Microcystis.*

The active principle is an endotoxin, *released from old or dying cells.* Actively growing cells washed free of pond water, or culture media, and swallowed, have little or no toxicity. In cells grown in culture, little of the toxin is found in the medium. The toxin was first thought to be an alkaloid, but it is now known to be a cyclic polypeptide which occurs in both the free and acid form. It is water and alcohol soluble, but insoluble in acetone and ether. The cyclic polypeptide has 10 amino acid units. Two units of glutamic acid, alanine and leucine, and one unit each of aspartic acid, *d*-serine, valine, and ornithine give it a molecular weight of 1200. The minimal lethal dose of the purified material is 0.5 mg to mice interperitoneally. Death occurs within the hour, and thus it is called a fast death factor.

The compound is a neurotoxin with an effect similar to eating the poisonous mushroom, *Amanita.* There is little blood clotting, liver cells begin to break down, the liver enlarges, and there is general congestion of the spleen. It is suspected that the toxicity could be due to the cyclic nature of the compound, or the *d* forms of two amino acids.

In addition to toxicity caused by *Microcystis,* several other blue-green algae have been responsible for death of domestic animals. *Aphanizomenon* is frequently found in bloom proportions, but some blooms give no toxicity at all. Some researchers believe that the "toxicity" of the organisms is attributable to a previous *Microcystis* bloom. It is well known that a bloom and crash of one blue-green population can be followed by massive growth of another. These cycles can repeat every few days. Sawyer and others have recently shown a specific *Aphanizomenon* toxicity. The effect is due to interference with Na^+ influx in nerve cells.

In both marine and freshwater habitats the filamentous *Lyngbya* has been implicated as a form producing a toxin. The organism has been reported as responsible for death of horses in India, as well as contributing to the poison of reef fish, which have the organism in their intestinal tracts.

Anabaena flos-aquae, found in shallow freshwater lakes, can produce a VFDF, measured in laboratory mice. First there are tremors, followed by mild convulsions, paralysis, and eventual death. Because it is a VFDF, the toxin cannot be similar to that of *Microcystis.* It is known that it is water soluble and the effect is similar in some respects to shellfish poisoning. The same VFDF can be detected with at least two species of *Anabaena,* using several types of laboratory animals. However, blooms of *Anabaena flos-aquae* contain both toxic and nontoxic strains. In nature the *Anabaena* toxin can affect ducks. Even with laboratory experiments with the cyclic polypeptide from *Microcystis* there was no effect on these birds. One *Anabaena* VFDF has been identified as 2.9 diacetyl-9-azabicyclo-[4.2.1] non-2,3-one. Another report with an *Anabaena* identified the toxin as an alkaloid with a molecular weight of less than 300.

One should remember that certain characteristics of the blue-green algae, whether they are toxic or not, make them a troublesome group. They grow rapidly in some bodies of water, especially those blue-greens that tolerate pollution. The abundant sheath material, which can cause the organisms to stick together or to other organisms, can provide an excellent bacterial substrate. Those blue-greens that have pseudovacuoles can float at the water surface, and thus they accumulate in large numbers. This group of algae can provide a great number of problems for the pollution biologist.

Toxicity in the Marine Environment

Prymnesium is a small brackish water, haptophycean flagellate. The organism occurs naturally in brackish water ponds used for fish culture. In Israel it has been responsible for massive fish kills in these ponds. The toxin itself inhibits the transfer of oxygen across gills and is a potent hemolytic agent. Although it has not been precisely identified, it is known to be thermolabile and saponinlike. Saponin is a glycoside commonly found in higher plants, which foams strongly when shaken in water. When injected, saponins are very powerful hemolytic agents, even in low concentrations—they destroy the red blood cells. Control of the *Prymnesium* in the fish tanks is by means of ammonium sulfate, which at the alkaline pH of brackish water causes lysis of the flagellate. Large populations cannot be formed in ponds treated with ammonium sulfate, and thus the toxin cannot become concentrated.

Red Tides

As with the use of any common name, there can be different interpretations as the term "red tide" is used in various localities. For some, the term is used to describe the massive growths of dinoflagellates responsible for fish kills, such as occur off the coasts of Florida and California. In addition, it has been used to describe the growth of other dinoflagellates responsible for the poison accumulated in shellfish. Organisms in the genera *Gonyaulax* and *Gymnodinium* are those involved in both problems. Some species appear to be specific to a given area, and others, such as *Gonyaulax polyedra,* form the poison only under specific conditions. One strain of *Gonyaulax,* known to be the cause of shellfish poisoning in nature, will not form the toxin in culture.

An increase in vertical stability, allowing reproduction at a faster rate, apparently triggers red tides. The change in stability can be brought about by heavy rains, the accompanying run-off (which occurs during calm, sunny days), and the meeting of dissimilar waters. Others say that the bloom of *Gymnodinium breve* on the west coast of Florida results from an increase in chelated iron.

Shellfish poisoning occurs along the northeastern coast of North America, as well as in the north Pacific. It has been reported from Norway, France, New Zealand, Great Britain, and Mexico, with some concentration at around the thirtieth latitude. In the United States the organisms are *Gonyaulax catenella* on the west coast, and *Gonyaulax excavata* on the east coast, where yearly outbreaks occur around Nova Scotia. In 1972 a *serious* "red tide" occurred in the greater Boston, Massachusetts, area. The bloom of the dinoflagellate usually lasts just a few weeks, and often it is safe to eat shellfish about two weeks after the red tide. However, if there has been a concentration of the toxin, longer delays are necessary before the shellfish are safe to eat.

The poisoning to humans comes from eating filter feeders, animals such as clams, scallops, or mussels, which filter the plankton from seawater as their source of food. The shellfish can accumulate poison when there are just 200 dinoflagellates per ml, hardly enough to give water a reddish color. The toxin is stored in the hepatopancreas. The toxin may be inactivated, or possibly secreted by the animal within two weeks. On the west coast it is sometimes dangerous to eat California mussels or Alaskan butter clams, unless one is certain of their origin. Butter clams store up to 80% of the toxin in the siphon. Detoxification may take one year!

After eating shellfish and accumulating sufficient toxin, the victim first experiences a numbing of the lips, tongue, and fingertips, usually within 30 min. The diaphragm is soon affected and in serious cases respiratory failure can result. One with respiratory failure can be kept alive by artificial respiration. Since the first symptoms are similar to overindulgence with alcohol, the poisoning is sometimes not taken seriously. If death does not occur within 24 to 36 hours, there is recovery with no permanent effects.

The compound from *G. catenella* has a molecular weight of 372, and is called saxitoxin. In the purified form it is dangerous to handle. It cannot be easily crystalized. It is a tetrahydropurine which interfers with Na^+ movement in nerve cells. Only 0.18 mg is sufficient to kill a 20 g mouse in 20 minutes.

The giant fish kills caused by "red tides" were first thought to result from loss of oxygen in waters. This low DO resulted when the masses of dinoflagellates began to decompose. Now we know that the death of fish is due to the release of a toxin from living dinoflagellates. This exotoxin is thus fundamentally unlike the endotoxin responsible for shellfish poisoning. The compound or compounds produced and released by *Gymnodinium breve* have been extracted in the laboratory and the concentrate immediately kills fish in controlled experiments. Perhaps the massive fish kills recorded when populations of red tide organisms die could be attributable to the release of all toxin a little prematurely. The material from *G. breve* is more soluble in lipid solvents than in water, and has a molecular weight of about 1500.

In addition to the death of fish, which can drastically affect the local economy, the red tide can cause eye discomfort as well as both throat and mouth irritations in humans. These symptoms come from the release of an irritant from the water. The tourist industry is directly affected when dead and decomposing fish remain on beaches for a period of days.

Massive fish kills have been reported in the Gulf of Mexico, California, Peru, Japan, Australia, India, Africa, and in some sections of Europe. The red tides in Florida are caused by *Gymnodinium* or *Gonyaulax,* while the California dinoflagellate, which causes some damage almost every summer, is a species of *Gonyaulax.* When there is a true "red tide" there can be from 10^6 to 10^8 cells per liter of seawater.

Medical Aspects

In the previous discussion of algal problems directly affecting humans, shellfish poisoning was considered. We shall see that algae cause other medical problems, but it is possible that we might be able to use algae or algal products for our benefit.

Prototheca is a colorless *Chlorella*-like organism which has been known for many years. Recently it has been found to be present in primary and secondary infections in humans. Although far from being a very widespread problem, it has caused extreme discomfort in reported cases. In some cases, the *Prototheca* has been confirmed as the causitive organism by means of EM studies.

Two occurrences have been reported in Africa. One farmer had a spreading lesion on his foot that did not respond to treatment with antibiotics; surgery also failed to correct the condition. A second case involved a man with a pimple-sized lesion on the side of his face which did not respond to any usual treatment. In fact, the lesion continued to grow and new ones also appeared. Other similar cases have been recorded but thus far, fortunately, they are not common occurrences.

A secondary infection of *Prototheca* has been confirmed in a terminal cancer patient. Although admittedly an insignificant problem when viewed on a worldwide scale, it is discouraging to recognize our inability to control growth of these organisms.

A species of *Lyngbya* has been implicated as a cause of swimmers' itch in Hawaiian waters.

In a number of countries it is not unusual to find that algae are used to treat certain ailments. For example, in Southeast Asia certain species of *Acetabularia* were known as a cure for gallstone problems. In Ireland, some European beauty spas, and in the United States, applications of *Chondrus* extracts have been useful in keeping the skin smooth. Originally it was noticed that those who gathered Irish Moss commercially had smooth skin. Some chemicals in the algae are absorbed by the skin and help slow down natural aging phenomena.

Also they keep the skin from drying, perhaps because of a natural water-holding capacity.

Algae alone, or mixed with other medicines, have been prescribed for consumption, lung diseases, diarrhea and also constipation (one might wish to provide bulk or, on the other hand, a gelling quality), goiter, and bladder disorders. They have been suggested as additives for diets, inasmuch as one can get a feeling of eating sufficiently, but have little digestible material.

For many years there have been searches for antibiotic properties of algae, and these have met with some success. However, no antibiotic has yet been developed on a commercial scale. Either the activity is not of sufficient potency or the production would be excessively costly. Antibiotic properties have been reported from several groups, including the red, brown, and green algae, and diatoms. In nature certain algae, *Sargassum,* for example, release extracellular products that are active against bacteria, and in the laboratory, organisms such as *Dasya, Halimeda,* and *Macrocystis* release antibiotics. In 1974, after examining several dozen British marine algae, Hornsey and Hide concluded that about one-third appeared to possess outstanding antibacterial properties.

Debromoaplysiatoxin, an anticancer compound, has been isolated from a *Lyngbya gracilis* population from the Marshall Islands in the Pacific. Chloroform extracts of the alga and several relatives from the same family display activity against a lymphocytic mouse leukemia.

Some species of the haptophycean alga, *Phaeocystis,* release acrylic acid when the algae are digested. In the Antarctic the intestinal tracts of some marine birds were found devoid of a bacterial flora. After investigation of field material, as well as extensive culture studies, investigators found that birds that had *Phaeocystis* in their food web showed the marked reduction in the numbers of bacteria. Although this production of acrylic acid does not affect us directly, it is an example of the production of an antibacterial substance by algae. This has stimulated others to look for algal antibiotics.

There have been many cases of toxicity of algae to humans because of touching, inhaling, or ingesting microscopic forms. They can cause skin disorders, itching, blistering, and even severe skin lesions. In these instances, unlike those reported earlier in the chapter, the organism is only temporarily associated with its host. Thus certain of these problems are contact phenomena and perhaps many are caused by the chemistry of the wall material of the particular alga. When inhaled they have caused respiratory problems ranging from congestion to acute bronchitis or even pulmonary edema, in which the lungs fill with fluid. In other instances where some water has been accidentally swallowed or when water with algal blooms has been used in the drinking supply, individuals may have cramps, pain, diarrhea, and vomiting.

Several algae in at least a half dozen classes have been listed as involved in taste and odor problems in domestic waters. These problems have existed for centuries. Although the problem can also be due to other organisms, *Synura* and certain diatoms and dinoflagellates can impart an objectionable taste, which has been described as metallic, fishy, or cucumberlike. With *Synura* n-heptanal is at least partly responsible for the odor. The problem must be attacked early by preventing blooms of troublesome algae, because treatment of a municipal water supply from a reservoir, river, or wells can be costly. While controlling algal growth, water can be obtained from another reservoir, or from a water layer that does not contain the organism and problem chemicals.

FERTILIZER

For centuries marine algae have been gathered for use as fertilizer (Fig. 20-1). In France, Spain, Great Britain, India, Australia, New Zealand, and the United States, the algae were recognized as a free source of nutrients. In colonial days in New England there was a holiday in the fall so that the people could gather the seaweed and haul it to the farms. In most areas, the labor costs and the availability of commercial fertilizers have eliminated the holiday as well as the need for a seaweed harvest.

The first records of the use of marine algae as a fertilizer are from the first century A.D. We know that the Chinese, Greeks, and later the Vikings used them as manure for their crops. Wet or dry algae were deposited on land, providing nitrogen, potash, trace elements, and some phosphorus. In addition, they provide organic matter, which increases the water-holding capacity and improves the physical condition, or crumb structure, of the soil. Inasmuch as certain of the nutrients can be lost during the drying process, use of wet seaweed is considered best.

When we compare the nutrient value of seaweed to manure, it is seen that there is more phosphate in manure, but there are equivalent amounts of nitrogen. However, unless the seaweed is used for a number of years, the full value of the nitrogen is not realized. Only after some months or years does the soil microbial community break down all the available nitrogen. Marine algae usually provide more trace elements than manure and are higher in potassium.

We know from experimentation with several marine algae that they contain growth hormones. When used as fertilizers, these compounds are then available to the cultivated plant and provide growth factors not present in other types of fertilizer. Some suggest that certain seaweeds can provide an immunity factor against crop disease. In some countries, coralline red algae, such as *Lithothamnium* and *Phymatholithon,* are harvested for use as a soil conditioner. On the Atlantic coast of Europe in areas with acid soils, they are a limy soil dressing. In France there are laws regulating the use of the available crop by coastal communities.

Other organisms utilized for their soil enrichment value include the brown algae *Fucus, Laminaria, Alaria, Sargassum, Macrocystis, Lessonia, Egregia,* and *Eisenia.* The green algae *Enteromorpha* and *Ulva* are gathered less frequently, for they provide significantly less bulk.

Certain soils in the Aran Islands, off the west coast of Ireland, are composed of sand and seaweed decomposition products. Formerly, little soil was available for crops in and among the rocky substrate. Over the years the mixture of sand and seaweed has built up a soil in which a good potato crop can be harvested.

In controlled tests with fruits, vegetables, and ornamentals, when seaweed fertilizers were used, relatively fast seed germination, increased growth, and increased fruit yield have been reported. When a seaweed preharvest spray was applied to peaches, the percentage of marketable fruit increased. However, most of these are isolated reports, and broad use of algae for these purposes is not the rule.

Nitrogen Fixation

Some attention has been paid to the use of nitrogen fixation in increasing the world food production. With nitrogen fixing bacteria this research has explored the possibility of association of bacteria with a crop plant, the transfer of the genetic material of nitrogen fixing bacteria (i.e., the nif operon) to other potentially more useful bacteria, and the possible associations of free-living nitrogen fixers and crop plants.

There have been some experiments with use of blue-green algae in rice paddies. In experiments in Japan with *Tolypothrix,* the rice yield increased when the blue-green alga was present with the rice plant. It would appear that selection of an alga that would persist through the years would provide a constant source of nitrogen. However, the nitrogen must be released from the alga in order to be supplied to the rice plant. One might select for an alga that would release a nitrogenous extracellular product, or find an organism that would grow during the rice off-season, but later be available, after decomposition, to the rice crop. The nitrogen would then be continually added to the rice paddy.

Minerals

As we have seen, the algae have been used as a source of elements when applied as a fertilizer. This would include not only N, P and K but also Cu, Co, Fe, Zn, Mn, Mo, and B. In seventeenth century France, algae were gathered because they are a source of soda or sodium carbonate. This material was necessary in making glass and pottery, for alum, and in the soap industry. In Great Britain in the eighteenth century tons of kelp ash provided by *Fucus, Ascophyllum,* and *Laminaria* were sold. (Kelp was formerly the term for the burned ash of brown algae, but now it is frequently used for the larger brown algae that are gathered for minerals, alginates, etc.) The best soda was obtained from the rockweeds *Fucus* and *Ascophyllum.* The techniques for preparation were quite involved. Use of seaweed decreased markedly in the late eighteenth century as new sources of soda were developed.

Although the same types of seaweeds could be used if one wanted a source of iodine, the *Laminarias* (Fig. 20–10) yielded three to nine times more iodine than the rockweeds. In the early 1800s there was a 30-year period during which the algae were the main source of iodine. Certain kelps can concentrate the element 30,000-fold over ocean levels. When iodine was discovered in Chile, seaweed utilization tapered off. During World War II the iodine supply in Japan was cut off, and industry again relied on the kelps. Various processes were used, but as much as 50% of the product could be lost during manufacture.

Potash, a crude form of potassium carbonate and the oxide of potassium, has been extracted from algae during various periods, particularly in some countries during time of war. The Pacific coast kelp industry, gathering *Macrocystis, Nereocystis,* and some *Pelagophycus,* was formed for harvesting algae as a source of potash. The seaweeds are harvested from large boats. New plants are formed from the base and thus the population can proliferate. One, and sometimes two, harvests per year are possible.

FOOD

With the increasing world population, algae have sometimes received attention as an alternate food source. Algae have been used for centuries as fodder. Areas in which marine macroalgae can be grown are limited, and macroalgae are not very attractive as a food to most of us. The microalgae are the only realistic algal food source. Harvesting the ocean has been suggested, using the algae to feed animals such as shellfish. Aquaculture, using up our waste nitrogen and phosphorus or harvesting organisms from sewage oxidation ponds, has also received some attention. All of these topics are interrelated. After discussing the gathering of marine algae for use as food for animals or for humans, we will discuss the conditions of culture that might be applied to most systems.

Fodder

In Great Britain, Scandinavian countries, France, Iceland, Canada, and the United States algae have been used as fodder, that is, coarse food for domesticated animals. In the United States and Great Britain businesses were developed for the manufacture of food for cattle, sheep, pigs, and poultry, mainly from brown algae. Few are still in operation.

In Iceland, sheep, cattle, and horses are allowed to browse along the coast. On some islands in Great Britain the coastal area is fenced and the sheep remain *outside* the fence, in the seaweed zone. Where the seaweed is gathered, local regulations have been enacted. Although varying somewhat from country to country, not much algal fodder is gathered today. During wartime, or years of recession, one finds more reliance on this available food, and in some areas it could be the prime winter feed.

The brown algae, including the rockweeds and kelps are most frequently used as fodder (Figs. 20-1 and 20-10). The reds *Rhodymenia* and *Chondrus* are considered of value, especially when the animals are browsing. At times *Enteromorpha* has been utilized. The freshwater blue-green, *Gloeocapsa,* mountain dulse, occasionally is gathered for fodder. One must use caution with cyanophytes because of their possible toxic nature (page 434).

Seaweeds can provide as much protein as a good crop of oats or hay, that is, at least 10% of the dry weight. Certain *Laminaria* species are among the best in percent *available* protein. In addition they provide vitamins and minerals that might otherwise be lacking in the diet. Some products are sold as supplements to the diet of domesticated animals, such as horses. It is claimed that, in general, the animals are in better health after using seaweed supplements for several weeks.

Have You Eaten Algae?

Marine algae collected from nature have been important to the diets of certain peoples for many years, and they have supplemented the diets of countless others. In some countries use of seaweeds has been a necessity at times, but when conditions improve, the same material is fed to livestock or used as a fertilizer. They were of more importance prior to the twentieth century, but even today in some countries of the Orient the harvest and use and even culture of seaweeds is most important. Japan is the principal user but at least some seaweeds are eaten in China, the Philippines, Hawaii, and many Pacific Islands, Australia, New Zealand, Europe, Iceland, North and South American countries, and Great Britain.

Representatives of the Chlorophyceae, such as *Ulva, Enteromorpha, Monostroma, Codium, Caulerpa,* and the Rhodophyceae, including *Rhodymenia, Porphyra, Ahnfeltia, Ceramium, Gracilaria, Hypnea, Iridea,* provide up to 20% of their wet weight as protein. The Phaeophyceae, for example, *Eisenia, Alaria, Dictyota, Durvillea* (Fig. 20-9), although utilized in some areas, provide the least protein of the three algal groups. However, the precise food value is frequently unknown because comparative studies are lacking. It would depend partly on the amount of the alga that one could digest. This would increase as the human intestinal flora adjusts to a significant portion of the diet being composed of algae. In one study, Schlichting reported that the protein quality of a mixture of several seaweeds, such as *Laminaria, Caulerpa,* and *Rhodymenia,* would be very good, but he did not obtain as favorable data from some cultured green algae, including *Pithophora, Pediastrum* and some desmids.

A great deal of work, especially with laboratory animals (mice, rabbits, chickens), has been done on the food value of cultured unicellular green algae. Many of these studies report that the food value of organisms such as *Chlorella* or *Scenedesmus* is high, and that algae can compete with milk and beef. Some claims are excessive, if one considers food habits that are soundly established, potential markets, and changes in industry. However, as Krauss pointed out, certain green algae could be excellent sources of protein, carbohydrate, and fat, as well as vitamins. Within limits, the ratios of carbohydrate:fat:protein can be regulated by the nutrients made available for growth. If certain conditions of growth produce, at a competitive price, a useful product that might contain *one* essential component of our diet, certainly this could be added as a supplement. It is not the food value of native or cultured algae, but the cost of production of the product, which limits production of algae.

Recipes

The favored way of eating green algae is as a supplement in a salad or as a component of soup. The brown algae are added to soups, bread, and are also candied, pickled, and roasted. Certain of the red algae are eaten raw. This would include *Rhodymenia* or dulse, commonly found in northeastern North America. (The same organism is seldom utilized in Japan, where it is also found!) Red algae can be used in soups, boiled, minced, fried, or lightly roasted, or added to salads.

In 1971 Pennelly compiled a number of recipes from the Orient, Europe, and America using green, brown, and red algae. These included pickled and seasoned seaweeds, soups, puddings, and main dishes. With vegetables, rice, pork, ham, and so forth, some fine dishes have been prepared. One issue (1975) of *Gourmet* magazine included a recipe for the pudding, Blanc Mange, prepared from Irish Moss.

Culturing Marine Algae

In many cases there have been attempts to culture certain marine algae because of their food value. Local populations were being depleted, and with the increased demand there was not a sufficient crop. Perhaps one of the best known and most successful operations has been the Japanese culturing of nori, that is, *Prophyra* (Fig. 20-12).

After publication of a 1949 study by Drew, considerable attention was devoted to the systematic culture of *Porphyra* in Japan. The marine alga had been harvested by the Japanese for several centuries with the organisms actually being cultivated on bamboo poles. In 1951 a tornado dealt a lethal blow to the nori industry. Working with the knowledge of

Figure 20-12. *Conchocelis* growing on shells suspended from rods. Part of the process for culturing nori or *Porphyra*. Courtesy of W. Johanson.

the life history as provided by Drew, the Japanese were able to reestablish a nori population quickly. First, *Porphyra* spores were liberated into sea water by stirring some of the alga very vigorously in a drum of freshly collected sea water. The water was then poured, while mixed (for the spores are heavy and sink) into shallow tanks. In the latter there were oyster shells. Upon germination the spores developed small filaments which then grew both in and on the shells. The water was changed periodically and several months later the *Conchocelis* stage (Fig. 20-12) had reached maturity.

At that time the shells were stirred again vigorously, and strings or netting were dipped into the container. Monospores (or alpha spores) from the *Conchocelis* stage attached to the strings or netting. The latter were then incubated in shallow bays and within two months the crop was ready for harvest.

Nori stripped from the strings and collected in pails is poured into 20 cm square frames with porous lower surfaces. After drying for a few hours in the sun, the paperthin squares can be stored in the dry state for long periods. Over 6 billion sheets per year have been harvested. More than 70,000 are employed and over 12,000 acres are under cultivation. Nori is eaten as a supplement in the diet, in soups, or used to wrap rice or other foods.

Recently the Chinese have been most successful in the culturing of kelp and in establishing an extensive kelp industry. More than 1500 years ago they imported kelp from Korea, and later from Japan. During World War II their supply was not available, and thus they soon became interested in establishing their own populations in the shallow bays on the Chinese shore. Fertile plants were imported from areas in the north, where they had been established in the 1920s from a Japanese *Laminaria* species. After gaining an understanding of ways to manipulate reproduction and the life history, the Chinese were able to establish cultures on stones, rafts, or in artificial tidal pools constructed by damming in the intertidal zone. The rafts are most successful; they are constructed of rope and bamboo. Rope is unravelled, juvenile plants are placed between strands, and the rope is allowed to ravel again, fixing the plant in position. After attaching many plants to one raft, an earthenware jar containing measured amounts of fertilizer is placed on the raft. The porous jar releases nutrients gradually.

Some bryzoans and ascidians become attached to the kelp, as well as certain amphipods which attack the holdfast and stipe. They are removed by weeding.

With a large labor force the Chinese are able to carry on this type of culture and harvest. The kelp is then used as a food, or to obtain minerals and alginates.

Freshwater Algae as Food

Freshly collected freshwater algae are not commonly utilized by humans. Most information concerning their food value comes from studies with cultured forms, which have been of some interest to plant physiologists because of rapid growth rates. Blue-green algae from nature are the forms commonly used as human food. In Mexico, South America, Africa, and China, *Nostoc, Spirulina, Phormidium,* and *Chroococcus* have been collected, sometimes cleansed of foreign material, dried, and used. The organisms could appear on the market in this state, or they might be ground up to be added as a flavoring. *Nostoc* balls can be found on the market in South America (Fig. 3-19). In Mexico, tecuitlatl, the local name for certain blue-greens from lakes in the valley of Mexico, has been harvested for centuries. Amomoxtli refers to *Nostoc,* and the name cocolin to a mixture of the blue-greens *Phormidium* and *Chroococcus,* along with certain invertebrates that grow in the blue-green mats.

Spirulina grows naturally in some Mexican and African alkaline lakes. Apparently as much as 65% of the dry weight can be protein. Since the organisms grow well in these lakes, and also in others subjected to wastes from mining operations, culture of this organism has attracted attention in the United States, Mexico, Japan, Israel, and some countries of Europe.

Culturing Microalgae For Food

Having found crops of algae most useful as foods, or food supplements for either humans or domesticated animals, there has been an increased interest in the culturing of algae as a source of food. We will see that microalgae are usually chosen as the most favorable organism for these studies. Unless one is interested in a specific macroalga (e.g., *Porphyra*) for a specific food or source of an economic product, the microalgae are easier to maintain in mass culture and provide us much better growth rates than the macroalgae.

In addition to their value as a potential source of food, the microalgae have been considered likely organisms for use in space, in aquaculture, and in the treatment of domestic wastes. Again, we rely on their rapid growth rates and the ease of manipulation of these cultures. They could be useful as photosynthetic gas exchangers, that is, to use up the carbon dioxide in a closed system while releasing photosynthetic oxygen. In aquaculture operations, algae remove large amounts of nitrogen, phosphorus, and other elements from our sewage. Whether we think of growing a marine or freshwater microalga, a *Chlorella,* a blue-green, diatoms, or a flagellate, the principles of culture are the same. Thus we will discuss some of the factors that must be considered whenever we propose to grow a microalga in large quantity.

1. *Selection of the Organism.* There are many organisms available in culture collections. Since a number have been utilized experimentally for years, there is a vast literature on their growth characteristics, with a doubling of cell number *possible* every two hours with at least some species. Some researchers might select marine organisms, others freshwater types. For certain experiments, cultures free of other organisms, axenic cultures, would be preferred. In outdoor culture, or sewage oxidation ponds, xenic cultures would be used, or one might simply rely on those species that occur naturally under such conditions. Axenic cultures would be necessary only when rigorous microbiological techniques are used at every step in the process.

2. *Culture Chamber.* Chambers can range from small models of glass or plastic that fit on a laboratory bench, to open tanks or ponds of many square meters area. In the laboratory if one wants to maintain sterility, a small chamber is preferred, while in large ponds one would be more interested in available sunlight, depth of the pond, and light penetration, as well as ease of harvesting a crop from the large pond.

3. *Nutrients.* In a sewage oxidation pond the nutrient solution would have been provided. One would have selected organisms that could most effectively remove the most troublesome nutrient or nutrients, probably first phosphorus and then nitrogen. Mineral requirements of microalgae are quite well known and dozens of formulations of media both for marine and freshwater laboratory culture are published. For culture of some organisms, organic requirements must be met. This could be an important cost factor. Usually organisms that must meet such requirements are not selected for mass culture. Nutrient replenishment can be achieved by initially providing large doses, or by periodic or continuous additions. If one is interested in nutrient removal, the run is terminated when nutrient levels are low.

4. *Light.* In outdoor culture natural sunlight provides the only illumination. In areas where the percent daily sunlight is below average, or on cloudy days in all areas, photosynthesis is reduced.

One must also consider light penetration. Light intensity at the surface could be too intense, whereas with a visibly green algal crop, light penetration is limited. Thus sewage oxidation tanks and ponds are never constructed very deep. Light penetration is improved by maintaining a more limited population as well as by reducing the turbidity caused by nonliving material in suspension. In the laboratory fluorescent illumination can provide sufficient light for excellent growth. One must consider whether to use a light-dark cycle and also take into account both light intensity and heat build up from illumination.

5. *Temperature.* Fortunately, with many algae the optimum growth rate is achieved at a temperature higher than room temperature. Many algae grow well in the 30 to 39°C range. For laboratory culture cost increases greatly when refrigeration must be employed to maintain a favorable temperature. In outdoor culture, cool night-time temperatures, and the size of the culture pond, maintain temperatures favorable for algal growth.

6. *Mixing.* The efficiency of the operation is reduced without adequate mixing. Some microalgae settle more slowly than others, but nutrient-cell contact and light penetration are not optimal if the organisms are in a layer settled on the bottom.

7. *Harvesting.* This can be the most difficult and frequently the most costly phase of the operation. Although at times filtration and sedimentation have been employed, and are still the preferred methods in some operations, the two main methods of getting food separation are flocculation or centrifugation.

A simple continuous centrifuge or a battery of centrifuges are standard. The material is pumped to the head of the centrifuge at a constant rate, with the efficiency of the centrifuge dependent on the flow rate and the centrifuge speed. Periodically, one must remove the concentrate from the head. The liquid is normally discarded, but in cultures water can be recycled for additional use, after appropriate addition of nutrients. With centrifugation there is a power cost, increasing greatly in recent years, as well as the high initial equipment cost.

Flocculation is achieved by adding chemicals, mixing, and allowing the resulting floc to settle. In order to get a floc, the negative charge on algal cells is neutralized so that aggregation and sedimentation will occur. The fluid is then drawn off the top and the algal crop is harvested from the bottom of the chamber. Alum, lime, and synthetic flocculants such as cationic polymers have been employed. Addition of small amounts of Fe will improve the activity of many flocculants. Most chemicals suggested are relatively inexpensive, but one might not want to contaminate the algal product with *any* chemicals. In some outdoor operations an autoflocculation is noted when the pH is near 10 and the organisms have been growing well on a hot sunny day. In the late afternoon investigators often take advantage of this phenomenon and harvest the crop. The time and space needed for these operations are disadvantages of this system.

Gravity sand filters can be used, even for the final drying. These are inexpensive, but the filters are easily clogged when a dense growth must be harvested, and sand in the product can be a serious problem.

In all these systems, after filtration, sedimentation, centrifugation of flocculation, the product must be dried, for at best it is only 4% (by weight) solids. The dewatering process, in which about 50 to 80% of the water is lost, is accomplished by drying in the sun, on sand beds, or in the oven. The final step, the drying process, which results in a product with about 10% water, is accomplished most efficiently with heated rotary drums. The material is spread on the drums and after drying is scraped into a storage bed. In some operations sand filters are used for all these steps, but although inexpensive, almost without exception the final product contains some of the filtering material.

Using this knowledge of the growth of microalgae, there have been numerous attempts to grow various microalgae for food, mostly on a small scale in laboratory pilot plants. The Japanese have successfully operated an outdoor plant for several years. A great deal of our information on the nutritional value of microalgae, frequently a *Chlorella* or close relative, has come from these studies.

Certain of the microalgae can be manipulated to produce a reasonable yield of protein. By modifying the level of nitrogen supplied to the alga, the ratio of carbohydrate to fat to protein can be altered. When selecting organisms, one of the high-temperature strains of *Chlorella* is often a candidate. Some strains can achieve a rate of 12 to 14 doublings per day. If the rate were reduced even to two doublings per day, an inoculum of 1 kg of cells would yield a 3 kg crop 24 hr later, *plus* the kilogram need for inoculation the next day. It is easy to understand why such algae are thought of as a potential food source.

The consumer can utilize most, if not all, of the alga itself. In contrast, with grain crop production, roots, stems, and leaves represent plant growth not used as food. With silage all but roots are utilized. If the producer can sell the entire algal crop, and there is no waste in the manufacture, algal production becomes even more attractive.

For a time there was some interest in using algae both as a source of food and a photosynthetic gas exchanger in the closed systems that we may one day have in space. All green plants are photosynthetic gas exchangers, utilizing carbon dioxide and producing oxygen. The rapidly growing algae are very efficient gas exchangers. Although this approach is technically feasible, thus far we have relied on chemical processes for gas exchange and new techniques in food preparation and packaging.

WASTE TREATMENT

With increased urbanization we have begun to hear more and more about our sewage treatment plants (STP). Many plants were not large enough to handle increased capacities, might not have been able to take care of flow during all parts of the year, or were providing only minimal treatment. Even in the 1960s most U.S. plants had only primary treatment. In primary treatment large material was removed by screens, and after reducing the flow rate, sedimentation of smaller particles was achieved. The resulting odoriferous, turbid liquid was chlorinated, at least during warmer months, and discharged. The organic load, biochemical oxygen demand (BOD), in the water after primary treatment may then be reduced by bacterial action, either in activated sludge or by use of trickling filters. In the first case the bacterial action is in an aerated tank and in the latter on the surfaces of the rock filter. These processes are called secondary treatment.

More and more stress was placed on bodies of fresh water receiving STP effluent. We initially emphasized reducing the organic load and thus added, or sometimes improved, secondary treatment. Rivers were "cleaned up" and many nuisance problems disappeared.

But now with the "clean" secondary effluent entering our rivers, streams, and lakes in increasing amounts, because of population growth, we can detect, especially in the area receiving discharge from cities, more inorganic chemicals, including N, P, and Na. Only small amounts of these chemicals are removed in primary and secondary treatment, and the remaining chemicals are readily available to microorganisms, especially algae. Obviously this calls for tertiary treatment and some such plants are already in operation. By precipitation and adsorption many of these chemicals can be removed.

Oswald and Golueke have experimented with using algae in sewage oxidation plants to remove the inorganic nutrients. Effluent from secondary treatment is allowed to flow into shallow tanks or pools and the naturally developing populations of algae and bacteria thrive. The bacteria break down remaining organic compounds and provide carbon dioxide for algal growth. Algae take in phosphorus, nitrogen, other elements, as well as CO_2, *and* release oxygen.

In a typical sewage oxidation pond operation there are shallow tanks or ponds with baffles to allow for circulation (Fig. 20-13). The long route traveled by the organisms makes it possible to provide mixing of materials and suspension of the alga. The important design considerations are the depth of the tank, the detention time, amount of circulation and recirculation, method of mixing, and the rate of waste application.

It is necessary to provide mixing so that organisms remain in the effective light intensity (or move rapidly in and out of the light) and so that nutrients become well distributed. Without a circulating system the organisms settle out prematurely or more easily attach to surfaces, and the system becomes less efficient. With STP effluent entering at a constant rate, the flow of incoming nutrients must balance the ability of the algae to remove nutrients. A very slow flow can be inefficient and costly. At the other extreme, with a rapid flow nutrient removal is not being achieved. When the population of organisms is reduced by dilution, and the algal growth rate is slow, the inoculum can be washed out.

With an efficiently operating oxidation pond, one has achieved:

1. waste treatment, or removal of inorganic elements, and some organic compounds,
2. water recovery, for use elsewhere,
3. algal production, and
4. gas exchange.

The water can be supersaturated with oxygen so that animal life could survive, and the algae can be harvested for use as animal food.

The problem of increased loads on our sewage treatment plants is not solely one of increased population size. During the 1960s there has been a tendency toward urban migration, not only of people, but also of certain types of agriculture. The industry for fattening and feeding livestock began to concentrate in large cities. Thus animal wastes that

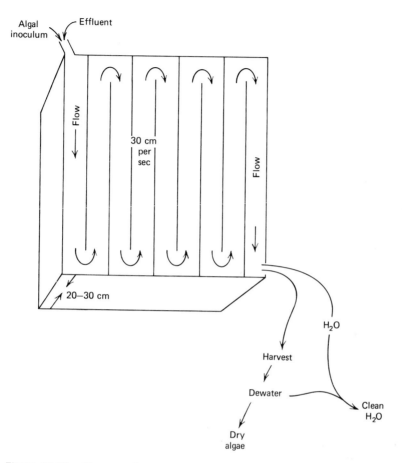

Figure 20-13. Diagrammatic presentation of the design of a simple chamber for growth of algae in sewage effluent. At the upper left the algal inoculum and the effluent to be treated enter the chamber. With a series of baffles the contact time in the apparatus is increased. By providing a gentle slope the desired flow can be achieved. These chambers are about 20–30 cm in depth, so that light penetration to the bottom is possible, or at least with some mixing, organisms receive an adequate supply of sunlight. After leaving the chamber the material is harvested, dewatered, and dried.

were deposited on the land, when animal feeding operations took place in rural settings, now more easily find their way into water near large cities. The recycling by placing wastes on the land, and using them for crop fertilizers, is most efficient.

In the United States the possible daily per capita water transmissible wastes are 2.5 kg from agriculture, 0.5 kg from food processing, and 0.1

kg as domestic wastes. Of the total, 3.1 kg per person per day, only 16% (or 0.5 kg) were delivered directly into water in the 1960s. At the same time 80% of domestic wastes was discharged into water. Projections to 1990 when animal feeding operations are centered in urban areas show that 50% of agricultural wastes will be in urban areas and transmitted to water.

When looking at the treatment of the effluent of a secondary treatment STP, operators might be interested only in the recovery of the water, without the nitrogen and phosphorus in it. Or the interest could be in the recovery of the nutrients, which could then be bagged and sold for fertilizer. With the use of algae in a sewage oxidation pond the algal product could be the greatest benefit for the whole system. Granted the water is now made "clean," and the nutrients are put in a form (bound up in organisms) that will not show up in our rivers, but of what value is the algal material?

In the United States, use of sewage oxidation ponds for algal growth could provide one-eighth of the daily protein we currently produce. The entire STPs and ponds would occupy about 10^9 square kilometers of land. In order to raise the same amount of protein in typical ways one would need 3.5×10^{11} sq km. We already use considerable land in our existing sewage treatment plants. Why not add a little more land and technology to the system and provide some protein, along with clean water?

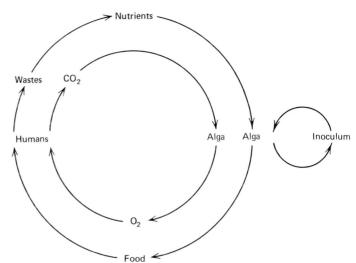

Figure 20-14. Cycling of nutrients and gases in the closed system of a photosynthetic gas exchanger. Such a system has been proposed for use by astronauts in space.

Considerable research with algae in closed systems has been conducted over the past two decades because of our interest in keeping humans in space for months at a time (Fig. 20-14).

AQUACULTURE

Recently there has been an interest in using our knowledge of growth characteristics of *both* algae *and* certain invertebrates to remove nutrients in an STP effluent. The proposal is to grow oysters or clams commercially on the crop of algae obtained from STP nutrients (Fig. 20-

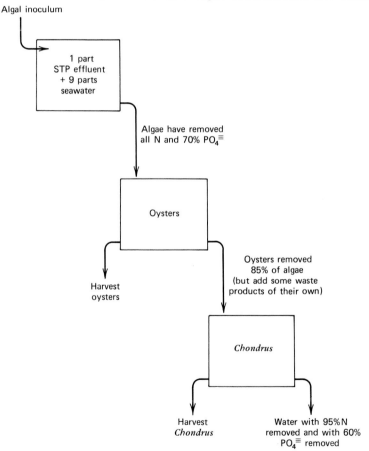

Figure 20-15. Model of a system used to remove nitrogen and phosphorus from sewage effluent, while growing a commercial product, in this case oysters.

15). At the Woods Hole Oceanographic Institute, Ryther and colleagues have been experimenting with such systems. Thus one would not only remove the nutrients polluting our streams, but also simultaneously provide a useful commercial product.

Effluent from a secondary STP is mixed with seawater at a 1:10 dilution so that the salinity is not greatly reduced. The unicellular algae, frequently diatoms from Martha's Vineyard Sound, grow in one tank (Fig. 20-16). When there is a healthy population, a portion of the crop is pumped (or fed by gravity flow) to the oyster tank. At the same time more effluent-seawater mix can be added to tank one. After the oysters have removed 85% of the algae, there is a significant increase in both N and P in the second tank, but this comes from oyster excreta. A third culture chamber, actually a run with *Chondrus* plants attached, reduces the levels of N and P. In sum, the system has removed most of the nitrogen and phosphorus *and* provided an oyster crop, as well as some growth of the commercially valuable *Chondrus*. Other organisms are being incorporated into this system, and experiments with growth of seaweed in seawater-effluent mixtures are underway.

Figure 20-16. Outdoor culture tanks and aquaculture laboratory at Woods Hole, Mass.

ARE THESE STEPS ECONOMICALLY FEASIBLE? IS THERE A MARKET FOR ALGAE?

In addition to the few ideas presented here, there have been many proposals to increase food production in the ocean. Some scientists have proposed that farms be established along our coasts, or that deep water, rich in nutrients, be pumped to the surface. We are all aware of the increased productivity in areas of upwellings. There is some debate concerning the amount of food that can be produced if we used even coastal ocean waters efficiently. Assuming there were no nutrient limitations in the ocean, Vischniac calculated that prior to 1970 we were harvesting about 20% of the productivity of the ocean. He based his figures on the amount of incident light that falls on the ocean as well as the efficiency of photosynthesis. Other researchers feel that these are conservative estimates.

The most serious limitation to using any mass culturing algal system at present appears to be our inability to control all important parameters when dealing with living organisms. Even when this control has been achieved, the cost of the system, especially harvesting the product, has been prohibitive. And although food costs in the early 1970s doubled in the United States, power costs have also increased very rapidly.

As yet, the biological systems we have discussed are not completely reliable, and certainly not thoroughly understood. But how much longer can we waste nutrients we deposit in streams or ignore available water rich in nutrients?

If we *had to remove* most N and P from our waste water, or *had to rely* on some algae for food, we probably have, or soon would develop, the necessary technology.

REFERENCES

Booth, E. 1965. The manurial value of seaweed. *Bot. Mar.* 8: 138–43.

Cheng, T. 1969. Production of kelp. A major aspect of China's exploitation of the sea. *Econ. Bot.* 23: 215–36.

Dawes, C. 1974. On the mariculture of the Florida seaweed, *Eucheuma isiforme.* Florida Sea Grant Program Rep. No. 5.10 p.

Dawes, C., J. Lawrence, D. Cheney, and A. Mathieson 1974. Ecological studies of the floridian *Eucheuma* (Rhodophyta, Gigartinales). III. Seasonal variation of carrageenan, total carbohydrate, protein and lipid. *Bull. Mar. Sci.* 24: 286–99.

Fryer, L. and D. Simmons. 1977. *Food Power from the Sea: The Seaweed Story.* Mason/Charter, New York. 220 p.

Goldman, J., K. Tenore, J. Ryther, and N. Corwin 1974. Inorganic nitrogen removal in a combined tertiary treatment-marine aquaculture system. I. Removal efficiencies. *Water Res.* 8: 45–54.

Golueke, C. and W. Oswald 1964. Role of plants in closed systems. *Ann. Rev. Plant Physiol.* 15: 387–408.

Harvey, M. and J. McLachlan 1973. *Chondrus crispus.* Nova Scotia Inst. Science, Halifax. 155 p.

Jensen, A. 1972. The nutritive value of seaweed meal for domestic animals. *Proc. VIIth Internat. Seaweed Symp.* 7-14.

Kadis, S., A. Ciegler, and S. Aj 1971. *Microbial Toxins.* Vol. VII. Algal and fungal toxins. Academic Press. 401 p.

Krauss, R. 1962. Mass culture of algae for food and other organic compounds. *Amer. J. Bot.* 49: 425-35.

Leeper, E. 1976. Seaweed: Resource of the 21st century? *Bioscience* 26: 357-8.

Levring, T., H. Hoppe, and O. Schmidt 1969. *Marine Algae: A Survey of Research and Utilization.* Cram. de Gruyter, Hamburg. 421 p.

Mackie, W. and R. Preston. 1974. Cell wall and intercellular region polysaccharides. In W. Stewart (Ed.) *Algal Physiology and Biochemistry.* Univ. California Press, Berkeley. pp. 40–85.

Mathieson, A. and R. Burns 1975. Ecological studies of economic red algae. V. Growth and reproduction of natural and harvested populations of *Chondrus crispus. J. Exp. Mar. Biol. Ecol.* 17: 137–56.

Miura, A. 1975. *Porphyra* cultivation in Japan. In J. Tokida and H. Hirose. (Eds.) *Advance in phycology in Japan.* Junk, the Hague. pp. 273–304.

Mynderse, J., R. Moore, M. Kashiwagi, and T. Norton. 1977. Antileukemia activity in the Oscillatoriaceae: Isolation of debromoaplysiatoxin from *Lyngbya. Science* 196: 538–40.

Othmer, D. and O. Roels. 1973. Power, fresh water, and food from cold, deep sea water. *Science* 182: 121-5.

Parker, H. 1974. The culture of the red algal genus *Eucheuma* in the Phillipines. *Aquaculture* 3: 425–39.

Pennelly, D. 1971. Food for thought. In: J. Rosowski, and B. Parker (Eds.) *Selected Papers in Phycology.* Dept. Botany, Univ. Nebraska, Lincoln. pp. 858–61.

Ryther, J., W. Dunstan, K. Tenore, and J. Huguenin 1972. Controlled eutrophication—increasing food production from the sea by recycling human wastes. *Bioscience* 22: 144–52.

Schantz, E. 1970. Algal toxins. In: J. Zajic. (Ed.) *Properties and Products of Algae.* Plenum Press, New York. pp. 83–96.

Schwimmer, M. and D. Schwimmer 1968. Medical aspects of phycology. In: D. Jackson (Ed.) *Algae, Man and the Environment.* Syracuse Univ. Press. pp. 279–358.

Volesky, B., J. Zajic, and E. Knettig. 1970. Algal products. In: J. Zajic (Ed.) *Properties and Products of Algae.* Plenum Press, New York. pp. 49–82.

Culturing algae

ISOLATION OF ORGANISMS

REFERENCES

INTRODUCTION

Over a century ago algae were first studied in culture. At that time it was proposed that we might determine the nutritional needs of an alga by measuring growth in solutions of mixtures of inorganic salts. Because chemicals were not pure, by present standards, success was achieved with addition of a minimum number of common salts. In early media formulae many of the elements we now know to be essential to algal growth were not present except probably as contaminants, for only carbon, nitrogen, hydrogen, oxygen, phosphorus, magnesium, sulfur, potassium, iron, and calcium were though to be required. With this knowledge, simple media were prepared and many cultures were established. Investigators obtained good growth and could study microorganisms in detail, especially unialgai cultures. They grew enough cells to obtain material for pigment analyses, and for observations of details of cell division and life histories.

It is not clear when the first "pure" (meaning bacteria-free) culture of an alga was obtained. At the turn of this century there were programs dealing with the culture of algae in laboratories in several countries. Beijerinck, in 1890, was certainly among the first to work with a culture of an alga free of all other organisms, now called an axenic culture. Pringsheim has described this era in some detail. Many of the early investigators devised their own combination of nutrient salts for growth media. Often the media still bear their names, although seldom are the media compounded in exactly the same way as when they were formulated. Changes include dilution, substitution of one or more salts, or the addition of a trace element mix. The composition of several of these older freshwater media, such as those of Beijerinck, Knop, and Bristol, is presented in Tables 21-1 and 21-2. Because of the modifications that have been made in recent years, it is not uncommon to find markedly different formulations for media bearing the same name. One recent account of Beijerinck's medium would reduce the level of the nitrogen and magnesium salts listed in Table 21-1, add an additional phosphate,

Table 21-1. FORMULATIONS OF FRESHWATER MEDIA DESIGNED
EARLY IN THE CULTURING OF ALGAE[a]

Salt	Medium (salts in mg l^{-1})		
	Beijerinck	Knop	Pringsheim
NH_4NO_3	1000	—	—
KNO_3	—	1000	200
$Ca(NO_3)_2$	—	100	—
$(NH_4)_2HPO_4$	—	—	20
K_2HPO_4	200	200	—
$MgSO_4 \cdot 7H_2O$	100	100	10
$FeCl_3$	1	1	0.5
$CaCl_2 \cdot 2H_2O$	—	—	0.5
Total inorganic salts	1301	1401	231

[a] Note that calcium could be lacking and that trace element requirements were not known. When these media are used today, a trace mix (Table 21–5) is added. In addition, the total salt levels are often reduced by dilution with distilled water.

increase the level of phosphates (for buffering), and, finally, add a trace element mix to the original medium. Beijerinck would not recognize his modified medium! As we shall see later, all of these changes have been carefully developed in attempts to improve the basic medium through application of recent knowledge.

A freshwater medium Bristol compounded in the 1920s has found wide acceptance in many laboratories. Naturally, it has been modified since then. One formulation has been made more dilute than when compounded. A trace element mix has also been added. These media have been called modified Bristol's and Bold's basal medium (BBM). If one were to attempt to duplicate work performed with Bristol's medium, or any medium, it would be wise to determine the exact modification used. Modified Bristol's, or BBM, come closest to being standard freshwater algal media. Comparative experimentation would be greatly simplified if there were more standard media.

In the development of the early media, and in presenting more recent adjustments, some consideration was given to the total salts concentration (and salinity, when applicable) (Tables 21-2, 21-3), source of nitrogen, pH, and buffer actions, and concentration of elements that are easily precipitated (Table 21-2).

Table 21-2.   COMPOSITION OF BRISTOL'S MEDIUM AND THREE MORE DILUTE MEDIA[a]

	Medium (salt in mg l^{-1})			
	modified Bristol's	Chu # 10	PAAP	3.07
$Ca(NO_3)_2$	—	40	—	—
$NaNO_3$	250	—	26	2
K_2HPO_4	75	10	1	0.03
KH_2PO_4	175	—	—	—
$MgSO_4 \cdot 7H_2O$	75	25	15	1
$MgCl_2$	—	—	6	—
$CaCl_2$	25	—	4	—
$NaCO_3$	—	20	—	—
$NaHCO_3$	—	—	15	—
Na_2SiO_3	—	25	—	—
$NaCl$	25	—	—	—
$FeCl_3$	—	0.8	—	—
Trace mix (Table 21-5)	No	Yes	Yes	Yes
TRIS	—	—	—	Yes

[a] Although there is no trace mix listed in Bristol's medium as it was originally designed, most current modifications employ one. The trace mix listed in Table 21-5 could be used. When that same mix is added to medium 3.07, it is diluted 1 : 1000. Chu # 10 medium and the PAAP medium each have their own trace mix, but that listed in Table 21-5 could be satisfactory for most purposes.

Table 21-3.   A COMPARISON OF SOME OF THE MAJOR IONS AND ELEMENTS IN FRESH WATERS, AND IN TWO FRESH-WATER MEDIA

		Percent Total Salts As:		
	mg l^{-1} total salts	$Ca^{2+}, Mg^{2+}, CO_3^{2-}, SO_4^{2-}$	N	P
Natural waters	c. 150	73	1–2	<1
Bristol's medium	625	7	7	8
Chu 10 medium	133	27	6	2

CHOOSING A MEDIUM

Studies dealing with the algae in culture have suffered because there is no widely used, standard, "packaged" laboratory medium for organisms from fresh water environments. However, there are such formulations for bacteria and fungi. When attempting to grow the latter, it is a simple procedure to weigh out a few grams, place the material in a volume of water, and sterilize. In contrast, Bristol's or one of its modifications are composed of several salts that must be weighed individually. This complication is sufficient to discourage many researchers. Some of the marine media discussed below are used in many laboratories, but preparation is even more time-consuming than for freshwater media. Certainly this detail, or the need for a particular chemical, or the lack of an adequate balance have also discouraged many phycologists from making at least an attempt at culturing.

Nevertheless there are several ways to approach the culturing of algae. It is possible to establish some cultures rather easily, especially if undefined media are used. Determination of the purpose for establishing a particular culture will make it easier to choose among the approaches examined below. Is the material to be cultured for teaching or for research? Can it be used satisfactorily in the vegetative state, or must it be reproductive? Is it sufficient to have a unialgal culture, or should it be axenic? Will the culture be maintained for a short time, or will a stock culture be kept for a number of years?

Undefined Media

Soil Water Bottles.   A most suitable medium for maintaining many algae, both marine and freshwater, can be prepared by placing a good garden or greenhouse soil in either seawater, tap water, or even distilled water. The soil should never have been in contact with pesticides or chemical fertilizers. A number of combinations have been suggested, but a one to ten mixture will work in most cases. A gram of good garden soil is placed in 10 ml of liquid, or one can work with multiples of the same. The combined soil and water are thoroughly mixed and sterilized. This can be accomplished by fractional sterilization, that is, boiling for an hour on three consecutive days, or under pressure in an autoclave. If a good soil has been chosen, for example, a garden soil that would support good growth of higher plants, and if the water does not contain toxic elements, say, copper from the piping, the resulting solution will contain all the inorganic elements needed for growth. Soil extract was once prepared only by fractional sterilization, that is, boiling for an hour on three successive days. Certainly the solution is different when the higher temperature of autoclaving is used, even if the time necessary for sterilization is only 20 minutes. Organic compounds, especially B vitamins, may be broken down by excessive heat. But because autoclaving is simpler than fractional sterilization, and the resulting solution has been successfully employed, that procedure is preferred in some laboratories.

The student could experiment with the method of autoclaving, the type of soil used, the proportion of soil to water, and the source of the water. Some researchers prefer to add a "pinch" of calcium carbonate to soil water bottles in order to provide a more beneficial pH. (The pinch is herein defined as about 0.1 g in 100 ml.) The soil could be collected from the edge of the body of water, or even the sediment in the body of water from which the culture originated. The water used in the soil water bottle could come from the same site. With cultures of freshwater algae some investigators prefer getting all nutrients from the soil, and thus use distilled water. For marine species, aged seawater is often preferred, but it is probably no better than a good freshly collected seawater.

The container in which the soil-water combination was prepared can be used for growth of the algae. Some researchers prefer using soft glass bottles, even pint milk bottles. Usually the bottle contains between 100 and 250 ml of solution, and is about half full. Bottles with straight sides take up less space.

Soil Extract.   This is the solution that results when soil water bottles are prepared. The same procedures are used as described above, but a larger quantity is prepared, on the order of a liter of soil extract. After autoclaving, the solution is clarified overnight, and then the soil extract is decanted. This procedure is repeated until a clear solution is obtained, or the fine particles in suspension can be filtered from solution. The soil extract is placed in smaller containers for storage, or used immediately. *But,* after filtration or settling, the solution should be autoclaved again.

Soil extract is sometimes used alone for a culture medium. The advantage over the soil water bottle is that bottom organisms are more easily sampled. It is also added to other media to provide trace elements or growth requirements (see later). This addition can be concentrated as one part soil extract to ten parts of medium or as dilute as one part in 200.

Both soil water bottles and soil extract have many successful applications not only because they are often easier to prepare than standard media, but also because they can provide an unknown growth factor. Often a culture will grow in soil extract, but not in an inorganic medium. It is usually found that the soil extract provides a vitamin requirement. Once this is known, the vitamin can be added to the defined medium.

Defined Media

A glance of Tables 21-1, 21-2 and 21-4 shows that in addition to C, H, and O, which plants obtain from carbon dioxide and water, 10 elements are listed for freshwater media. Early plant physiologists knew some of the elements that were essential for growth of land plants. Soon phycologists or microbiologists found ways of putting these elements in solution so that algal growth was promoted. To simplify the preparation of media, they attempted to provide two essential elements as one salt, utilizing both anion and cation. Thus potassium phosphate provided both K and P.

Of what value are Si, Cl, and Na, which are provided in the media in these tables? For diatom glass walls silica is essential, but it is not necessary for many other algae. Chlorine is needed by some algae in minute amounts, but there is still much to be learned about its role in algal nutrition. It is often convenient to provide freshwater algae an essential cation as a chloride and thus avoid some problems of solubility. Sodium is essential for blue-green algal growth. Both Na and Cl are

Table 21-4. FORMULATIONS OF SEVERAL MEDIA USED IN THE CULTURING OF ALGAE[a]

| | (milligrams l^{-1}) | | |
	Tamiya	Ishiura and Iwasa	Sager and Granick
NH_4NO_3	—	—	300
KNO_3	5000	303	—
K_2HPO_4	—	697	100
KH_2PO_4	1250	680	100
$MgSO_4{\cdot}7H_2O$	2500	370	300
$FeSO_4{\cdot}7H_2O$	3	—	—
$CaCl_2{\cdot}2H_2O$	—	30	—
Na_3citrate	—	—	500
Trace mix (Table 21–5)	yes	yes	yes
Total salts	8750+	2080+	1300+

[a] Note that the total salt levels do not include the trace mix. These media are more concentrated than the average medium in use.

provided in larger amounts to marine and brackish water species because salinity must be maintained.

The major salts of many media are listed in Tables 21-1, 21-2, 21-4, 21-8, and 21-9. Many other formulations may be found in the literature. If many batches of any medium, either freshwater or marine, are to be made, it is convenient to prepare stock solutions. With certain salts used in large quantities, such as NaCl for marine media, or when solubility of the salt is low, stock solutions might not be prepared.

It is often convenient to prepare stocks by making a hundred-fold concentration of individual salts. Thus for Bristol's medium (Table 21-2) 25 grams of $NaNO_3$ can be added to a liter of distilled water (or 2.5 g in 100 ml). Volumetric flasks should be used for the preparation of stock solutions. When it is time to prepare the medium, 10 ml containing 250 mg $NaNO_3$ can be pipetted from the concentrate. If all six salts of Bristol's medium and the trace mix were prepared as $100 \times$ stock solutions, 10 ml of each could be delivered to 930 ml of distilled water. Then the final concentrations would be as listed in Table 21-2. Similar procedures would be followed with other media.

The selection of a medium is easier if the investigator wishes to grow an organism that has been cultured previously. In starting new isolates the best choice can be made with some knowledge of the habitat, or of requirements of similar organisms already in culture. The first step is to select a widely used freshwater or marine medium, such as one described in these tables. A trace mix (see below and Table 21-5) should be added to the major salts, but some researchers prefer a few milliliters of soil extract. The soil extract could also satisfy a requirement for organic compounds, demonstrated by many algae in culture. The B vitamins are the compounds which are most frequently needed. These could be added to any of the media listed (Table 21-6) at a level of tenths of a milligram for thiamine, and tenths of a microgram for biotin and B_{12}. Many organisms that have organic growth requirements are not isolated in some laboratories because, in order to cut down on bacterial and mold growth during isolation, no organics are put into the isolation medium. Later in this chapter, we will consider other organic growth requirements as well as utilization of organic substrates available in the environment.

It is thus clear that a medium can be chosen, or one can be defined, that will select for a particular alga—the organism that the investigator wishes to culture. As the organism then grows in culture, utilizing chemicals, some elements could become limiting. In order to avoid

Table 21-5. A TRACE ELEMENT MIX, PROPOSED BY WATT AND FOGG, FOUND USEFUL WITH MANY FRESHWATER ALGAL MEDIA[a]

	mg l^{-1}
$FeCl_3$	5.0
$CaCl_2 \cdot 2H_2O$	26.5
$MnCl_2 \cdot 4H_2O$	0.3
$CoCl_2 \cdot 6H_2O$	0.02
$CuSO_4$	0.01
$ZnSO_4 \cdot 7H_2O$	0.04
$Na_2MoO_4 \cdot 2H_2O$	0.02
Na_2EDTA	6.5
Total salts	38.39

[a] When preparing this mix, it is convenient to make a 100 X solution. Then 10 ml of the concentrate can be added to a 990 ml volume of the medium being prepared. If this trace mix is added to some older media, as presented in Table 21–2, it is wise to add elements such as Fe or Ca only in the trace mix. It is in this way that modifications of the original medium are often made.

deficiencies, and to keep the culture actively growing for some time, the media used by investigators 50 years ago became more concentrated. Algal weeds, organisms found under a variety of conditions in nature, thrived in these media.

Early investigators did not realize that some of the salts were precipitating. Thus additions of phosphates to some media often went directly into the precipitate. There must also be considered the contantly changing nutrient conditions in a large flask of algae growing in a

Table 21-6. LEVELS OF B VITAMINS FREQUENTLY USE IN BOTH MARINE AND FRESHWATER MEDIA[a]

	($\mu g\ l^{-1}$)	
Vitamin	Marine Media	Freshwater Media
B_{12}	0.5–1	0.1
biotin	0.5–15	0.1
thiamine	100 –2000	100

[a] The vitamins are often prepared in a single stock solution. The vitamin mix can then be added to defined media. However, such a vitamin mix is also often part of the enrichment provided when natural water is the base of the culture medium.

concentrated medium. In addition to the organisms taking up salts, there is the release of extracellular products into the flask. Usually there is a pH alteration, most commonly toward the alkaline range, unless there is strong buffering.

Early in the development of media, buffer action was achieved by use of both monobasic and dibasic phosphates. But unless the latter are supplied in sufficient amounts, they are not effective for long-range experimentation. Note the high levels of phosphate in some freshwater media (Table 21-4). Phosphates may precipate during media preparation, especially in the alkaline pH range. How much phosphate is then in solution? Which cations have also precipitated? Because answers to these questions are not usually known, such media are not useful for nutritional studies. Some media are buffered with carbonates. More recently organic buffers have been added to media, especially those which can not be metabolized. TRIS (Tris hydroxymethylaminomethane) is very useful in the alkaline range, and glycylglycine has been used in some marine media. However, some microorganisms can metabolize glycylglycine. Other organic buffers which are effective for specific cases have been proposed.

When an element is lacking or the supply is exhausted, organisms survive by parcelling out what is available *in* the cell. Eventually a deficiency symptom appears, such as a chlorosis without adequate Mg, or a switch from protein synthesis when the cell is under nitrogen stress. Under nutrient limitation algae frequently have enlarged cells, with many granules and thickened walls.

Trace Elements and Chelators. Early investigators were able to culture algae with just the major elements given in the previous tables, because the salts used were not pure. We now know that other elements are essential for growth (Tables 21-5, 21-10, and 21-11). C, H, O, P, K, N. S, Ca, Fe, and Mg are known as *major* elements. They are often supplied in milligram per liter amounts. The additional requirements are the *minor* or *trace* elements. The trace elements are required in microgram per liter quantities, or even in some cases in lesser amounts. Although there is considerable Na in marine media, it is not required by most algae, and therefore for them sodium would not be in either category. For blue-green algae it is probably a trace element. At present we are learning more about precise levels needed for individual algae, as well as ways to keep them in solution and available to the organism. EDTA, ethylenediaminetetraacetic acid or editic acid, is commonly used for chelation, that is, complexing the element so that it is available to the alga. Trace elements needed by algae include Fe, Ca, Co, Cu, Mn, Zn, Mo, and B. In addition, V, I, Sr and Li are supplied to some algae (Table 21-11). Fe can be either a major or minor element.

Since an element needed at a very low level could be present as a contaminant, the requirement for a trace element can be difficult to establish. When a medium formulation lacking an essential trace nutrient is used successfully, can one be certain that the element is available as a contaminant?

Freshwater Media.   Much of what has already been discussed deals with freshwater media. It is clear that there are many types of media in use, some developed many decades ago. In many cases any one of several media could be substituted for the one in use in a particular laboratory.

With freshwater organisms there has been little use of natural waters as media, or of enrichments of natural waters. However, there has been wide acceptance of soil extract, or of additions of soil extract to defined media. The novice should experiment with either Bristol's or one of the dilute media currently being developed.

Dilute Media.   Why are total salt levels in many media so high? There are several reasons. With high levels of nutrients organisms will grow in culture for weeks, even months. In addition there is a historical reason, for early investigators were unknowingly providing trace elements as impurities with the major salts. If one is interested solely in the culture of large quantities of cells, concentrated media are useful, and often more growth can be obtained by increasing salt levels further.

Chu proposed that media could be selected because they possess characteristics that are similar to the water in which the algae are growing. First, concern about the total salt concentrations affects choice of media, and, as a result, dilutions of many media have been successfully employed. With marine media the salinity can be reduced only slightly, but when enrichments are proposed it is possible to significantly change the level of an individual ion. In the 1930s Chu looked at the composition of fresh waters and attempted to formulate media with ions in the same relative amounts as found in nature. He also wanted total salt levels comparable to those of natural waters. Compare his medium (Table 21-2) with Bristol's or Tamiya's media (Tables 21-2 and 21-4). Although Chu medium number 10 has been successfully used for culture of many organisms, the levels of many ions are considerably higher than those found in nature where, for example, there would be about a milligram per liter or less of available nitrogen. However, with regard to total salt level, and amount of nitrate and phosphate, Chu's medium is an improvement over Bristol's (Table 21-3). Further supporting data for reducing the amount of nitrogen are presented in Table 21-7. Some dilute freshwater media and certain marine media have a level of nitrogen lower than that proposed by Chu. When experiments were designed to support more and more cells, or to grow larger and larger organisms, the level of nitrogen in a medium would be increased (Table 21-2).

Medium 3.07, with the lowest level of inorganic salts of any freshwater medium in Table 21-2, has all inorganic salts at or *below* levels in nature. It can be used in the laboratory if the cell number is kept lower than usual for culture studies. Thus it will support growth of about 10^5–10^6 cells per ml, whereas 10 to 1000 times more cells can be cultured in other media. When this dilute medium is selected, the organisms are transferred often, even daily, so that there is nutrient replenishment. Some minor adjustments of nutrients in medium 3.07 are essential if the investigator wishes to culture a variety of freshwater organisms. The medium was designed for a specific study of one organism, and it should be expected that other freshwater algae might have slightly different growth requirements. If available, a continuous culture apparatus could be utilized. It would give a continuous nutrient replenishment without requiring the daily transfer of organisms.

Table 21-7. THE LEVELS OF NITROGEN FOUND IN SEVERAL MARINE AND FRESHWATER MEDIA

Medium	mg N l^{-1}
Freshwater	
Tamiya et al.	693
Sager and Granick	105
Ishiura and Iwasa	42
Bristol	41
Chu 10	7
P A A P	4
3.07	0.33
Marine, artificial	
ASP-6	5
Müller, modified	16.7
Marine, enriched seawater	
Erd Schreiber	16.7
f/2	13

Historically, marine media have been designed to be more like natural waters, but still the levels of specific elements can be far higher than those encountered in nature. With low levels of N and P in many marine media it is necessary to transfer the cultures more frequently than the average freshwater culture.

Marine Media.   Researchers involved with the culture of marine organisms often use natural seawater, or more frequently, the seawater with enrichments (Table 21-8). Seawater is far more uniform in chemical composition than fresh water and can become a useful medium with just a few additions such as N, P, and Fe, or even soil extract. There are a number of defined media; two of the more successful formulations are presented in Table 21-9. Trace element mixes are listed in Tables 21-10 and 21-11.

With marine organisms there are additional media, that is, commercial preparations. These are sold in bulk quantities and are synthetic mixes which are quite simple to prepare. There is some difficulty in duplicating the medium each time it is prepared, although companies do give fairly accurate formulations when a supply is shipped. These media are preferred when large quantities of media are needed.

Because of the high salt concentrations, and the alkaline pH, there are often complications in the preparation or manipulation of marine media. At times some problems, such as solubility of a salt, can be solved by

Table 21-8. TWO MARINE MEDIA PREPARED BY ENRICHING SEAWATER

	Erd Schreiber[a]	f/2[b]
$NaNO_3$	100 mg	75 mg
$Na_2HPO_4 \cdot 12 H_2O$	20 mg	—
NaH_2PO_4	—	5 mg
Na_2SiO_3	—	7-13 mg
Trace mix (Tables 21-10 and 21-11)	—	+
Vitamins	—	+
Soil extract	50 ml	—
Sea water	to a liter	to a liter

[a] Although the early formulation for Erd Schreiber was simple, it can be modified by addition of a trace mix or a vitamin mix (or both).
[b] With medium f/2 buffering is sometimes provided with 500 mg of either TRIS or glycylglycine.

Table 21-9. THE COMPOSITION, WITH AMOUNTS OF MAJOR SALTS, OF TWO SYNTHETIC MARINE MEDIA[a]

| | mg l^{-1} | |
	ASP-6	Modified Müller
NaCl	24000	26700
KCl	700	730
MgSO$_4$·7 H$_2$O	8000	1560
MgCl$_2$·6 H$_2$O	—	1020
CaCl$_2$	420	1130
NaHCO$_3$	—	200
NaNO$_3$	300	100
Na$_2$HPO$_4$	—	20
glycerophosphate	100	—
KBr	—	22
Na$_2$SiO$_3$	70	20
TRIS	1000	—
Trace Mix (Table 21–10)	+	
(Table 21–11)		+
Vitamins	±	±

[a] After Provasoli.

Table 21-10. TRACE ELEMENTS FOR CERTAIN MARINE MEDIA, AS PROPOSED BY PROVASOLI[a]

	mg l^{-1}
P II mix	
EDTA	10
Fe	0.1
B	2.0
Mn	0.4
Zn	0.05
Co	0.01
S II mix	
Br	10.0
Sr	2.0
Rb	0.2
Li	0.2
I	0.01
Mo	0.5

[a] Two mixes are prepared, usually as 100 X stock solutions, which are then added to the marine medium being compounded. The elements are added to the trace mix as chlorides, acids, etc.

Table 21-11. A TRACE ELEMENT MIX, MODIFIED AFTER MÜLLER, USED WITH SOME MARINE MEDIA

	$mg\ l^{-1}$
$SrCl_2$	2.26
$ZnSO_4$	1.29
H_3BO_4	2.0
$MnSO_4$	0.57
$FeCl_3$	0.12
$NaMoO_4$	0.17
RbCl	0.016
$AlCl_3$	0.012
KI	0.02
$CoSO_4$	0.0034
LiCl	0.006
$CuSO_4$	0.00083
Na_2EDTA	20.0

lowering the salinity merely by dilution with distilled water. It is suggested that the directions given in publications for the preparation of a particular medium be *carefully* followed. Sterilization can be by filtration rather than by autoclaving. Provasoli, McLaughlin, and Droop have presented a history of the development of artificial media, and have provided many formulae for media currently used.

Carbon Dioxide and Aeration.   With all the types of cultures discussed thus far, be certain to provide air exchange to the culture so that CO_2 is available. This can be merely by diffusion through cotton plugs, the stopper most commonly used. Loose fitting plastic or stainless steel enclosures are becoming more popular because they are easier to manipulate and can be used repeatedly. With screw-top containers be certain that the caps fit loosely during autoclaving as well as incubation. Media may be stored with the cap fitting tightly.

Carbon dioxide may be bubbled into cultures by using a mixture of 1 to 5% CO_2 in air. However, special care must be taken to keep the culture sterile and to provide moisture to the dry air mix, lest the culture dry out, or salts become more concentrated in a rapidly evaporating medium.

Medium Sterilization.   The earliest procedure used was that of fractional sterilization (see discussion above under soil water bottles).

Now autoclaving, the use of steam under pressure, for about 20 minutes is the widely used sterilization method. However, as noted earlier, when all the salts of a medium are put together there can be a precipitate during autoclaving. Some researchers have avoided this problem by autoclaving some components separately. Organic compounds may be degraded by the heat, and thus are frequently sterilized separately, often by filtration.

Now it is becoming more and more common to sterilize certain media by filtration. Glass or stainless steel apparatus can be purchased and easily sterilized. The filter may be sintered glass or a membrane type with a pore size sufficient to trap bacteria. There are several kinds of membrane filters but all are sterilized separately in moist heat and assembled in the sterile filterng device aseptically. Sterilization by filtration is very effective, but often slow to complete because of the small pore size of filters. This procedure is preferred when working with natural waters or when organic compounds are to be added to the medium.

NUTRITIONAL TYPES

So far we have emphasized those algae that have inorganic nutrient requirements, or perhaps an additional requirement for a vitamin. But many algae either require an organic compound, or can utilize the compound if it is supplied. Photosynthetic forms that grow in the absence of organic compounds are sometimes simply called autotrophs, while other algae that utilize organic sources have been labeled heterotrophs. It is now known that these terms are not specific enough, and do not take into consideration other possibilities of utilization. Thus, following the recent review of uptake and utilization of organic carbon by Nielson and Lewin, a more complete terminology is presented.

Photolithotrophs

Algae called photolithotrophs get their energy from light, use water as a reductant, and increase the level of cell carbon by the reduction of carbon dioxide. These organisms are frequently called photoautotrophs or simply autotrophs. Some are obligate photolithotrophs, whereas others might grow during some periods in this manner, but can use an alternate system (see below). Those that use alternative systems are facultative photolithotrophs.

Heterotrophs
Heterotrophs are algae capable of sustained growth and cell division in the dark. Both the carbon and the energy are derived from the organic substrate. Many of these same algae are also capable of photolithotrophy. Thus if they are in sunlight, and they do not have any organic matter available, they are photolithotrophs. If placed in complete darkness, as in the sediment or in a laboratory experiment, they can exist as heterotrophs (see discussion below about substrate concentrations). In some cases in the presence of light the utilization of the organic compound can actually be suppressed.

Mixotrophs
Although at present few of these organisms are known, they *require* organic matter in addition to that carbon reduced in photosynthesis. Apparently there is a loss of photosynthetic capacity. Mixotrophs are successful organisms because they have been able to use simple organic compounds such as acetate. Some call these organisms photolithotrophic heterotrophs because they are a combination of the first and second groups.

Photoassimilators
In the light a number of algae, even some that are not capable of heterotrophic growth, can utilize an organic substrate when it is available in sufficient concentration. The amount of carbon that comes from the organic fraction will vary with the organism, as well as the carbon source available. These organisms have often been called photoheterotrophs or the process labeled photometabolism. Some organisms can break down an organic compound in respiration, and then use the carbon dioxide in photosynthesis.

Under standard conditions there is a limit on the growth that can be achieved by a particular alga. If the system is saturated by means of the reduced carbon from photosynthesis, then there will be no photoassimilation.

Auxotrophs
This term would be applied to any alga that required an organic compound, if the addition of this substrate did not add *significantly* to the carbon in the cell. Vitamins are usually the compounds included under this term, for they are often needed only in microgram per liter amounts.

Phagotrophs
Only a few algae, such as *Ochromonas,* are capable of using particulate food. These are the phagotrophs. They ingest food particles or small organisms, digest them, and pass the digested compounds from the vacuole into the cell. Algae that are phagotrophs are often quite similar to nonphotosynthetic protists even in other ways.

Discussion
Because natural waters have a measurable level of organic compounds, there has been some interest in the role of algae in their utilization. Early culture studies with the algae concentrated on high levels of simple compounds such as sugars. A number of algae were isolated which would utilize one or more organics as a source of carbon in the dark (heterotrophy) or as a supplement to photosynthesis in the light (photoassimilation). Unfortunately the levels of individual compounds were in grams per liter. In nature it is not often that even a milligram per liter of an *individual sugar* would be found. Thus, with early laboratory studies of heterotrophy or photoassimilation, it was impossible to extrapolate to growth in the ocean, a stream, river, or pond.

Now studies of uptake of organics have shown that algae in general obtain a sugar only by diffusion of the compound into the cell, whereas bacteria have transport mechanisms capable of taking in the compound at low substrate levels. In selected cases one can demonstrate transport of some compounds by algae. Thus, normally the bacteria keep those compounds that can be metabolized at levels below which simple diffusion into algae can occur. Why then are 1 to 3 milligrams of carbon per liter as dissolved organic compounds reported from nature? Apparently much of this matter is of large molecular weight, is broken down much more slowly by bacteria than are most substrates, and is not metabolized by the algae. The age of some of these compounds has been measured as greater than 200 years in some parts of the ocean.

The algae themselves are sources of much matter in solution in nature. A number of studies have shown that, at some point in their growth, a large percentage of the photosynthate can be excreted by particular algae. Some researchers have theorized that the organism is utilizing the environs as a storage reservoir. However, unless the organism would have a mechanism for keeping bacteria from the area, this would be a poor mechanism for *storage.* If high concentrations of a particular substrate could be maintained, the organism might use the organic later, for example, during the dark period.

In some environments where there is a large concentration of organics, for example, in a sewage oxidation pond or a farm pond, some algae utilize these in addition to relying on photosynthesis. Some organisms have developed a high degree of specificity and thus will metabolize only one organic compound. Others might require a certain light-dark cycle and will not demonstrate any heterotrophic capabilities under long-day conditions. And yet other organisms appear to be able to use a broad spectrum of compounds heterotrophically, and in photoassimilation. Eventually we will learn that the ability of many algae to utilize organic matter is useful, or potentially useful, in many environments in which they grow.

CONDITIONS FOR GROWTH

Organisms are conveniently grown in test tubes or flasks, but any container that permits light to penetrate, surrounds the organism so that it can be maintained in sterile conditions, and does not release toxic materials can be employed. For precise experimentation, rigorously cleaned pyrex glassware must be used. Pyrex glassware withstands the heat of sterilization better than soft glass and does not easily lose chemicals by leaching.

Apparatus

Organisms can be maintained on any shelving material that will not interfere with the quality and quantity of light (see below). In constructing such shelving some consideration has to be given to the effect that even incomplete enclosure will have on air movement and thus temperature. It is not at all uncommon to have quite a temperature range in a "constant temperature" room.

Organisms can be grown in stationary culture, but often some apparatus that will agitate the algae is utilized. Agitation can be accomplished by bubbling in air, which also provides additional carbon dioxide. More frequently, culture vessels are swirled, rotated, or shaken by placement on a laboratory shaker or rotator. The investigator can easily manipulate frequency and degree of agitation so that optimal conditions are achieved. For marine organisms tidal simulators can be constructed that will bathe the seaweed in nutrient solution and then periodically expose it to air.

Illumination

Natural illumination is not often used in laboratories; it is too intense, with over 100,000 lux falling on the culture surface at noon on a sunny day. In addition, within the confines of a flask or test tube there is a buildup of heat, often above the lethal temperature. However, some good work has been performed in the diffuse light of a north window. The student might try this and learn that adequate growth can be obtained under such simple conditions, without the use of expensive equipment. Natural illumination is not the choice when the researcher or teacher designs experiments that can be repeated in the same laboratory or elsewhere. How can the investigator duplicate cloud cover, the reduction in intensity caused by an adjacent building, or how can he (she) be certain of the length of daylight? Thus artificial illumination is almost always used, and fluorescent light is the most popular choice. At about 30 to 60 cm from two 120 cm fluorescent bulbs, it is possible to obtain between 4000 and 6000 lux illumination, an illuminance more than adequate for most experimentation. Inasmuch as light quality can be altered when different types of bulbs are used, most laboratories have now narrowed the choice to either cool white or daylight fluorescent bulbs. Although adverse effects of continuous illumination are difficult to prove with most organisms in culture, it is wise also to purchase an inexpensive timer. The two most commonly used cycles are either 15 hr light and 9 hr darkness or a 12 L: 12 D cycle. With day length varying with the seasons, it is difficult to decide which is the most "natural" cycle. Few experiments deal with cultures subjected to seasonal changes in light dark cycles.

If incandescent lamps, or other types of illumination are used, remember that temperatures can become quite high within an enclosed flask or test tube. When incubators are used (see below), these lamps, and even the ballasts from fluorescent lights, can raise the temperature considerably. Ballasts can be mounted outside the incubator, and cooling jackets can be supplied for incandescent lamps.

Temperature

During the previous discussions there has been mention of temperatures favorable for growth. Most culture facilities are maintained at a constant temperature of 20° or 25°C. By arbitrarily selecting a temperature in this range, there is also selection for organisms that will survive under these conditions, and many will. But such organisms are often grown in the laboratory at temperatures that are at least 5° to 10°C warmer than the water from which they were collected. Is use of such a constant temperature natural? Certainly not, and many cold water forms will not tolerate such "heat." Thus refrigeration must be provided when the cold water forms are cultured. If thermophilic species are grown, incubators, even with temperatures up to 60° to 70°C, must be used. It is interesting that the optimum temperature in nature for some forms appears to be some degrees lower than the laboratory optima. It is thus difficult at times to extrapolate from laboratory data to the field, but, on the other hand, an organism has demonstrated that it has that temperature tolerance. Perhaps in another locality the organism will be found growing well at the higher temperature.

In many laboratories a specially designed constant temperature room is available. For large-scale culturing, walk-in incubators can be purchased. Smaller, refrigerator-size incubators are available. These incubators are easily assembled by making a few adjustments and modifications to an existing refrigerator.

The temperature variation in large incubators and constant temperature rooms is not less than ± one degree celsius. Even in small incubators it is difficult to do better with temperature control. For accurate control, a water bath is a must. The control can be maintained at ±0.1°C by using a thermoregulator. However, because of the cost involved, it is only practical to use water for temperature control with experimental material. Maintenance cultures and culture collections grow well under the slight temperature variation of constant temperature rooms.

TYPES OF ALGAL CULTURES

Maintenance Cultures
Many organisms are kept in culture collections to be used for either teaching or research. These stock cultures are kept in liquid or on agar slants. (One and one-half percent agar is used to solidify any liquid medium.) They are grown in dim light at a favorable temperature in containers that retard water loss. Screw top test tubes or bottles can be closed, but not tightly, to prevent rapid drying. Soil water bottles are a favored maintenance medium for some organisms, and soil extract is sometimes solidified.

Cultures Of Limited Volume
Often experiments are conducted in flasks or test tubes that are filled to about half the liquid capacity. The vessel is inoculated with a small number of cells and incubated under favorable conditions, which may

include agitation, for days or even weeks. There is a lag period, a period of accelerating growth and then a stationary phase. Samples may be taken during various stages, or the culture may be sacrificed for experimentation at any time.

These cultures of limited volume, terminology used by Fogg, are the type of culture most often used in the laboratory. They can be sampled frequently during morphological and taxonomic studies and can yield large quantities of cells for physiological experiments. It is with such cultures that concentrated media were designed.

Mass Cultures

Whenever organisms are grown in bulk in numerous vessels, or in chambers larger than the normal laboratory glassware, the term mass culture might be applied. Sometimes there is not a clear distinction between some cultures of limited volume and certain mass cultures. These scaled-up experiments could be used to produce enough organisms for a pigment analysis, isolation of an enzyme, or an extracellular product.

Because rapid growth rates can be maintained indefinitely by some microalgae, there have been suggestions for mass culturing algae for food (chapter 20).

When considering algae for such mass cultures, the high-temperature strains, those that maintain optimal growth rates at around 40°C, are most often selected. The best growth rates among the algae are found in this group. Under optimal growth conditions, that is, favorable temperature, light saturation, nutrient and gas replenishment, a doubling of cell number every 2 hr can be maintained with one species of *Chlorella.* (Thermal algae, found in hot springs, grow at temperatures about 40°C, even up to 75°C, but do not have exceptional growth rates.) It is less expensive and in some ways easier to maintain cultures at 40°C than at controlled room temperature. The heat given off by lamps can be used, and only in warmer climates or during the summer would cooling be necessary.

At present technical problems make mass culturing less than satisfactory. Maintaining uniformity of product and a culture free of other organisms is not an easy task. Rapid growth of an organism at the next trophic level can wipe out an algal culture overnight. When outside tanks or lagoons are used, a series of cloudy days or cold winter temperatures mean little output for that period. On the other hand, with a plentiful supply of nutrients in many natural waters and good growing seasons in many parts of the world, technological advances should bring increased utilization of mass cultures.

Continuous Culture

Cultures of limited volume, described previously, have as their principal disadvantage the ever-changing nutrient conditions within the culture vessel. Early in the experiment the nutrients are in excess, but later the stationary phase may be brought about by nutrient depletion. These fluctuations in nutrient level can be eliminated by the technique of continuous culture. Theoretically, the organism remains in the logarithmic phase of growth.

Often continuous cultures have been suggested as a most useful approach to mass culturing. The culture is initiated by preparing for sterilization a growth chamber or reaction vessel. A tube connects the growth chamber to a large container of sterile culture medium. After inoculation of the culture chamber, a population of cells will soon be established. Then new medium is gradually added to the culture chamber (Fig. 21-1). Old medium with cells for harvest leaves the reaction vessel by way of an overflow valve. This continuous harvest can

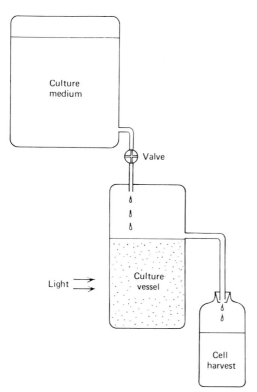

Figure 21-1. Diagram of a continuous culture apparatus. The algae in the culture vessel may be agitated by aeration or a stirring device in the bottom. The flow of culture medium is controlled by a valve. The latter may be set for a certain flow, or it may be activated when the cell density becomes too high.

be collected in a number of ways. The rate of addition of new medium must be adjusted so that the culture will not be flushed from the reaction vessel. This rate is determined in part by the growth rate of the alga.

The rate of addition of new medium can best be controlled by a metering pump, or electronically by way of a reading of the density of the culture. In the latter case, when the cell density is above the selected reading, a valve allows enough new medium to enter to dilute the culture.

When there is frequent transfer of cultures, a practice necessary when dilute media or natural waters are employed, the investigator is using a procedure intermediate between those used with cultures of limited volume and continuous cultures.

Not only are continuous cultures more expensive than typical stationary cultures, but they are also subject to numerous mechanical problems. They require some attention, especially in attempting to maintain sterility. Nevertheless, continuous cultures are used in industry for growth of other microorganisms, and are well worth considering for algal research.

Algal Assays

Several types of algal assays are used both in the laboratory and in the field. For example, radioactive carbon can be added to light and dark bottles incubated in the field. The bottles contain the algal population of the water studied and grow, utilizing the carbon. Measurements of carbon utilization are simple and accurate. This assay gives us some measurement of productivity. In addition, some organisms are routinely used in the laboratory and in industry to assay for vitamin level. *Euglena gracilis* is the assay organism for vitamin B_{12}.

Assay for the Carrying Capacity of Natural Waters. For several years most of us have been aware of the increased quantity of nutrients that find their way into all bodies of water. As we are faced with the enormous problems of monitoring bodies of water receiving these nutrients, and of quickly gathering useful information for effective action, some attention has been focused on the use of algae as assay organisms. Laboratory procedures are suggested for a quick, reliable, and uncomplicated measurement of the amount of algal growth a body of water can support (its carrying capacity). Previous exhaustive examination of physical, chemical, and biological parameters at many stations are quite useful, but are very time consuming.

Natural waters, with around 1 mg N l^{-1} and levels of P measured in micrograms, are poor media for growth of algae, especially when usual laboratory procedures are followed and organisms are incubated for days. Nutrients are quickly exhausted, and much of the incubation period is spent in the stationary phase of growth. Good growth occurs in nature because of the cyling of nutrients. It is for this reason that continuous cultures are sometimes suggested for algal assays. However, continuous cultures, while potentially the most useful, can be the most demanding.

Several key requirements have been suggested for any algal assay test: simplicity, low cost, reproducibility, and means of applying results to field conditions. Stationary cultures would not be the best choice for reasons outlined above. It is not wise to use moribund cells for the assay, but that is what develops in natural waters after a few days of laboratory culture in stationary flasks. Spiking natural waters (separate additions of elements) would provide a means of detecting limiting nutrients, and this could be successfully done in stationary culture.

On the other hand, a laboratory procedure utilizing bacteriological methods and supplies could easily meet requirements of cost, easy standardization, and reproducibility. With frequent transfer of liquid cultures, thereby reducing cell number and replenishing nutrients, certain advantages of continuous cultures are realized.

Selenastrum has been proposed by a government-industry study as a national test organism for any freshwater algal assays. It is grown in a dilute culture medium so that the danger of nutrient carryover, that is, nutrients in the inoculum, is minimized. (Even when using washed cells, there can be sufficient nutrients within cells for growth of several generations.) After active growth in the dilute medium, it is inoculated into sterile-filtered river water. Organisms are grown at a constant temperature in fluorescent illumination on a roller tube rotator. The gentle agitation provides a motion similar to that in a stream with a moderate flow. Organisms are transferred daily, reestablishing a population of 4×10^5 cells ml^{-1} each day. Growth can be determined in a number of ways. When the daily growth rate is measured, clear differences in the carrying capacity of natural waters from location to location, or during different seasons, can be shown.

ISOLATION OF ORGANISMS

Perhaps the first goal of many people when they begin culturing is to maintain an organism in the laboratory, and thus establish at least a crude culture. The organism, along with attached biota and other organisms in the accompanying solution, is usually picked up and placed in a large container. The "culture" should then be placed in a favorable temperature with adequate illumination. Organisms can be placed in a north window where diffuse light will be adequate and where there would be little chance of high, and injurious, temperatures. For maintaining cultures for teaching purposes, or for some types of experiments with life history studies of larger algae, this type of "culture" might be sufficient.

However, for growing organisms for experimentation, many investigators now feel that it is important to establish an axenic culture. Xenic cultures, with one or more bacteria in addition to the alga, are commonly used for some investigations, but serious nutritional studies are nearly impossible to attack with the xenic culture because nutrient changes might affect the other microbes and not the alga. In the past, pure culture has been used to denote both axenic or just unialgal cultures, and thus one must be cautious about its meaning.

A clone is a culture established from one individual and thus there is genetic homogeneity. A gone is established from the germination product of a reproductive structure, i.e., the progeny or a zygote. Axenic cultures are almost of necessity always clonal because it is highly unlikely that an investigator could isolate two or more individuals of one species simultaneously *without* bacteria. Isolation techniques center around procedures that will make it easier to establish clonal cultures.

If the organisms are of sufficient size, individuals can be easily manipulated through the steps of isolation, even without the aid of a microscope. But with these larger forms it is always possible that some miscroscopic forms are attached to the large form, or come along during isolation, suspended in the liquid. Other organisms, such as large colonies, can be manipulated easily under the dissecting microscope, and smaller forms can be isolated under the compound microscope. The difficulty of isolating small forms is compounded because of the reduced working distance between objective and specimen, as well as because directions are reversed with such magnification.

When individual organisms are isolated under a dissecting microscope the following procedure is used. First, find the organism and draw it into a small pipette. (The pipette can be drawn to size from sterile soft glass tubing.) Place the organism in a sterile liquid medium in a spot plate and transfer from one depression to the next, diluting out any associated microorganisms. Finally, place the organism in a sterile vessel and incubate. To obtain an axenic culture it is often necessary to repeat this procedure once the initial culture is actively growing.

If the organism cannot be picked with any ease under any of these conditions, it might be treated as one manipulates bacteria, and thus streaked or plated on a solidified surface. (The medium selected for growth would be solidified with 1.5% agar.)

At present a spraying technique is quite popular (Fig. 21-2), with the collection of microscopic forms drawn up into a micropipette. Then compressed air is blown *across* the micropipette tip, atomizing the suspension of algae. The small drops are then evenly distributed on the surface of agar (Fig. 21-2). Petri plates inoculated by spraying or by streaking with a bacteriological loop are then incubated under artificial illumination at a constant temperature. After several days of growth the individual cells isolated on the agar surface have divided several times and a microscopic aggregation or population has developed. These populations can be easily isolated by using a sterile micropipette. The tip of the micropipette is touched to the surface of the algal population and hundreds of cells are then drawn in by capillary action. Larger algae, in smaller numbers, or individually, can be isolated by using a slightly larger micropipette. Quickly, the contents of the micropipette can be transferred to a sterile tube of liquid medium by breaking the tip, with the alga, into a test tube (Fig. 21-2).

With the microalgae one assumes that most of the individual algal populations, which develop on the agar surface after streaking or spraying, are unialgal. After the isolate has been placed in a small tube and allowed to grow, miscroscopic examination will be used to determine if there are two or more species of algae present. Bacteria, if present, often can be seen under phase microscopy, but a check for growth in a bacteriological medium is quite simple. If necessary, as is most often the case, the actively growing algal culture is again streaked, sprayed, or plated in order to separate the actively growing alga and associated bacteria. When this plate is incubated in the light, the alga grows and quickly divides. Again algal populations soon become visible. Actively growing algae can frequently be separated from all other organisms by a few such serial isolations, and the repeated isolations from populations started by individual organisms usually result in an axenic culture, which is also clonal.

When the culture with bacteria, the xenic culture, grows well, but the axenic culture cannot be established, consider the possibility that the bacteria are providing one or more growth requirements. For successful culture in the absence of bacteria the growth factor could be supplied to the isolation medium. In addition, during isolation it might be wise to consider some alterations of the medium or the culture conditions to retard bacterial growth. Bacteria could be interfering with the alga in other ways. It is for these reasons that organic compounds are used sparingly. When they are supplied in high concentrations, isolation media or agar plates are frequently overrun with extensive growth of molds and bacteria. When active growth and repeated isolations do not separate algal and bacterial forms, ultraviolet light or antibiotics can be used in an attempt to kill the bacteria selectively or to reduce their numbers. In this way the next plating or spraying is more advantageous for the alga.

If there is difficulty in establishing an axenic culture, some knowledge of the organism would be most helpful. When larger cells have bacterial associations and smaller ones do not, gradient centrifugation is most helpful. If the alga can release flagellated asexual cells, one might concentrate on isolation procedures that would favor growth of zoospores. Since these cells are formed within a parent cell or sporangium wall, upon release they are free of other microorganisms. Zoospores can sometimes be formed when a population from nature is put into a medium in the laboratory, or by transferring an older culture to new medium. Transfer from agar to liquid medium stimulates abundant zoospore formation in many unicellular forms. Transfer to light after a lengthy period in darkness (more than overnight) produces zoospores in some forms, whereas a shift from a higher to a lower temperature might be successful in other cases. Some flagellated cells respond to light and aggregate at one side of a vessel where they can be collected.

Marine microalgae can be treated exactly as freshwater forms. Culture of some marine marcoalgae has been difficult, for they are too large to be simply collected and placed in large containers. Portions of the plant body are most often selected for culture, or only reproductive bodies and spores are used, with the hope that the juvenile stages of the next generation will survive in culture. However, great strides have been made in the last decade and it is to be expected that those interested in marine organisms will continue to add names to the list of marine species in culture.

Figure 21-2. Diagrammatic presentation of some procedures used to obtain a culture of a microalga. A solidified substrate is provided by adding 1.5% agar to the nutrient solution. The algae from nature are isolated on the agar surface by streaking with a bacteriological loop, or by spraying. After incubation, macroscopic populations originating from individual cells are isolated with a capillary pipette. The algae in the tip of the pipette are quickly placed in a test tube by breaking the tip from the pipette. The culture can then be grown by incubation under controlled conditions.

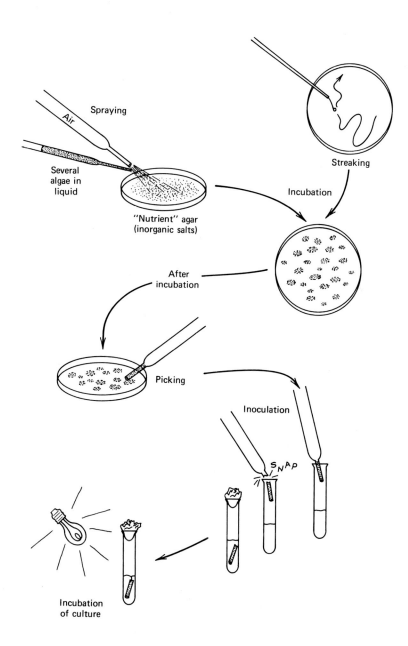

Spraying

Air

Several
algae in
liquid

"Nutrient" agar
(inorganic salts)

Streaking

Incubation

After
incubation

Picking

Inoculation

SNAP

Incubation
of culture

REFERENCES

Aaronson, S. 1970. *Experimental Microbial Ecology*. Academic Press, New York, 236 p.

Allen, H. and J. Kramer. 1972. *Nutrients in Natural Water*. Wiley-Interscience, New York. 457 p.

Benoit, R. 1969. Geochemistry of eutrophication. In: *Eutrophication: Causes, Consequences, Correctives*. Nat. Acad. Sci./Nat. Res. Council. Publ. 1700. pp. 614–30.

Chu, S. 1942. The influence of the mineral composition of the medium on the growth of planktonic algae. I. Methods and culture media. *J. Ecol.* 30: 284–325.

Carlucci, A. and P. Bowes. 1970. Vitamin production and utilization by phytoplankton in mixed culture. *J. Phycol.* 6: 393–400.

Fogg, G. 1965. *Algal Cultures and Phytoplankton Ecology*. Univ. Wisconsin Press, Madison. 126 p.

Fries, L. 1973. Requirements for organic substances in seaweeds. *Bot. Mar.* 16: 19–31.

Gerhart, D. and G. Likens. 1975. Enrichment experiments for determinating nutrient limitation: Four methods compared. *Limnol. Oceanogr.* 20: 649–53.

O'Brien, W. 1972. Limiting factors in phytoplankton algae: Their meaning and measurement. *Science.* 178: 616–7.

O'Kelley, J. 1974. Inorganic nutrients. In W. Stewart (Ed.) *Algal Physiology and Biochemistry*. Univ. Calif. Press, Berkeley. pp. 610–35.

Pringsheim, E. 1946. *Pure Cultures of Algae*. Cambridge Univ. Press. 119 p.

Provasoli, L. 1963. Growing Marine Seaweeds. *Proc. 4th Internat. Seaweed Symp.* pp. 9–17.

Provasoli, L., J. McLaughlin, and M. Droop. 1957. The development of artificial media for marine algae. *Arch. Mikrobiol.* 25: 392–428.

Starr. R. 1971. Culture collection of algae at Indiana University. In: J. Rosowski and B. Parker (Eds.) *Selected Papers in Phycology*. Dept. Botany, Univ. Nebraska, Lincoln. pp. 817–43.

Stein J. 1973. *Handbook of Phycological Methods. Culture Methods and Growth Measurements*. Cambridge Univ. Press. 448 p.

Tamiya, H. 1958. Mass Culture of algae. *Ann. Rev. Plant Physiol.* 8: 309–34.

APPENDIX

Often in the rush of the semester the undergraduate may feel that there is "not sufficient time" to search for more information about the subject matter being studied, however interesting. The lecture notes and text book serve as the sole sources. It should be emphasized that a great deal more information on the biology of the algae, including topics such as their ecology, life histories, and physiology, will be found in the current and past literature.

For an exhaustive search of literature on a particular subject you should consult abstracting journals such as *Biological Abstracts*. Review articles are also helpful. Your reference librarian would help you find additional sources. To get a feeling for the research areas currently receiving some attention, you might wish to peruse at least some of the following:

American Journal of Botany
Annual Review of Microbiology
Annual Review of Plant Physiology
Archives of Microbiology
Botanica Marina
Botanical Magazine, Tokyo
Botanical Review
Botaniska Notiser
British Phycological Journal
Bulletin of the Torrey Botanical Club
Canadian Journal of Botany
Ecology
Freshwater Biology
Journal of Experimental Marine Biology and Ecology
Journal of Phycology
Journal of Protozoology
Journal of the Marine Biological Association of the United Kingdom
Limnology and Oceanography
Nature
Nova Hedwigia
Pacific Naturalist
Phycologia
Physiologia Plantarum
Planta
Plant Physiology

Protoplasma
Revue Algologique
Science
Taxon
Transactions of the American Microscopical Society
University of California Publications in Botany

Do not overlook the fine research, past and present, done with the algae in many other countries, especially France, Germany, India, Japan, and Sweden. Some of this work is published in local journals in the native language, providing more evidence that we all must be familiar with a language other than our native tongue.

With some regularity marine phycologists publish a series of papers in the various proceedings of Seaweed Symposia. The ninth symposium was held late in the summer of 1977; a publication can be expected within a couple of years of that date.

Many reference volumes offer additional material. Some are assemblages of related (Stein, 1973; Whitton, 1975) or unrelated (Parker and Brown, 1971) papers. These chapters or papers would not have been published previously. Others are collections of selected papers already published elsewhere (Rosowski and Parker, 1971). The latter can become dated quickly, but a wise selection of outstanding papers, as well as the selection of additional references, can prolong the usefulness of such volumes.

REFERENCES

Parker, B. and R. Brown. 1971. *Contributions in Phycology*. Allen Press, Inc. Lawrence, Kansas. 196 p.
Rosowski, J. and B. Parker. 1971. *Selected Papers in Phycology*. Dept. Botany, Univ. Nebraska, Lincoln. 876 p.
Stein, J. 1973. *Handbook of Phycological Methods*. I. Culture Methods and Growth Measurements. Combridge Univ. Press. 448 p.
Whitton, B. 1975. *River Ecology*. Univ. California Press, Berkeley. 725 p.

GLOSSARY

Aerobic   Requiring oxygen for respiration.

Agar   Gelling compound (polysaccharide) derived from certain red algal walls. Used in the preparation of solidified culture media.

Akinete   Reproductive cell of a blue-green alga. Larger than a vegetative cell and thick walled.

Alga   (Plural—algae) Unicellular or multicellular, usually photosynthetic, motile or nonmotile primitive plants lacking vascular tissue. Marine, freshwater, and soil forms are known.

Algicide   One of many chemicals or agents used to kill algae, especially troublesome forms.

Algin, alginic acid   Wall component of certain kelps (brown algae) used as a gelling agent or a stabilizer (as in dairy products).

Allophycocyanin   One of the water soluble phycobilin pigments located in phycobilisomes of blue-green and red algae.

Amylopectin   The branched component of the starch molecule. Turns red with iodine potassium iodide (I_2KI).

Amylose   The unbranched component of the starch molecule. Turns blue with iodine potassium iodide.

Anaerobic   Living in the absence of oxygen.

Anaphase   Stage in nuclear division, following metaphase, in which the new chromosomes have moved away (in two groups) from the equator of the division figure.

Androspore   Flagellated stage of a few filamentous green algae. It germinates and attaches near the egg cell and a dwarf male filament results.

Annual   An organism that lives and reproduces during one growing season.

Antheridium   Male reproductive structure producing reproductive cells that are usually motile.

Anticlinal   Perpendicular to a surface of a structure or organism, as a periclinal division.

Apical At the tip or apex.

Apical cell A cell at the tip of a structure or organism that produces derivatives that form the plant body.

Apical growth Nuclear and cytoplasmic divisions, enlargement and differentiation at the apex of an organism.

Apical plane A surface achieved by sectioning through a pennate diatom from the bow to the stern.

Aplanospore A nonmotile reproductive cell in certain algae that also have the potential for producing motile reproductive cells or planospores (zoospores).

Aquaculture The science of growing aquatic organisms in the laboratory or field under artificial conditions, often utilizing human waste products, in liquid form, for nutrients.

Archegonium A female gametangium of green plants other than algae, with the egg surrounded by a jacket of nonreproductive cells.

Areola Although the term should probably be used for the margin of silica around a hole, it describes types of passages through the silica wall. These passages have a perforate membrane or velum in the passage itself, or even two membranes. Considerable terminology can result, especially if observations are made with the electron microscope. In at least one *Coscinodiscus* the type of areola is a loculus with a cribrum (one type of velum) at one surface and a foramen at the other!

Armored Possessing wall components, or platelike structures, adjacent to the plasmalemma. They may be ornate, as in some dinoflagellates.

Asexual Pertaining to the phase of an algal life history that has reproduction without benefit of sexual processes; *or,* certain reproductive structures that are involved in such a life history.

Assay Growth experiment with algae aimed at determining the amount of primary production in a body of water, its carrying capacity, the vitamin content of water, etc.

Autogamy Union of gametes produced by one individual.

Autoplasmolysis The process of reduction in size of the protoplast by self-regulation of water loss from the vacuole.

Autotroph   Organism capable of synthesizing protoplasm strictly from inorganic substances, e.g., most photosynthetic algae.

Auxospore   The reproductive cell of diatoms formed when there is diameter increase. Often produced when there is fusion of egg and sperm; thus also a diatom zygote.

Auxotroph   Organism requiring an organic compound, as a vitamin. This substrate does not add *significantly* to the cell carbon.

Axenic   Culture containing organisms of only one species, and thus unialgal, but also free of all other organisms.

Bangiophyte   Common name for one of two groups of red algae. These are the primitive forms which may be unicellular, filamentous, or parenchymatous.

Basal body   The portion of the flagellum base just inside the plasmalemma. In cross section the microtubules and connecting fibers have a complex orientation.

Benthic   At or on the bottom of a body of water, often referring to attached algae along the seashore.

Binucleate   Containing two nuclei, as in some dinoflagellate cells.

Bipartite   Composed of two parts.

Bipartition   Division into two equal or nearly equal parts.

Bladder   Inflated bulbous portion of a plant body, as in some brown algae.

Blade   Flattened, elongate thallus found in some red, brown, and green algae.

Black zone   The spray zone of some marine coastal areas, colored by blue-green algae and/or lichens.

Bloom   The abundant growth of microalgae in natural waters, often producing visible coloration to the water. It may be composed of one or more species.

Calcium carbonate   Chemical compound, $CaCO_3$, deposited by some algae exterior to the plant body, but adherrent.

Carotene   Class of pigments, yellow and orange, found in all algae.

Carotenoid   Class of pigments that includes both carotenes and xanthophylls.

Carpogonium   Female cell of red algae.

Carposporangium   A red algal cell that contains a carpospore. The terminal cell of a gonimoblast filament of florideophytes.

Carpospore   Reproductive cell of red algae produced in the carpogonium some time after sexual reproduction. Develops into a tetrasporophyte.

Carposporophyte   The entire third phase of a red algal life history from zygote to carpospore. This phase remains attached to the female gametophyte.

Carrageenan   Sulfated polysaccharide produced by red algae and deposited in walls of a few species; utilized as a stabilizer or gelling agent.

Cell plate   The partition developed between newly dividing protoplasts, positioned between the nuclei.

Cellulose   Carbohydrate found in the walls of many algae.

Cell wall   The material, outside the plasmalemma, covering many algal cells, such as the glass walls of diatoms.

Central cell   In the red algae, the cell positioned along the axis, surrounded by a group of pericentral cells.

Centric   Those diatoms with radial symmetry.

Centroplasm   In the blue-green algae the inner part of the protoplasm containing DNA. Surrounded by chromoplasm.

Charophyte   The line of green algae that gave rise to higher plants. Includes organisms that have a persistent spindle apparatus at the time of nuclear division. Formerly used for just *Chara* and close relatives with no consideration of ultrastructural detail.

Chitin   A material best known as the covering of certain animals, as arthropods. One form, chitan, is found in a few diatoms.

Chlorophyll   The pigment complex found in photosynthetic plants. All plants have chlorophyll *a*. Chlorophylls *b, c,* and *d* are found in one or more groups of algae.

Chlorophyte   Common name for the group of green algae that is considered the most primitive, when ultrastructural features are examined. At nuclear division there is early collapse of the spindle apparatus.

Chloroplast   The membrane-bound organelle in which photosynthetic pigments are located.

Chromoplasm   In the blue-green algae the outer layer of protoplasm in which thylakoids, with their photosynthetic pigments, are located.

Chrysolaminarin   A food reserve of certain algae, e.g., diatoms. It is a beta 1:3 linked glucan with 1:6 linked branches.

Class   The category used in classification ranked below division or phylum. It is at the class level that there is the most agreement among phycologists for positioning algae.

Cleavage   The partitioning of the cytoplasm after nuclear division.

Clone   A population derived from one individual.

Closed spindle   The microtubules involved in nuclear division are formed within a persistent nuclear envelope.

Coccolithophorid   A haptophyte that usually possesses mineralized scales on the cell surface.

Coenocyte   A multinucleate organism lacking cross walls, except when reproductive cells are formed.

Colony   An aggregation or close association of motile or nonmotile cells of an organism in two or three dimensions. Does not include filaments.

Conceptacle   A cavity in the plant body in which reproductive cells are formed; the cavities in which *Fucus* gametes develop.

Conjugation tube   The hollow cylinder, an extension of the cell wall between adjacent gametes, through which a gamete migrates prior to fusion with another gamete.

Contractile vacuole   Microscopic pulsing vesicle found in several microflagellates, used for osmoregulation.

Cortex   The tissue just inside the epidermis, but not in the center or medulla.

Cortications   Cells or filaments, developing at a node, covering an axial cell, or the exterior of a plant body, producing a cortex.

Costa   The internal or external, frequently elongate, thickening of a diatom wall.

Crustose   A growth form adhering closely to the substrate.

Cryophiles   Organisms inhabiting cold water, snow, or ice.

Cryptophytes   Common name for members of the Crytophyceae.

Cryptostomate   A pit or opening on fucoid thalli from which hairs project; not associated with reproduction.

Cyanelle   The endophyte in a host-endophyte relationship.

Cyanome   The two-membered complex in a syncyanosis or symbiosis between a host and a blue-green algal cell.

Cyanophyte   Common name for a member of the Cyanophyceae.

Cystocarp   The aggregation of gonimoblast filaments and sterile covering (pericarp) in a red alga. The carposporophyte and pericarp.

Cytokinesis   Cytoplasmic division, which may or may not immediately follow nuclear division.

Cytoplasmic strand   A thread of cytoplasm through a vacuole, connecting portions of the cytoplasm, or the bulk of the cytoplasm with a central nucleus.

Cytoskeleton   The structural framework of a microalga or a zoospore, composed of microtubules. Observed with the electron microscope.

Deciduous   The capacity to fall off.

Dendroid   In the form of a tree.

Desiccation   The capacity to dry up.

Desmid   Common name for advanced, freshwater, unicellular or filamentous green algae which lack motility, or motile reproductive cells.

Diffuse   The capacity for cell divisions and growth taking place throughout the plant body.

Dinoflagellate   Common name for a member of the Dinophyceae. Some members are red tide organisms.

Diploid   Possessing a complete set of paired chromosomes; 2 n.

Dioecious   Having the male reproductive organs on one individual and the female reproductive organs on a second individual.

Division   The primary grouping or category of organisms in which those of common descent are placed.

DO   Dissolved oxygen.

Dome cell   One of the cap cells on the female reproductive structure of *Chara*.

Ecophene   An ecological variant.

Electron microscope   (EM) The optical instrument that obtains an image by use of a beam of electrons focused magnetically.

Endophyte   An organism living within another plant.

Endoplasmic reticulum   The system of membranes within the cytoplasm, excluding those of cell organelles.

Endospore   A spore formed within the parent wall in a select group of blue-green algae.

Endosymbiont   In an association of two organisms, the member that lives within the host cell.

Epilimnion   The circulating layer of water above the thermocline.

Epiphyte   An organism growing upon a plant.

ER   Endoplasmic reticulum.

Euglenoid   Common name for a member of the Euglenophyceae.

Eukaryote   Any organism possessing membrane-bound organelles, thus including most plants and animals. Contrast with prokaryote.

Eustigmatophyte   Common name for a member of the Eustigmatophyceae.

Exospore   Spore produced by certain blue-green algae at the tip of an elongate cell, utilizing a portion of the parent cell wall.

Extracellular product   Product of metabolism released from an organism into the environment.

False branching   The "branch" formed in some blue-green algae when a filament ruptures through the sheath and produces an appendage toward the side. Not a true branch because there are no cell divisions in a second plane.

Fibrillar   Composed of fibers.

Filament   A threadlike growth form or linear arrangement of cells, composed of a group of cells in a series.

Flagellate   Cell or organism capable of motion through flagellar activity.

Flagellum   An elongate cell appendage or organelle that enables microorganisms and reproductive cells to move about freely.

Flagellum complex The assemblage of flagella, basal bodies, and cytoskeleton.

Flora The plant life of a region, or perhaps even a different era.

Florideophyte Common name for the advanced red algae that have apical growth of the filamentous or pseudoparenchymatous plant bodies.

Flotation An act or state of floating.

Foliose Resembling a leaf.

Fractional sterilization The rendering of a solution sterile by boiling for one hour on three consecutive days.

Fragmentation The mode of reproduction in which new individuals are formed when the original thallus separates into segments.

Fusiform Shaped with a taper at both ends, as a spindle.

Gametangium Structure that encloses the gametes.

Gamete A reproductive cell, often recognizable as male or female, that fuses with a partner to form a zygote.

Gametogenesis Process of the formation of gametes.

Gametophyte The plant that produces gametes. It is often haploid, as in organisms with sporic and zygotic meiosis, but diploid gametophytes are found where there is gametic meiosis.

Gas vacuole Small, gas-filled vesicle in blue-green algae which provides some buoyancy to the organism.

Gelatinous Possessing a jellylike consistency.

Girdle The area about the midline of a diatom or dinoflagellate.

Girdling lamella The membrane or thylakoid, often in groups of three, encircling the chloroplasts of organisms in several classes of algae.

Gliding A motion, without benefit of flagella or cilia, common to some filaments in the blue-green algae and to some diatoms.

Globule The male reproductive structure of *Chara*.

Glycogen A starchlike compound that is a principal storage material in animals.

Golgi or Golgi apparatus   The cell organelle, composed of a series of flattened sacs, which is involved in the formation of many wall components, such as scales. Also secretory.

Gonidium   An asexual reproductive cell of *Volvox*.

Gonimoblast filament   The filamentous outgrowth from a red algal zygote, as well as from other fusion products such as the placental cell. Carposporangia develop at the tips of these filaments.

Granum   The chloroplast substructure formed by repeated stacking of thylakoids.

Hair   An elongate outgrowth from the cell.

Haploid   Possessing a complete set of unpaired chromosomes; *n*. With a single complement of chromosomes.

Haptonema   The flagellumlike organelle of one class of algae. It may be straight or coiled and it functions in part in the attachment of cells.

Haptophyte   Common name for organisms in the class Haptophyceae.

Heterocyst   The reproductive cell of blue-green algae, usually not larger than a vegetative cell, which possesses polar nodules at the points of attachment to adjacent cells.

Heterogamy   Grade of sexuality in which reproductive cells are flagellated and morphologically dissimilar.

Heteromorphic   Different morphologies.

Heterothallic   Different thalli. Self incompatibility; thus requiring two individuals to complete a sexual phase.

Heterotrichous   Growth in which there are both prostrate and erect portions of the plant body.

Heterotroph   Organism capable of sustained growth on an organic substrate in the dark.

Holdfast   A portion of the plant body specialized for attachment, as the basal portion of a *Laminaria* plant body.

Homothallic   Self compatibility. Thus requiring only one individual to complete a sexual phase.

Hormogonium   A segment of a blue-green algal filament, formed by the loss of one cell of a parent filament. A form of fragmentation.

Hormospore   Reproductive structure formed by the production of a thickened wall around a series of cells in a blue-green filament.

Horn   A process extending from the surface of a cell, such as a dinoflagellate.

Hypolimnion   The relatively undisturbed layer of lake water below the thermocline.

Intercalary   The positioning or occurrence of a structure or process at a point other than at the base or apex. At times localized at a specific point, as the intercalary meristem of *Laminaria*.

Intercalary band   One of the several extra girdle bands between the valves of an elongate centric diatom.

Internode   The part of the axis between two nodes.

Inversion   The process of turning inside out.

Isogamy   A grade of sexuality in which gametes are identical in size and morphology.

Isomorphic   The same morphology, as in a life history in which both generations are identical in appearance.

Karyogamy   Fusion of nuclei and association of their chromosomes.

Karyokinesis   Nuclear division.

Keel   The longitudinal extension from the valve of a pennate diatom, as in the keel of a ship. However, in diatoms the keel may be positioned at the midline or at the edge of the valve.

Kelp   One of a number of brown algae which has a large thallus, as a relative of *Laminaria*.

Lamella   One of the layers or membranes observed with the electron microscope, including photosynthetic lamellae or thylakoids.

Lamellate   Composed of layers, *or* in the form of a leaf.

Leaf   A flattened outgrowth from the stem of a plant with conduction through xylem and phloem. Without the phloem tissues algae cannot have true leaves.

Leucosin   *See* chrysolaminarin. A storage product of the Chrysophyceae; a beta 1:3 linked glucan with 1:6 linked branches.

Life history   The complete picture of the reproduction of an organism, including all possible phases such as gametophyte and sporophyte.

Littoral   The coastal region from the spray zone to the edge of the water at low tide; the area covered by a tide.

Lorica   A cell covering that is not completely in contact with the protoplast.

Luminescent   Capable of producing light at low temperature by way of physiological and cellular processes.

Macroalga   Organism that can be seen with the naked eye, such as benthic seaweeds.

Macrandrous   Lacking the ability to produce dwarf filaments, as in an *Oedogonium*.

Mating type   Terminology used in sexual reproduction when sex differences, males and females, are not clear. Compatible strains may be called plus and minus mating types.

Medium   A solution or formulation, often defined, on or in which an organism will grow, since all essential chemical requirements are available.

Medulla   The center or core of a structure or organism, surrounded by a cortex.

Meiosis   Reduction division. The type of nuclear division in which the diploid number of chromosomes is reduced to the haploid number.

Membrane   A general term that is most often used to indicate the unit structure around the cell surface, and around and in cell organelles.

Meristoderm   A dividing tissue (meristem) at the periphery of a brown algal stipe, enabling the stipe to grow secondarily and become thicker.

Meristem   A dividing tissue.

Mesokaryotic   The nuclear type found in dinoflagellates. Chromosomes are usually seen in their condensed form.

Metaboly   Cellular movement (without flagella), as the changes in overall shape possible with euglenoids.

Metaphase   The stage in nuclear division in which the chromosomes are lined up at the cell equator.

Microalga   Any one of thousands of microscopic algae that are both motile and nonmotile and occur as unicells, colonies, and filaments.

Microfibrillar   Composed of layers of fibrils as seen under the electron microscope. Wall structures are sometimes seen as groups of fibrils.

Micron μm. The unit of measurement that is 1/1000 of a millimeter.

Microtubules   Structural elements of nuclear division, components of the core of a flagellum, or cytoskeleton elements of a zoospore. In cross section they are hollow.

Minus   One of two designations for mating type, when maleness and femaleness are not apparent. *See* plus.

Mitochondrion   The organelle associated with intracellular respiration.

Mitosis   Nuclear division in which the resulting nuclei have the same chromosome number as the parent nucleus.

Mixotroph   An organism that requires organic matter, in addition to the reduced carbon from photosynthetic processes.

Monad   The flagellated cell of the chrysophytes.

Monera   One name for the kingdom containing prokaryotes.

Monoecious   One household. Male and female reproductive cells are formed on one individual.

Monophyletic   Developed from a single stock or parent-type.

Monospore   In the red algae a spore produced by the alteration of a single vegetative cell.

Morphogenesis   The development of form.

Mucilage   A general term for the carbohydrate complex often found associated with plant walls, reproductive structures, etc. The material is not firm.

Multiaxial   Composed of several central strands or axes.

Multinucleate   With many nuclei.

Multiseriate   Composed of several rows of cells, as in a filament several cells thick.

n   The haploid number of chromosomes.

Naked   Lacking a wall.

Nannandrous   Capable of producing dwarf male filaments, e.g., in certain species of *Oedogonium*.

Nanoplankton or nannoplankton   Floating or passively moving organisms in the 10 to 60 μm size range.

Naviculoid   In the form of a boat.

Nitrogen fixation   Ability to utilize gaseous nitrogen as the source of cell nitrogen. Common in blue-green algae.

Node   The place or location on an axis where branches or leaves develop.

Nodule   A thickening of the wall of a naviculoid diatom along the midline, either at the center or at the poles.

Nucleolus   The dense spherical body in the nucleus, composed of protein and RNA.

Nucule   The oogonium and associated sterile cells in the charophytes.

Oogamy   Grade of sexuality in which the male cell is flagellated and the female cell is both larger and nonmotile.

Oogonium   The single-celled female gametangium.

Open spindle   In nuclear division the microtubules are not surrounded by a persistent nuclear envelope.

Organelle   A component of the cell, such as a chloroplast, mitochondrion, nucleus, golgi.

Pair   The male and female, or plus-minus, sex cells attached and beginning to fuse.

Palmelloid   Possessing a common, thick, mucilaginous wall material.

Parallel evolution   Evolution of similar structures independently in two separate lines.

Paramylon   The food reserve of euglenoids.

Paraphysis   Sterile cell or filament growing with and around sex organs or sporangia, perhaps offering some protection to the reproductive structures.

Parenchyma   Simple tissue composed of similar thin-walled cells, which function not only in photosynthesis, but also in storage. In the algae parenchyma cells are the components of structure and form in certain species.

Parietal   Toward the margin, or near the periphery.

Parthenogenesis   Development of a new individual without fertilization of the egg.

Pectin   Any of a group of wall compounds, complex polysaccharides, which, when extracted, would produce gels.

Pellicle   Cell covering in certain organisms, such as that found under the plasmalemma in euglenoids.

Pennate   Diatoms that are typically boat-shaped and bilaterally symmetrical. Not centric.

Pericarp   The outer layer of the cystocarp. The urn or covering around the gonimoblast filaments of a red alga.

Pericentral cell   In certain red algae, the ring of cells that develops around the axial cell. The cells may be visible from the exterior, or themselves covered by a cortex.

Periclinal   Around the axis, parallel to the surface, such as periclinal divisions.

Periphyton   Algae, usually microalgae, attached to many types of substrates, including other organisms. The term may be used to refer only to organisms growing on plants.

Periplast   A type of cell covering of the protoplast, often including other cell covering materials. Sometimes used in the cryptophytes, and also as a synonym for pellicle.

Perennial   Lasting for years, such as plants that continue to live from year to year.

pH   A symbol used to denote the acidity (below 7) or alkalinity (above 7) of a solution. A logarithmic scale.

Phagotroph   Organism capable of using particulate food.

Photic zone   The upper region in a water column in which there is sufficient light for photosynthesis.

Photoassimilator   Organism capable of using organic matter in the light, while photosynthesizing.

Photoheterotrophy   Photoassimilation.

Photolithotroph   Organism that derives its energy from light, using water as a reductant and thus reduces carbon dioxide.

Phragmoplast   The structure, composed of microtubules extended between recently divided nuclei, prior to cell plate formation.

Phycobilin   A water-soluble pigment group found in three classes of algae. They are found in or on thylakoids, associated with protein.

Phycobilisome   The particle on the thylakoid surface of cyanophytes and rhodophytes containing the phycobilins.

Phycobiont   The algal partner in lichens.

Phycocyanin   The blue phycobilin.

Phycoerythrin   The red phycobilin.

Phycology   The science or study of the algae.

Phycoplast   The microtubule assemblage in nuclear division in one group of green algae. The microtubules are arranged parallel to the division plane of the dividing protoplast between two closely positioned nuclei.

Phycovirus   A virus that will infect and usually destroy a specific alga.

Phytoplankton   The floating and passively moving photosynthetic organisms in marine and fresh waters.

Pigments   In the algae colored chemicals that are normally associated with photosynthesis. An exception would be the stigma pigmentation.

Pit connection   The link between cells, especially in the red algae in which the opening is blocked by a conspicuous plug.

Placental cell   The product that results from fusion of the zygote with several cells in some advanced red algae.

Planospore   A motile spore of an organism that can also produce nonmotile spores in the same type of sporangium.

Plant body   Thallus.

Plasmalemma   The membrane that delimits the outer boundary of the cytoplasm.

Plasmogamy   Fusion of protoplasts, often followed by nuclear fusion.

Plastid   The organelle that contains the photosynthetic pigments. Chloroplast.

Plurilocular   Reproductive structure of brown algae. It is divided into numerous separate compartments, each of which releases a reproductive cell.

Plus   One of two designations for mating type, when maleness and femaleness are not apparent. *See* minus.

Polar cap   The small portion of an *Oedogonium* parent cell wall which after division becomes part of one of the newly formed cells. Caps are formed in a series when several divisions take place at the same site.

Polar nodule   The plug formed in a heterocyst at the point of connection with vegetative cells.

Polyhedral body   In the blue-green algae, an organelle that resembles the bacterial carboxysome.

Polymorphism   Many morphologies. Used in connection with organisms that are capable of forming several distinct thallus types.

Polyphyletic   Developed from many parent stocks.

Polysaccharide   A compound, such as starch, composed of more than three sugar units per molecule.

Potash   Potassium carbonate. Often a crude compound produced by burning plants, seaweed mulch, etc.

ppt   Parts per thousand or grams per liter.

ppm   Parts per million or milligrams per liter.

Prasinophyte   Common name for a group of green algae with numerous types of scales, some of which are on flagella.

Primary production   The weight of new organic material produced by a group of plants over a period of time.

Procarp   The early developmental stage of a cystocarp.

Progressive cleavage   Division of the protoplasm of a multinucleate cell, with the formation of a number of uninucleate cells taking place simultaneously.

Prokaryote   Organism which lacks membrane bound organelles. The bacteria and blue-green algae.

Prophase   Initial stage of nuclear division in which condensed chromosomes are seen.

Prostrate   Lying flat, often attached to the substrate.

Protist   A kingdom of organisms containing a diverse group of primitive, often unicellular individuals. Kingdom formed once higher plants, higher animals, and prokaryotes are clearly defined as in individual kingdoms.

Protoplast   The bulk of the cell, exclusive of the cell covering or cell wall.

Pseudodichotomous   A growth form, initially not a dichotomy, but with additional growth of one arm, now clearly a fork.

Pseudoparenchyma   Resembling a parenchyma, but in reality of filamentous origin with an aggregation and intertwining of filaments.

Pseudoraphe   Clear line along the midline of a pennate diatom, formed by the absence of ornamentation. It is the space between the ends of adjacent striae or costae.

Punctum   A thin area, depression, or hole in the glass wall.

Pure culture   A poorly defined term which means axenic to some, but unialgal to others.

Pyrenoid   The proteinaceous structure in the plastid about which starch is formed in some algae.

Quadriflagellate   With four flagella.

Raphe   A linear slit in the face of the valve, either marginal or median.

RD   Reduction division or meiosis.

Receptacle   A fertile area on which sporangia and gametangia develop.

Red eyespot   Stigma. Pigmented body involved in movement of a flagellated cell toward light.

Reduction division   Meiosis. The type of nuclear division in which the diploid number of chromosomes is reduced to the haploid number.

Reticulate   In the form of a net.

Repeated bipartition   Division of the protoplast after each nuclear division.

Rhizoidal   In the form of a rhizoid, a colorless filamentous appendage which attaches some algae to a substrate. The rhizoid may be a cell extension, or composed of a few cells.

Ribosome   Cell organelle involved in protein synthesis.

Root   Higher plant structure which functions in anchoring and in absorbing water and nutrients. Since a root has xylem and phloem algae cannot have true roots.

Scalariform   In the form of a ladder.

Scales   Thin, flat plates on the cell surface of many microorganisms. Scales may themselves be ornamented and bear spines.

Schizophyta   The division for the bacteria.

Secchi disc   A circular white disc, attached to a cord, suspended in a body of water in order to determine transparency.

SEM   Scanning electron microscopy.

Septate   Possessing cross walls or septa. In diatoms, with internal incomplete partitions.

Septum   An internal partition, composed of silica, attached to the wall of a diatom.

Seta   A hairlike growth which is a distinct cell.

Sexual   Capable of producing egg and sperm. Also the plant body or generation which will form gametes.

Sheath   The wall material, a cylinder of polysaccharides, in which a chain of blue-green algal cells is located. Some move within the sheath.

Silicoflagellate   Common name for certain chrysophytes with an external silica cytoskeleton.

Siphonous   Tubular and multinucleate, without cross walls. Sometimes broadly used for all coenocytes.

Soil extract   Solution formed by sterilizing garden soil and water (1 g in each 10 ml). Used for culturing some algae, or as an addition to other media.

Spermatangium   Structure in which sperm are formed.

Spermatia   Nonmotile male cells of the red algae.

Spine   Extension of the cell wall of certain algae.

Sporangium   Structure in which spores are developed.

Spore   Asexual reproductive cell. May be haploid or diploid.

Sporophyll   A leaf that bears spores.

Sporophyte   The spore-producing phase in an algal life history. Often diploid, but many haploid phases also are capable of reproducing with spores.

Sporopollenin   A wall component, most resistant to decomposition, found in some algae, as well as in pollen.

Spray zone   The black zone. Along the coast the zone above the littoral, subject to spray from breaking waves.

SR   Sexual reproduction.

Stem   Upright supporting structure of higher plants. With xylem and phloem in stems, algae cannot have true stems.

Stigma   Red eyespot. Pigmented body of motile cells associated with response to light.

Stipe   Stemlike part of an advanced brown alga.

Stp   Sewage treatment plant.

Stria   A linear close arrangement of puncta.

Structured granule   The cyanophycin granule of blue-green algae.

Subapical   Below the apex.

Sublittoral   The coastal zone below the level of the low tide. In this zone benthic organisms are always covered by the tides.

Sulcus   A furrow or groove.

Sulphated galactan   One of the complex carbohydrates found in red algal walls, e.g., carrageenan.

Swarmer   Flagellated microalga, especially a reproductive cell.

Symbiont   An organism associated with another in symbiosis.

Symbiosis   A mutually beneficial association between two organisms.

Syncyanosis   The association of a blue-green alga with a host cell.

Synzoospore   A complex zoospore formed by the failure of an original group of many zoospores to separate; now the organism always produces the complex form.

Telophase   Stage in nuclear division in which chromosomes have moved from the metaphase plate and two new nuclei have formed.

TEM   Transmission electron microscopy.

Teratology The science of the study of abnormal forms or monsters.

Tetrapyrrole A chemical ring structure that is the base form of several algal pigments.

Tetraspore In the red algae a spore formed with others in groups of four. Spore formed after reduction division in red algae.

Tetrasporophyte The diploid phase of the life history of a red alga, the form producing tetraspores.

Thallus The plant body.

Theca Wall covering of organisms in some classes of algae, such as the combination of plates in the wall of one group of dinoflagellates.

Thermocline Layer of water, between the epilimnion and hypolimnion, in which the water temperature drops $1\,°C$ for each meter in depth.

Thermophile Organism that lives and grows in warm or hot waters.

Thylakoid Photosynthetic membrane, found in plastids in eukaryotes and in the chromoplasm of prokaryotes, that contains the photosynthetic pigments.

Tinsel flagellum Pleuronematic type, or flagellum with numerous lateral hairs.

Tissue An aggregation of cells of similar origin and type.

Trace element Minor element. Chemical required for plant growth in small amounts, micrograms per liter, e.g., Co and Mo.

Transapical plane In diatoms a transverse plane.

Trichoblast A uniseriate branch of certain red algae on which reproductive cells may develop; often deciduous.

Trichocyst Organelle, as in the cryptophytes, that can eject hairlike components into the environment.

Trichogyne Female hair. The projection from the surface of the carpogonium of red algae to which spermatia attach.

Trichome Linear series of cells of a blue-green alga surrounded by the sheath.

Trichothallic Intercalary growth located at the base of a hair.

2 *n* Possessing a complete set of paired chromosomes.

Ultraplankton Free floating or weakly swimming organisms in the 0.5 to 10 μm size range.

Ultrastructure Form seen with the aid of the electron microscope.

Uniaxial One series of cells along the center line or axis.

Unilocular Reproductive structure, with one chamber, in which reduction division takes place and brown algal spores are formed.

Uninucleate Possessing just one nucleus.

Uniseriate One row of cells in a series.

Unit membrane In cells examined with the electron microscope the basic membrane form observed.

Vacuole A space or cavity in the cell usually filled with water and some chemicals, such as those in excess or in storage. A portion of the protoplasm, limited by a membrane, that does not contain in its boundaries other organelles.

Valve Diatom wall component; theca; frustule.

Vascular plant Higher plant with conducting tissue, xylem, and phloem.

Vegetative Not reproductive. Capable of providing structure, growing, photosynthesizing.

Vegetative cell division Cell division, increasing the size of the plant body, incorporating part of the parent cell wall into those walls of the newly formed cells.

Vesicle A small, spherical, membrane-bound structure in cells, functioning in release of products from cells. Scales and other wall components are formed in vesicles.

Volvocalean Referring to organisms that lack vegetative cell division. It is a colonial line limited in its evolution because of the method of division.

Whiplash Flagellum type that lacks hairlike lateral appendages.

Whorl An arrangement of similar parts in a circle, as attached to the stem around the node.

Xanthophyll Pigment category; hydrocarbons, located in plastids, yellow, orange, and red in color. Function in photosynthesis.

Xanthophytes Common name for members of the Xanthophyceae.

Zoospore Flagellated asexual reproductive cell.

Zooxanthellae Symbiotic algae, not including the chlorophytes, associated with animals. Dinoflagellates and cryptophytes are members of the zooxanthellae.

Zoochlorellae Symbiotic green algae.

Zygote Diploid cell resulting from fusion of egg and sperm.

Index